JOHN S. SCOTT

A Dictionary of Building

Illustrated by Clifford Bayliss

Cross-referenced to
A Dictionary of Civil Engineering

SECOND EDITION

Penguin Books

Penguin Books Ltd, Harmondsworth, Middlesex, England
Penguin Books Inc., 7110 Ambassador Road, Baltimore, Maryland 21207, U.S.A.
Penguin Books Australia Ltd, Ringwood, Victoria, Australia
Penguin Books Canada Ltd, 41 Steelcase Road West, Markham, Ontario, Canada

—

First published 1964
Reprinted 1969
Second edition 1974

—

Copyright © John S. Scott, 1964, 1974

—

Made and printed in Great Britain
by Hazell Watson & Viney Ltd
Aylesbury, Bucks
Set in Monotype Times

PENGUIN REFERENCE BOOKS

A DICTIONARY OF BUILDING

John S. Scott, who was born in 1915, is a chartered structural engineer, a certificated colliery manager, and a qualified linguist. He has worked in Austria, France, Saudi Arabia and Romania and has spent ten years in civil engineering. He now works as a mining and civil engineering writer and also translates technical texts from French, German and Russian. He is married, has six children and lives in London.

ABBREVIATIONS

(1) Subjects

carp. = carpentry
d.o. = drawing office practice
elec. = electrical
joi. = joinery
mech. = mechanical engineering
pai. = painting

pla. = plastering, floor, and
    wall tiling
plu. = plumbing
q.s. = quantity surveying
tim. = conversion and working
    of timber

The same sense of a term may be used in several of the subjects listed above. For the scope of each subject, see its definition in the text. Bricklaying, masonry, building drainage, and other general building terms are included but not classified under an abbreviation.

(2) Dimensions

For abbreviations of units, see the list of conversion factors on page 7, or the Penguin *Dictionary of Civil Engineering*.

(3) Authorities

ASHVE = American Society of Heating and Ventilation Engineers
ASTM = American Society for Testing and Materials
BS (or BSS) = British Standard (Specification)
BSCP = British Standard Code of Practice
BSI = British Standards Institution

CROSS REFERENCES

Cross references are indicated by italic type, those to the *Dictionary of Civil Engineering* by the letter *C*.

The great bulk of the terms used in building and civil engineering has forced the publisher to divide the material into two volumes, a *Dictionary of Building* and a *Dictionary of Civil Engineering*. Since many terms have two or more senses, these senses are likely to be listed separately in the two volumes. It will help the reader if he will first study the list of subject abbreviations at the beginning of each book, which is a rough guide to its contents. Some other subjects, less easily classified are shown in the list.

ACKNOWLEDGEMENTS

Dictionary of Civil Engineering	*Dictionary of Building*
concrete, whether mass, reinforced, prestressed, or lightweight; cast iron, wrought iron, steels	building trades, their tools and materials, including plastics
welding	brazing, soldering, capillary joints, silver brazing
underpinning, bridges, space frames, tension structures, timber preservation, timbering of excavations, concrete formwork	carpentry, timber houses, shoring of buildings
scientific terms such as dewpoint, electro-osmosis, terotechnology, tribology	contracts, bills of quantities, quantity surveying
sewerage and sewage disposal	house drainage
non-destructive testing, geophysics applied to civil engineering, hydrology	heating and ventilation, insulating materials and techniques, degree days

In this (1974) edition all architectural terms, including those describing the shapes of mouldings, have been deleted because they are better dealt with in the Penguin *Dictionary of Architecture*. This has made more space for modern building information.

ACKNOWLEDGEMENTS

Extracts from British Standards and Codes of Practice are reproduced by permission of the British Standards Institution, 2 Park Street, London W1A 2BS. The frequent, numerous, and cheap Building Research Digests (HMSO) and the annual Specification (Architectural Press) are highly informative and up to date.

There are two recent dictionaries of painting and decorating: J. Snelling, *The Painter and Decorator's Book of Facts*, 1972, and J. H. Goodier, *Dictionary of Painting and Decorating Trade Terms*, 1961. J. B. Taylor's *Plastering*, 1970, is highly informative, so is P. L. Marks's *Practical Building Terms*, 1953. W. Kinniburgh's *Dictionary of Building Materials*, 1966, is most helpful on timbers, trade names, paints and materials, especially their chemistry.

The author would be grateful to be told of mistakes and possible improvements for the next edition.

IMPERIAL-TO-METRIC (MAINLY SI) CONVERSION FACTORS AND ABBREVIATIONS OF UNITS

The 500 pages of BS 350 (*Conversion Factors and Tables*) are the best source of conversion factors, but those given below are the few that builders really need. Others are printed in the Penguin *Dictionary of Civil Engineering*.

LENGTH
: 39·37 inches (in.) = 3·281 feet (ft) = 1 metre (m)
25·4 millimetres (mm) = 1 in. = 2·54 centimetres (cm)
1 mile = 1760 yards = 5280 ft = 63 360 in.
= 1·6093 kilometre (km)

AREA
: 1 in^2 = 6·45 cm^2 = 645 mm^2
10·764 ft^2 = 1 m^2; 100 m^2 = 1 are = 119·6 yard^2
10 000 m^2 = 100 ares = 1 hectare (ha) = 2·47 acres
4840 yard^2 = 1 acre = 0·4047 ha; 1 mile^2 = 2·59 km^2

VOLUME
: 1 imperial gallon contains 4·546 litres = 10 lb of water
1 United States wet gallon contains 3·785 litres
1 m^3 = 1000 litres = 35·315 ft^3 = 1 tonne of water
1 ft^3 = 28.32 litres

TEMPERATURE
: 9 deg. Fahrenheit (F.) = 5 deg. Kelvin, Celsius or Centigrade (K. or C.), therefore
temperature, deg. K. or C. = $\frac{5}{9}$(temperature °F. −32)
0°C. = 32°F.

CONVERSION FACTORS

WEIGHT or
MASS

16 ounces (oz) = 1 pound (lb) = 454 grams (g)
2240 lb = 1 long ton = 1016 kilograms (kg)
1 kg = 2·2046 lb, so 1000 kg or 1 tonne = 2204·6 lb

FORCE

1 pound force (lb f) = 4·448 newton (N) and
1 N = 0·2248 lb f
1 kilogram force = 2·2046 lb f = 9·807 N

PRESSURE or
STRESS

1 lb per inch^2 (psi) = 6894·8 newton/m^2 (N/m^2)
$\qquad$ = 6·8948 kilonewtons/m^2 (kN/m^2)
1000 kN/m^2 = 1 newton/millimetre^2 (N/mm^2)
1 meganewton/square metre (MN/m^2) = 1 N/mm^2
= 145 psi
1 standard atmosphere = 101·325 kN/m^2
= 0·101 325 N/mm^2
1 bar = 100 kN/m^2 = 100 Kilopascal (kPa)
NOTE The pascal (Pa) is the SI unit for pressure or
force per unit area, adopted in 1971 by the Conférence
Générale des Poids et Mesures. Thus
1 pascal (Pa) = 1 N/m^2 and 6·895 kPa = 1 psi
1 kilopascal (kPa) = 1 kN/m^2 = 0·001 N/mm^2
1 megapascal (MPa) = 1 MN/m^2 = 1 N/mm^2

HEAT

1 British thermal unit (Btu) = 1055 joule (J)
= 1·055 kilojoule (kJ)
= 252 gram-calories (c)

On the Continent the kilogram-calorie or large calorie of 1000 gram-calories has been used but as these are not SI units they will probably drop out in favour of the joule (J) and kilojoule (kJ) so that the main conversion will be:

1 Btu = 1055 joule = 1·055 kJ
1000 Btu = 1·055 megajoule (MJ)
1 Btu/hour = 0·293 J/second = 0·293 watt (W)
3415 Btu/hour = 1 kilowatt (kW)

THERMAL CONDUCTIVITY

or K-value: 1 Btu in./ft² hour deg. F. = 0·144 W/m deg. C.

THERMAL CONDUCTANCE and AIR-TO-AIR TRANSMITTANCE or U-VALUE

1 Btu/ft²hour degree F. = 5·678 W/m² degree C. Though they use the same unit, the U-value and the conductance are different. The air-to-air transmittance or U-value gives a direct measure of the insulating usefulness of a wall, roof, etc. The conductance is numerically slightly larger (therefore less insulating) because it does not include the air resistance each side of the wall or roof and is therefore less immediately useful to the builder than the U-value.

A

abatement [carp.] The reduction in size of timber necessarily involved in sawing and planing it.

abrasion resistance [pai.] The resistance of a *finish* to wear, by the finger or by even softer surfaces.

abrasives Hard materials (usually powders, grits, or stones) in various forms such as paste, *glasspaper*, emery wheels, or water-resisting emery cloth, used for rubbing down paint, wood, or other building materials, or for sharpening tools. The best known abrasives are quartz or flint sand, powdered glass, carborundum, emery (corundum), garnet, *diamond* (*C*). For the greatest effect, an abrasive should be only slightly harder than the surface being abraded. The usual bonding materials for abrasives are *shellac*, rubber, synthetic *resins*, and other *plastics. See* **hardness** (*C*).

ABS Acrylonitrile-butadiene-styrene, a *plastics* from which pipe is made.

absorption (1) The water absorbed by a brick, expressed as a percentage of the weight of the brick. Engineering bricks in Britain are boiled for five hours to measure their absorption. *See* **saturation coefficient.**

(2) In acoustics, that property of a material which reduces echoes (reverberation) within the room. It has little effect on the passing of sound through a wall or floor, except in so far as it reduces the sound within the room. The absorption of sound by the materials of partitions or wall coverings is made use of to improve the acoustical properties of rooms. The *reverberation period* of a room is reduced when it is filled with such absorbent objects as soft furnishings or human beings, or by cutting openings, particularly windows, in the walls. *See* **audiograph, sabin.**

absorption rate or **suction rate** or **initial rate of absorption** The amount of water absorbed by a brick when it is partly immersed in water for one minute. This property is used in USA for describing bricks. *See* **water retentivity,** *and above.*

absorptive power or **emissivity** The relative rate of emission or absorption of heat from a body, compared with that from a similarly shaped black body in the same conditions. *See* **eupatheoscope.**

abstracting [q.s.] The process, before drawing up a *bill of quantities*, of assembling and adding similar tasks in the *contract* which will be paid for at the same price under one *item*. The items are arranged in trades to make billing easier.

abut To meet at one end, usually in conversation called 'butt'.

abutment An intersection between a roof surface and a wall rising above it. *See also C.*

abutment piece [carp.] (USA) A *sill* or *sole plate.*

11

abutting tenons [carp.] Two *tenons* entering from opposite sides and meeting in the centre of a *mortise*.

accelerated weathering or **accelerated ageing** The exposure in testing of *paints* or other building materials to severer cycles of heat, frost, wetness, and dryness than are naturally experienced.

accelerator (1) A substance such as calcium chloride ($CaCl_2$), added in small quantities (up to 2%) to *cement* or to a *concrete* mix to hasten its hardening rate, the opposite of a *retarder*. It usually also accelerates the *set* (3). *See* **admixture**.

(2) An inorganic salt (alum, K_2SO_4, or $ZnSO_4$) added to anhydrous calcium sulphate to make it a useful wall plaster.

(3) A hardener or catalyst; a substance which, when mixed with a *synthetic resin*, increases its hardening rate. For *glues* it may be mixed either in the pot or by *separate application* of resin and hardener to each side of the joint. It is usually an organic acid. These acids are also used in paints based on synthetic resins.

(4) A pump for circulating the water through a heating system. *Small-bore* systems almost invariably include a pump, unlike the older systems which used thicker pipes.

access An approach to a building, room, or building service for maintenance or repair. A maintenance access is usually a removable panel in a duct, chase, or conduit, so arranged that pipes or cables can be repaired.

access eye or **inspection fitting** or (USA) **cleanout** An opening in a drain closed by a plate held on to the opening by a bolt or wedge. It is often provided at pipe bends to enable the pipe to be rodded. It is therefore sometimes called a *rodding* (*C*) eye.

acetone [pai.] A highly inflammable, powerful, organic *solvent* used for removing paint or finger-nail varnish. It is also part of many aeroplane *dopes* and *lacquers*.

aciding Lightly *etching* a surface of *cast stone*.

acoustical fibre building board Low density *fibre board*, sometimes perforated to increase its sound *absorption*.

acoustical reduction factor The reciprocal of the *acoustical transmission factor*. *See* **sound-reduction factor**.

acoustical transmission factor Of a surface, partition, or device, at a given frequency and under specified conditions, the ratio which the sound energy transmitted through and beyond the surface, partition, or device, bears to that incident upon it (BS 661). The *sound-reduction factor* is its reciprocal.

acoustic clip *See* **floor clip**.

acoustic construction Any building method aimed at reducing the sound entering or leaving a room, for example *discontinuous construction*.

acoustic plaster Wall plaster having a high sound *absorption*, often containing aluminium powder which evolves gas on contact with water, This gas remains in the set plaster, giving it a honeycombed structure.

especially with a porous slag aggregate. *Vermiculite* is also used as an aggregate in acoustic plaster. Lime plaster, though its absorption is not high, has twice the absorption of ordinary gypsum plasters, which are *hard plasters* and poorly absorptive.

acoustics The science of sound. The sound *absorption* and transmission of walls, and their shape, control the acoustic properties of rooms.

acoustic tile Square blocks, often perforated, made of soft, sound-absorbent material such as *cork, insulating fibre board*, or plaster, instead of the hard ceramic materials of which most tiles are made, which have very low sound *absorption*.

adamant plaster Quick-hardening *gypsum plaster*, sometimes with sand or sawdust, for floating coats.

adaptor [elec.] An electrical fitting for taking power from either a *lampholder* or a *socket outlet* in a way for which it was not designed. An adaptor is therefore either a lampholder adaptor (or lampholder plug) or a plug adaptor (or *socket outlet* adaptor).

additive An *admixture*.

adds [q.s.] Overall quantities, before *deducts* such as window openings are subtracted from them.

adhesion (1) or **bond** The sticking together of structural parts by mechanical or chemical bonding, using a *cement* or *glue*. Timber parts are stuck with glue, bricks are bonded in *mortar*, and steel is bonded to concrete by its adhesion with the cement. *See* **specific adhesion** (*C*).

(2) [pai.] The attachment of a *paint* or *varnish* film to its *ground*.

adhesives (1) [tim.] *Glues*.

(2) *Bitumen* (*C*), *resins*, *mastics*, *cements*, and so on for fixing *wall tiles*, floor blocks, etc.

admixture or **additive** In *concrete* or *plaster*, a substance other than *aggregate*, *cement*, or water, added in small quantities to the *mix* (*C*) to alter its properties or those of the hard concrete. *Accelerators*, *plasticizers*, and *air-entraining agents* (*C*) are admixtures. BS 4049:1966 (*Plastering Glossary*) states that an admixture is added to a mix, but an additive is added to a binder.

adobe Earth used for making unburnt, sunbaked *bricks* or *blocks* about $45 \times 30 \times 10$ cm ($18 \times 12 \times 4$ in.), containing straw chopped to about 15 cm (6 in.) length as reinforcement, in Central America, southwest USA, Australia, and other semi-arid regions. Lima Cathedral is built of adobe blocks. *See* **cob**.

adze or (USA) **adz** [carp.] A tool like an axe, used for roughly surfacing timber. It was the predecessor of the *plane*. The cutting blade of an adze is perpendicular to the shaft, while that of an axe is in the same plane. The adze strikes the timber towards the operator, who must take care not to hurt himself. This very ancient, obsolescent tool is coming into its own in the *cutter block* used in many modern joinery machines. *See* **railway tools** (*C* illus.).

adze-eye hammer [carp.] A *claw hammer* in which the head has a socket which extends farther up the handle than usual. The term is American; but these hammers are also much used by *carpenters* in Britain because the socket supports the handle firmly and enables a strong *withdrawal* force to be given by the hammer when pulling nails.

aerated concrete or **cellular concrete** An *insulating material* which can be one of two types: (1) a sand-cement mix aerated by special chemicals and whisking machines, weighing at least 1040 kg/m³ (65 lb/ft³), with a *conductivity* of 0·29 W/m deg. C. (2·0 Btu in./ft²hour deg. F.); (2) a neat cement mixed with a foaming agent, which weighs at least 320 kg/m³ (20 lb/ft³), with a conductivity as low as 0·079 W/m deg. C. Aerated concrete is nailable and easy to saw, but like many other insulators, must not be allowed to get wet, or its insulation value is impaired. It has much more *moisture movement* (*C*) than ordinary concrete. *See* **lightweight concrete, gas concrete.**

aerograph [pai.] A *spray gun* for paint.

African mahogany [tim.] The best-known timbers of this name imported to Britain are from West Africa, generally Nigeria or Ghana, of the Khaya family. (Trees of many different species are also exported from East and South Africa as mahogany.) Khaya is generally similar to *Honduras mahogany* but is less uniform and sometimes spongy or with *shake*. Makore and gaboon are other 'mahoganies' imported to Britain from West Africa.

after-flush [plu.] The small quantity of water remaining in the *cistern* after a w.c. pan is *flushed*. It trickles slowly down and remakes the *seal*.

after-tack [pai.] The defect of a paint film which has been *tack* free and then becomes *tacky*.

ageing or (USA) **aging** [pai.] The storage of *varnishes* during or after manufacture to improve their *gloss* and reduce *pinholing*, *crawling*, and *lining*. *See also C*.

agent In a civil engineering or large building *contract*, the person who legally represents the *contractor* and acts for him on all occasions at the site. He is often a civil engineer.

aggregate Broken stone, slag, gravel, sand, or similar inert material which forms a substantial part of such materials as *plaster*, *concrete*, asphalt, or tarmacadam. Aggregate is described as coarse if it stays on a screen with 5 mm (3/16 in.) square holes, and as sand or fine aggregate if it passes through them.

aging (USA) *Ageing*.

Agrément Board An organization set up by the Government in 1966, with offices in Hemel Hempstead, to encourage innovation in the building industry by issuing certificates of appraisal of inventions and practices. It was based on the working of an agrément board founded in France in 1958. The scope of its assessments was recently widened to include traditional products with an export potential. It is a member of the European Union of Agrément (UEAtc) which exists to enable agrément appraisals in one country to be used in another. *See* **National Building Agency.**

air bottle A vessel connected to high points of a hot water system to collect air from them.

air brick A perforated block built into a wall to ventilate a room or the underside of a wooden floor.

air brush [pai.] A small *spray gun.*

air change A quantity of fresh air equal to the volume of the room being ventilated. The standard of ventilation is described by the number of air changes per hour. Hot underground kitchens may need up to 60 changes, boiler houses, laundries, and smoking-rooms 10 to 20, classrooms 6, teashops 4, library reading-rooms and typists' offices 2, storerooms 1½.

air conditioning The bringing of air in a building to a desired temperature, purity, and humidity throughout the year, often by washing with cold water and then heating or cooling the air, which is blown or sucked in as required. *See* **air change, air washer, plenum, modulated control.**

air-cooled slag Dense *blast-furnace slag*, often suitable for concrete aggregate or roadstone.

air-dry [tim.] A description of *timber* which has a *moisture content* in equilibrium with the surrounding air. In Britain, air-dry timber varies from 19 % to 23 % in moisture content, according to the species and the season of the year.

air duct An air passage, usually formed in sheet metal, plastics, etc. for ventilating a building.

air flue A small duct, built to withdraw bad air from a room.

air grating A perforated metal plate across an air duct where it enters a room. *See* **register.**

air gun A *spray gun.*

air house, air dome, air structure, balloon structure, inflatable, pneumatic architecture, pressurized structure or **radome** This subject, which is young so far as actual structures are concerned, will certainly develop fast because it already has more than 100 years of development behind it in the form of its parent the pneumatic tyre, patented in 1845. The first design for an air structure was for a field hospital in 1917 yet the first structures (if we except the sailing boat's sail), the radomes of the USA, were not inflated until 1949. With an internal pressure of 7 cm (2·8 in.) water gauge, these 15 m (49 ft) dia. 'balloons' can withstand winds of 240 kph (155 mph). Inflated structures have also been used by the British Army for making short-span bridges, for saving life in inflatable dinghies or life rafts and for ensuring the safe splashdown of the Apollo astronauts by the buoyancy rings and balloons on the spacecraft. Pneumatic *formwork*

(*C*) is used for bending plywood, for casting domes of concrete or plastics, and pneumatic lifting bags are used for salvaging aircraft. Even pneumatic jibs are used for cranes. Civil engineers have long known of Fabridams, inflated sausages (containing water) that can dam a river. Balloons of 30 m (98 ft) dia., inflated after they have been blasted from the Earth, have been used as passive communications satellites, merely to act as reflectors of radar waves by the thin deposit of aluminium on their surface. The Freyssinet *flat jack* (*C*) made of mild steel is probably the humblest but now one of the most indispensable examples of the pressurized structure. The largest radome, at Andover, Maine, is 64 m (210 ft) dia. and 49 m (160 ft) high.

There are two main types of pressurized structure but really unambiguous descriptions of them do not yet exist: (a) those that are held up by ribs or inflated 'tyres' and therefore do not need an air lock to allow inward or outward access; (b) those that do need an air lock, have no inflated ribs and therefore must be held up by an internal air pressure slightly above atmospheric, often 1·5 to 2·5 cm (0·6 to 1 in.) water gauge. The ribbed structures are called 'air inflated' and the non-ribbed ones 'air supported'. The American 'airmat' is a ribbed structure, costly because of its high inflation pressure of 20 to 80 kN/m² (3 to 13 psi) which necessitates a strong, expensive skin. The air-supported structures, with pressures below one thousandth of these levels, are consequently cheaper and capable of spanning larger distances but suffer a greater fire risk. They are often used as temporary shelter for growing crops in cold climates, as storehouses, to help winter building and so on. The pressure comes from a fan or blower, always working, because if it is stopped the roof will collapse sooner or later. The edges must be weighted, or buried in soil which comes to the same thing, so as to counterbalance the upthrust on the roof skin of the small excess pressure. If enough weight is not provided, an edge will lift and the roof will collapse. Collapse also occurs when fire breaks through the roof, so the designer has to choose the roof sheet and the door locations with great care. The doors are protected from air loss by air locks. (*See: Principles of Pneumatic Architecture* by R. Dent, 1971.)

air lock (1) (USA) *Weather strip*.

(2) [plu.] A stoppage of flow in a pipe due to a bubble of air being trapped in it.

air shaft or **light well** An unroofed space within a large building which admits some light and a little bad air to windows facing on it. Modern architects try to avoid light wells by open planning.

air slaking Chemical absorption, by *quicklime* or *cement*, of moisture from the air. Air slaking may so damage these *binders* as to make them useless.

air termination network Those parts of a lightning protective system at the roof, which collect the electrical discharges from the air, and lead them to the *lightning conductors*.

air test [plu.] Using a *screw plug* to block the top of a drain length (sometimes also a plug at its lower end) it is possible to pump air into a sound length of drain, and to test it by a *U-gauge* connected to the upper screw plug. This upper screw plug can by connected at the top of the *anti-siphonage pipe*, on the roof. Only about 8 cm (3 in.) of water pressure is possible because this is the maximum which the seals will stand, but a small U-gauge is extremely sensitive to small leaks. The test pressure is 4 cm (1·5 in.). The U-gauge is also used for testing the gas pipes fixed in a house, but the test pressure is then about 30 cm (12 in.). *See* **drain tests**.

air-to-air heat-transmission coefficient The *U-value*.

air-to-air resistance The resistance (R) to the passage of heat provided by the wall of a building, the reciprocal of the *air-to-air heat-transmission coefficient*. $R = \dfrac{1}{U}$.

air washer In *air conditioning*, a chamber through which air passes into water sprays or over wet plates. The water cleans the air of dust and smells, and brings it to nearly its own temperature. If heated later, the air can have a very low *relative humidity* (*C*).

alburnum [tim.] *Sapwood*.

alder [tim.] (Alnus glutinosa (British), Alnus rubra (American, Pacific coast)) A hardwood which fades to pale brown when seasoned. Very high *shrinkage* and suitable for conditions which are either dry or wet but do not alternate between the two. Suitable for piling, for *joinery*, and for making *plywood*.

alizarin pigments [pai.] Red, violet, blue, or orange *lakes* obtained from madder root or, synthetically, from coal tar.

alkali resistance [pai.] Paints which are used in bathrooms or laundries or on new concrete, brickwork, or plaster must be durable in contact with *lime*, *cement*, or soaps; that is, they must be alkali resistant. Many synthetic *resins* have this property. *Pigments* also must be carefully chosen, since some, e.g. Prussian blue, and *aluminium powder*, must not be used in alkaline situations. *See* **saponification**.

alkyd resin [pai.] A *synthetic resin* derived from an alcohol, such as glycerin, with an organic acid, such as phthalic acid. Alkyds are usually resistant to the weather, and are used as *binders* for *emulsion paints*. The name is derived from the first syllable of alcohol with the last of acid.

alligatoring [pai.] *Crocodiling*.

alligator wrench [plu.] A *pipe wrench*.

all-rowlock wall A wall built in *rat-trap bond*.

aluminium foil Aluminium *sheet* which is thinner than 0·15 mm (0·006 in.). It reflects both visible light and infra-red (heat) rays. It is therefore used as an *insulator*. Backed with Kraft paper, one sheet in the centre of 4 cm (1½ in.) cavity has a thermal *conductance* of 1·31 W/m² deg. C. (0·23 Btu/ft²h deg. F.) and a negligible weight. It is also used as an insulator on *sarking felt*.

aluminium paint [pai.] Paint made from aluminium powder. It reduces corrosion on steel owing to its *leafing* properties. It has been used effectively on asbestos cement and as a *sealer* over creosote as well as on metals, though each paint would need a different *vehicle*.

aluminium powder Flakes of aluminium obtained by putting *aluminium foil* through a stamp battery or *ball mill* (*C*). The highest metallic lustre is obtained in paints made with the variety called *leafing* aluminium powder.

amber [pai.] A fossil *resin* found in East Prussia, derived from extinct pine trees, and at one time, when it was much cheaper, used for making *varnishes*.

American bond Brick walling in which all courses are *stretchers* except the fifth, sixth, or seventh course, which is a *header* course. In USA it is also known as common bond, and in Britain as English garden-wall or Scotch bond.

American Society for Testing and Materials One of the organizations in USA which correspond to the *British Standards Institution* in publishing standards.

amyl acetate or **banana oil** [pai.] A solvent used in cellulose lacquers, *bronzing fluids*, and metallic paints. Its strong banana-like smell is not always pleasant.

anchor Any means whereby a building part is held by a masonry mass. It usually consists of some sort of bolt, or other metal part, *grouted* (*C*) into the masonry or held into it by expansion within a hole in the masonry. *See* **Rawlplug**, *also C*.

anchor block (USA) A wood *fixing brick*.

anchor plate A cast-iron plate, about $300 \times 300 \times 3$ mm ($12 \times 12 \times \frac{1}{8}$ in.), used as a *flooring tile* in factories. It is bedded in a plastic mortar *screed* so that *lugs* which project downwards from it hold it into the screed.

anchor strip A *lining plate* of thermoplastic laminated sheet.

angiosperms [tim.] A large group of flowering plants, not including pine trees, but including *deciduous* trees. All *hardwoods* are therefore angiosperms.

angle bead or **corner bead** A rounded corner, originally of *hard plaster*, but now often a protective, permanent *screed* for plasterers to work to at salient corners, made of a small, round *moulding* of wood or metal. *See* **angle staff, square staff**.

angle block or **glue block** [carp.] A small block, usually shaped like a right-angled triangle. It is glued and tacked into the corner of a frame to stiffen it, particularly under stairs.

angle board [carp.] A board used as a gauge so as to plane a timber face to a definite angle.

angle bonds The bonding of brickwork at the corners of footings, sometimes with metal ties or special bricks.

angle brace [carp.] or **angle tie** A bar fixed across an angle in a frame to stiffen it.

angle bracket [carp.] A *bracket*, particularly one projecting from a wall to hold up a shelf.

angle closer A brick cut specially to complete, i.e. close, the bond at the corner of a wall.

angle divider [carp.] A *bevel* which bisects angles and can also be used as a try *square*.

angle-drafted margin A *drafted margin* round two faces of a stone at a corner.

angle float [pla.] A plasterer's tool for shaping an internal corner.

angle gauge A *template* made for setting out or checking angles on a building site.

angle joint [carp.] A joint between two pieces of timber at a corner, as opposed to a *lengthening joint*. *See* **bridle, cocking, dovetail, end-lap joint, halving, housing, mitre, mortise and tenon**.

angle of saw-tooth [carp.] The angle between the *face* and the *back* (4) of a *saw-tooth*. It varies between 40° and 70°.

angle rafter or **angle ridge** [carp.] A *hip rafter*.

angle staff or **angle shaft** A wooden or pressed steel *angle bead*, sometimes enriched.

angle tie [carp.] (1) or **dragon tie** A horizontal timber which carries one end of the *dragon beam* and ties the *wall plates* together at a corner of the building.

(2) An *angle brace*.

angle tile or **arris tile** A *plain tile* moulded to a right angle for covering a *hip* or *ridge*, or, in *tile-hanging*, a corner of a building.

angle trowel or **twitcher** [pla.] a small rectangular *trowel* with upturned edges for working internal angles, or a vee-shaped trowel for working external angles (BS 4049).

angling or **angled** A description of a part which is set at an angle, usually not a right angle.

angular hip tile or **angular ridge tile** An *angle tile*.

anhydrite ($CaSO_4$) A mineral from which anhydrite plaster is made by mixing the ground mineral with *accelerators* of set. It differs from anhydrous *gypsum plaster*.

anhydrous Matter containing no water. It particularly applies to certain *gypsum plasters* which have been heated more strongly than *plaster of Paris*. These plasters do not react quickly with water and an *accelerator* of set is added to make them usable in building (*Keene's cement*).

anhydrous gypsum plaster Plasters such as *Keene's cement*.

animal black [pai.] Black *pigment* made by calcining such animal products as bone or ivory chippings. Drop black, bone black, and ivory black are three varieties of it. *Compare* **mineral black.**

animal glue or **Scotch glue** [carp.] *Glue* from the bones, sinews, hides, horn, or skin of animals, sold in cake or grain form, prepared by soaking the glue overnight and then heating to 60° C. (140° F.) (not boiling, which spoils the glue) and applying to the warm joint. The glue is strong but has no resistance to water. *Fish glue* has similar properties.

annual ring or **growth ring** or **year ring** [tim.] One ring of *springwood* with *summerwood*, added to the growing tree each year. Since the form of the growth rings is affected by the rainfall, it is theoretically possible to fix the date of an old timber in a house by comparing the sequences of its rings with other timbers grown in the same district which (if of the same date) will have rings in similar sequence.

annular bit [carp.] A *hole saw*.

annunciator A signal board which shows which of a number of bell buttons was pushed to operate the signal bell. It is also used in electrical fire-alarms. *See* **drop annunciator.**

anodizing *See* C.

anti-actinic glass A *heat-absorbing glass*.

anti-condensation paints Paints with a very rough surface which do not always prevent *condensation* but make it less likely that condensed water will form drops and flow down in streaks. Cork grains and similar material are sometimes mixed with such paint to form a rough absorbent surface which is also slightly insulating. This insulation may keep the surface warm and thus reduce condensation, but is a very expensive way of *insulating* a room.

anti-corrosive paints Paints which contain *inhibiting pigments* and delay corrosion of metal surfaces. Most *priming coats* for steel are anti-corrosive.

anti-fouling composition [pai.] A paint applied to ships' bottoms and marine structures to reduce the attachment of barnacles, seaweed, and other molluscs.

antimony oxide [pai.] (Sb_2O_3) A *white pigment* with *hiding power* second only to *titanium dioxide*. It is brilliant white with a faint pink under-tone and will not *feed*.

anti-noise paint A paint which is rough surfaced like *anti-condensation paint* and therefore has high sound *absorption*.

anti-siphon pipe [plu.] A pipe which admits air to the downstream side of the water *seal* of all the *traps* to which it is connected, and thus helps to ventilate the drains through its top end, the *ventilation pipe*. Its main function is to prevent the water seals being sucked away by water passing down the *soil pipe* or drain. *See* **single-stack system.**

anti-siphon trap [plu.] A *waste* trap which can lose its *seal* with difficulty or not at all. This is often done by increasing the volume of water in the seal or by an enlargement of the lower part of the U. The adoption of these traps in the *single-stack system* of plumbing may save the expense of an *anti-siphonage pipe*. *See* **deep-seal trap.**

anti-skinning agent [pai.] A substance such as *pine oil* added to a *paint* or *varnish* to prevent skin forming at the surface during storage. *Skinning* can also be prevented by filling the empty part of the tin with nitrogen or another inert gas.

anti-slip paint A paint containing sand, cork dust, asbestos fibre, or similar material, used for finishing wood floors or decks.

apartment (mainly USA) A dwelling unit for one family in a building containing more than two such units, usually called a *flat* in Britain.

apartment house A building containing three or more dwelling units (flats or apartments) served by a common entrance hall. (Mainly North American; known as a block of flats in Britain.)

appraisal A judgment of quality; for the building industry this generally means an estimate of the suitability of a building technique for a particular use. Appraisal certificates are issued by the *Agrément Board*, by the *National Building Agency*, and for the limited matter of timber frames, by the Timber Research and Development Association (*TRADA*).

apprentice A youth who agrees to work under a skilled master for small pay for a number of years on condition that the master teaches him to be a *tradesman*. All crafts were studied in this way in the Middle Ages, but technical schools are displacing the system. *See* **skilled man.**

apprenticeship An agreement between a youth and his employer, that the latter shall teach him a *craft*. The agreement is sometimes called an indenture.

approximate quantities [q.s.] A preliminary estimate of the *building components* embodied in a project before the processes of architectural and structural design have fixed the details of floors, walls, *columns*, stanchions, *roof*, and so on. If often involves *cubing*.

apron (1) A panel or board below a *window board*. It projects slightly into the room. *See below and C.*

(2) Vertical asphalt on a fascia or overhang of a roof.

apron eaves piece or **T-plate** A T-shaped section formed by bending and folding an edge of zinc roofing sheet. It fixes the eave of the zinc sheet and acts as a *flashing*.

apron flashing A one-piece *flashing*, such as is used at the lower side of a chimney penetrating a sloping roof. *Compare* **soaker, stepped flashing.**

apron lining or (Scotland) **breastplate** [joi.] Horizontal boards forming a vertical face to a stair *well*, and covering the apron piece.

apron moulding [joi.] A *moulding* on a *lock rail* of a door.

apron piece [carp.] A *pitching piece.*

apron rail [joi.] A *lock rail* of a door.

apron wall (USA) A *spandrel* in a multi-storey building.

arbitration [q.s.] In the final payments of building *contracts*, when a dispute has reached a deadlock, both parties may agree to refer the dispute to arbitration by an individual who may be mentioned in the contract. Most contracts stipulate that disputes shall be referred to arbitration, also the manner in which this shall be done. *Building surveyors* are sometimes arbitrators.

arcade A roofed passage of any sort with shops on one or both sides.

arch bar A steel bar of rectangular section under a brick *flat arch* to hold it in place, particularly a fireplace arch, a method of construction that is obsolescent in industrial countries, reinforced concrete lintels usually being preferred.

arch brace (1) A sloping strut in a timber *truss* near its support, and shaped like an arch. (2) A curved timber in a *frame*.

arch-brick or **voussoir** A wedge-shaped brick for building an arch (or lining a well).

arched construction Building by arches, that is carrying floor or wall loads on arches instead of on *lintels* or *beams* (*C*).

architect One who designs and supervises the construction of buildings. His main duties are preparing designs, plans and specifications, inspecting sites, obtaining *tenders* for work and the legal negotiations needed before building can start. His functions now extend into town planning and the study of the social and work activities that need buildings. The Architects (Registration) Acts 1931 to 1969 protect the title 'architect' in the United Kingdom so that only those who are registered with the Architects Registration Council may practise as architects. To qualify for registration a person must pass an architectural examination of university degree level as well as one in professional practice.

architectural assistant [d.o.] or **architectural draughtsman** A man or woman working for an architect, normally in the drawing office but sometimes on a site. He often is a registered architect.

architectural sections or **drawn sections** Drawn or *extruded sections* in copper, aluminium, or their alloys used as *trim* round the outside of windows, the porches of monumental buildings, and so on.

architrave [joi.] or **lining** *Trim* which is *planted* to cover the joint between the frame within an opening and the wall finish, particularly plaster, at a door or window.

architrave block [joi.] or **plinth block** or **skirting block** A block at the foot of an *architrave* into which the skirting board fits.

architrave jamb A *moulding* along the side of a door or window opening. *See* jamb.

arch-stone or **voussoir** A wedge-shaped stone in a stone arch.

arcuated construction *See* arched construction.

area or (USA) **areaway** A space separating the *basement* of a building from the surrounding earth. The basement is thus kept dry and naturally lit. *See* **dry area.**

armoured cable [elec.] Cable protected by strips of sheet steel wound round it, often used in buildings; called BX cable in USA. *See also C.*

armoured plywood [tim.] *Plymetal.*

armour-plate glass *Plate glass* which has been toughened.

arris (1) A sharp edge of a *brick*, of *plaster*, or of other *building elements*.
(2) The upper edge, sharp or rounded, of *asphalt* (*C*) *skirtings.*

arris fillet A *tilting fillet.*

arris gutter [carp.] The V-shaped wooden *Yankee gutter.*

arris rail [carp.] A triangular fence *rail* fixed to fence posts with an *arris* uppermost.

arris tiles *Angle tiles.*

arris-wise or **arris-ways** A description of the laying of bricks, slates, or tiles diagonally, or the sawing of timber diagonally.

artificial aggregates These include nearly all *lightweight aggregates* except pumice, but also *blast-furnace slag*, and *clinker*, sometimes regarded as a lightweight aggregate.

artificial marble *Scagliola* or *marezzo marble*, both made with *gypsum plaster.*

artificial stone High-grade *concrete* made from chosen sand or crushed stone, and coloured or pigmented cement. *Compare* **cast stone.**

artisan An old term for *tradesman.*

asbestine [pai.] Talc; an *extender.*

asbestos A mineral crystal, consisting of thin, tough fibres like textile, which can withstand high temperatures without change when pure. As a building *insulator*, loose asbestos weighs 144 kg/m³ (9 lb/ft³) and has a thermal *conductivity* of 0·05 W/m deg. C. (0·35 Btu in./ft²h deg. F). It is, however, more often used as *asbestos-cement* sheets, which have a conductivity of 0·2 to 0·6 W/m deg. C. (1·4 to 4·2 Btu in./ft²h deg. F.). As the thickness is small, the *U-value* is fairly high – for 6 mm (¼ in.) about 7·9 W/m² deg. C. (1·4 Btu/ft²h deg. F.). *See* **sprayed asbestos.**

asbestos blanket [plu.] A small blanket wrapped round pipes being welded or annealed or brazed, to keep the heat in.

asbestos cement Cement mixed with *asbestos* fibre to make cheap roof- and wall-*cladding* sheets, and stove pipes and pipes for carrying water under pressures as high as 250 m (800 ft) head. Not all asbestos cement is tough enough to withstand fire, particularly the common type which contains only about 15% asbestos. It can be cut with a fine-toothed saw or hacksaw, preferably wetted.

asbestos-diatomaceous earth or **asbestos-diatomite** Insulating, fire-resisting *building blocks* or boards made of asbestos fibre, diatomaceous earth, and a *binder.*

asbestos plaster An incombustible insulator for pipes, consisting of *asbestos-diatomite.*

asbestos roofing Roof covering of *asbestos-cement* sheets or *diamond slates*.

asbestos rope [plu.] *See* **joint runner.**

asbestos sheeting Corrugated, reeded, or otherwise patterned or plain sheets for roofing and wall *cladding*, made of *asbestos cement*.

asbestos wallboard and **asbestos wood** Sheets with a higher proportion of asbestos, consequently a higher fire resistance and better insulating value, than *asbestos-cement* sheet. These sheets can be nailed close to the edge without damaging the sheet, and are often used for lining rooms and workshops.

ash [tim.] (Fraxinus excelsior) English or European ash is a pale *hardwood* from white to light brown in colour, much used for making handles of striking tools (navvy's or miner's picks or sledge hammers). It can be easily bent when steamed. It is not used for building work but figured varieties are used as *veneers*.

ash dump (USA) A labour-saving device in houses, consisting of a container beneath the fireplace, into which the ashes fall, to be removed later through a *cleanout* door in the cellar or outside the house.

ashlar or **ashler** (1) A square-hewn stone.

  (2) Stone walls or facings finely dressed to given dimensions, laid in *courses* with thin joints, about 3 mm ($\frac{1}{8}$ in.) thick. The facing may be plain, vermiculated, with *drafted margin*, etc. It was used by the Egyptians in 3000 B C. *Compare* **rubble walls.**

  (3) In USA ashlar also includes walling of burnt clay blocks larger than brick size and would therefore include what is called *terra cotta* in Britain and sometimes also brickwork.

  (4) A *stud* in *ashlering*, a vertical timber which joins *ceiling joists* to *rafters*.

ashlar brick (USA) A brick hacked to resemble stone.

ashlaring Setting *ashlars*.

ashlering or **ashlaring** or **ashler pieces** [carp.] Vertical timbers (studs) about 1 m long, fixed in an attic from *joists* to *rafters* to carry the plaster or other facing of the *partition* which cuts off the acute angle where the roof meets the floor. *Clinker-block* partitions are now more often used in Britain since they are cheaper.

ASHVE The American Society of Heating and Ventilation Engineers.

asphalt roofing Roofing with *bitumen felt* or with *mastic asphalt* (*C*) laid in two or three coats.

asphalt shingles (USA) *Strip slates*.

asphaltum [pai.] US paint industry's term for *asphalts* (*C*) distilled from crude mineral oil as well as for some natural asphalts. In Britain the term is reserved for natural asphalts.

assembly gluing or **constructional gluing** or **secondary gluing** [tim.] Gluing together *plywood* or timber components to make wooden aircraft or similar structures. *See* **primary gluing.**

Association of Building Technicians A trade union for all design and supervisory staffs in building, from *architects* and *civil engineers* to *clerks of works* and *foremen*. It is a specialist section of the Union of Construction, Allied Trades and Technicians, covering non-manual workers (UCATT).

ASTM The American Society for Testing and Materials.

astragal [joi.] Scots for *glazing bar*.

atomization [pai.] The breaking up of fluid *paint*, *lacquer*, or *varnish* into fine drops by a compressed-air blast before it is ejected from a *spray gun*.

audiograph An *acoustic* instrument which measures the rate at which sound dies away in a room. It thus gives a measure of the sound *absorption* in a room.

auger [carp.] or **Scotch eyed auger** A drilling tool shaped like a corkscrew and used for boring holes in wood. A loop (eye) is forged in the top end, through which a wooden handle is passed. The auger can thus be twisted without a *brace*. Each auger can drill a hole of only one diameter but of considerable depth.

auger bit [carp.] A *bit* of *auger* shape which lacks the eye of an auger and can thus be inserted in a *brace*. Unlike the *twist drill* it is not used for drilling holes in metal. *See* **carpenter's tools** (illus.).

autoclave A pressure vessel in which (for example) *sand-lime bricks* are cured at high temperature and pressure in live steam.

automatic flushing cistern [plu.] A water *cistern* which periodically flushes a urinal or a drain of inadequate slope.

automatic immersion heater [elec.] An electric tube- or blade-shaped heater which is submerged in a hot water *cylinder* or tank and controlled by a *thermostat* built into it or in contact with the water.

autumn wood [tim.] Another name for *summerwood*.

awl or **scribe awl** or **scratch awl** [carp.] A wooden-handled tool with a steel point for marking or scribing wood or piercing hardboard or thin plywood. The point is parallel to the handle. *Compare* **gimlet**.

axe (1) [carp.] A tool for rough-dressing timber.

(2) A *bricklayer's hammer*.

axed Description of the surface of stone which has been struck with a *bush hammer* (*C*) or axe. Once-axed is rough, fine-axed is smooth.

axed bricks or **rough-axed bricks** Bricks cut to shape with *axe* or *bolster*, not rubbed, therefore laid with joints which are thicker than in *gauged brickwork*. The work is nevertheless skilled, the final trimming being done with a *scutch*.

axed work Stone faced with the axe, showing axe-marks.

axle pulley or **sash pulley** In the *sash window*, the axle pulley is a wheel placed at the top of the window (one each side of it), carrying the *sash cords*.

B

back (1) The *extrados* of an arch.

(2) [pla.] Of *gypsum wallboard*, a surface designed to be plastered, the opposite surface being the *face* (4).

(3) [joi.] The top edge of a saw.

(4) [joi.] The following edge of a *saw-tooth*, behind the *face* (3).

(5) [tim.] The unexposed surface of a sheet of plywood, which may be less beautiful than the face, but is equally strong and thick.

(6) The upper face of a *slate*. Compare **bed.**

back boiler A *boiler* fitted at the back of the hearth of an open fire or *roomheater* for providing hot water in a small house. Large back boilers may also heat water for space heating downstairs or upstairs.

back boxing [joi.] *Back lining.*

back drop or **drop connection** or **drop manhole** [plu.] A connection at a manhole, at which the branch drain enters by a vertical pipe.

back edging Cutting off a length of glazed earthenware (or other ceramic) pipe by first chipping through the glaze all round, then gradually chipping the earthenware below. until the pipe falls in two.

back filling or **back fill** (1) Earth, stone, or hard rubbish used for filling excavations after *foundations* (*C*) have been laid in them.

(2) (USA) *Backing* brickwork or *bricknogging.*

back-flap or **back fold** or **back shutter** [joi.] The part of a boxing shutter which folds behind the rest and is not seen when the shutter is folded away.

back-flap hinge [joi.] A door hinge screwed on to the face of a thin door and its frame. It looks like a *butt hinge* but is wider, and is used when the door is too thin to be carried by a butt.

background or **backing** or **base** [pla.] The surface on which the first coat of plaster is laid.

background heating The cheapest possible room heating, the opposite of full central heating; it has to be supplemented by electric, paraffin, coal or gas fires in cold weather.

back gutter or **chimney gutter** A gutter between the 'uphill' side of a chimney and a sloping roof, usually lined with *flexible metal*. See **cricket.**

backing (1) The bricks in a wall which are hidden by the *facing bricks.*

(2) Stone which can be used as random rubble.

(3) Coursed masonry built over an *extrados.*

(4) Fitting small wooden *furring* pieces on to *joists* to make a level surface for floorboards.

(5) The shaping of the top of a hip *rafter* to suit the roof slopes.

(6) Hollowing out the back of lining boards to ensure that they lie flat and do not rock.

(7) A *backup strip.*

backing coat [pla.] Any coat other than the finishing coat.

backings [carp.] *Furring* strips on joists.

backing up Using cheaper bricks for *backing* than for *facing* the wall.

back iron or **cap iron** or **cover iron** or **break iron** [carp.] The steel plate which stiffens the *cutting iron* of a *plane* and breaks up the shavings. In a metal plane it is held down by the *lever cap*. Compare **block plane**. See **carpenter's tools** (illus.).

back lining or **back boxing** [joi.] A thin strip closing a *jamb* of a *cased frame* of a *sash window*. It prevents the sash weights from rubbing against the brickwork.

back lintel The lintel supporting the *backing* of a wall, and not seen on the face.

back mortaring (USA) *See* **pargetting** (3).

back nut [plu.] A nut, on the long thread of a *connector*, which helps to make a tight joint.

back plastering (USA) Rendering or plastering the part of a wall which is not seen. It may mean putting an additional plaster skin between the outside *sheathing* and inner plaster in *frame construction* to make two air spaces instead of one, or rendering the inner face of the outer leaf of a cavity wall. See **pargetting**.

back putty *Bed putty*.

back saw or **backsaw** [joi.] A saw stiffened by a heavy fold of steel or brass along its back, such as a tenon saw or dovetail saw. See **carpenter's tools** (illus.).

back sawing [tim.] An Australian method of *flat sawing* timber which enables the heart to be *boxed*.

back shore or **jack shore** [carp.] An outer support in *raking shores*. It gives support to the foot of the *rider shore*.

back shutter [joi.] A *back-flap*.

backup (USA) The *backing* bricks or stones in a wall.

backup strip or **lathing board** or **backing** (USA) A narrow strip of wood nailed at the angle of a *partition* as a base on which to nail lathing.

back vent (USA) A *fresh-air inlet*.

badger (1) [joi.] A large wooden *rabbet plane*.

(2) A tool which is sent through a drain to clean the jointing mortar out from inside it.

badigeon [pai.] (USA) A *filler* laid on to wood or stone.

bag moulding [tim.] See **flexible-bag moulding**.

bag plug [plu.] or **expanding plug** An inflatable drain plug which blocks the lower end of a length of drain. See **drain tests**.

bakelite One of the earliest *plastics*, it was produced by Dr Baekeland in 1916. It is used for making electrical fittings, door handles, drawer pulls, and so on.

baking [pai.] A term used in USA for *stoving*.

balanced construction [tim.] A description of *plywood* which has an equal thickness of wood in both directions of the grain, is symmetrical about its centre line, and has an odd number of plies. Thus, in balanced *three-ply* the face and back *veneers* are of equal thickness and the core is twice as thick as either of them.

balanced-flue appliance [plu.] A *room-sealed appliance* so designed that it draws its combustion air from a point immediately next to that where it discharges its products of combustion, the inlet and outlet being together in a windproof terminal outside the room.

balanced sash [joi.] A *sash window. Compare* **sliding sash**.

balanced step or **dancing step** A *winder* so built that its narrow end is little narrower than the parallel treads of the straight part of the stairs. It is therefore more comfortable to walk on than a winder in which the *nosings* radiate from a common centre.

balancing valve *See* **lockshield valve**.

balection moulding [joi.] *See* **bolection moulding**.

Bales catch A type of *ball catch*.

bale tack A *lead tack*.

balk [tim.] *See* **baulk**.

ballast Unscreened gravel containing sand, grit, and stones of less than boulder size. *See also C*.

ball catch or **bullet catch** [joi.] An automatic door fastening in which a spring-loaded ball, projecting slightly from a *mortise* in the door, engages with a hole in a *striking plate* on the *door frame*.

ball cock or **float valve** [plu.] An automatic arrangement in flushing or other *cisterns* whereby a floating copper or plastic ball opens the cock when the cistern is empty, closing it as it floats up while the cistern fills. The ball is connected to a lever which opens or closes the valve. *See* **float switch** (*C*).

balloon framing [carp.] Modern American timber house construction in which the *studs* run to the roof *plate* past the floor *joists* which are nailed to them. *See* **braced frame, ribbon board**.

balloon structure *See* **air house**.

ball-peen hammer [mech.] An engineer's *hammer* with a round *peen*.

balsa wood [tim.] A *hardwood* which is as soft as *cork* and lighter. It is used for making structural models and lightweight insulating *coreboards*. The wood is light because it is mainly *parenchyma*, a light pithy material which exists in other woods in the *heart centre*. It weighs 110 to 160 kg/m³ (7 to 10 lb/ft³). It cannot be nailed or screwed but is easily glued. It was used for insulating the 'Methane Pioneer', which sailed in 1959 from the USA to Britain with the first cargo of liquid methane ever to cross the Atlantic. This cargo must be kept at about −160° C.

baluster [joi.] A post in a balustrade of a bridge or *flight* of *stairs*. Wooden turned balusters were first used in England in Elizabethan times and were then about 8 cm (3 in.) in diameter.

balustrade Collective name to the whole infilling from handrail down to floor level at the edge of a *stair*, bridge, etc.

bamboo houses Most houses are built of bamboo in countries where this fast-growing plant thrives. In 1972, according to the United Nations, more people lived in bamboo houses than in any other type of house, and they usually built their houses themselves.

band Scots for *bond*.

band-and-hook hinge or **strap hinge** or **band-and-gudgeon hinge** or **hook-and-ride hinge** A gate hinge made of a heavy wrought iron strip (the band), which drops on to a pin fixed in to a wall, often a *gate hook*.

banding or **railing** or **lipping** or **edging** [tim.] Inlays or strips to cover edges of *veneer* or ends of cores. On *flush* doors the *shutting stile* is always banded (slamming strip or clashing strip) and, in the best work, also the other edges.

band saw [tim.] A mechanical saw for cutting intricate shapes or for converting timber, consisting of an endless steel belt with teeth at one edge, running on two pulleys. The saw loop is closed by either brazing the ends together or by hooking them with hooks and notches built into the blade. Band saws move at about 30 m/s (100 ft/s).

banister (Scotland and elsewhere) A *baluster*.

banker (1) A board on which *concrete*, *mortar*, or *plaster* is mixed by shovel. It measures about $1·8 \times 1·3$ m (6×4 ft) with boards 23 cm. (9 in.) high on three sides.

(2) A mason's workbench, of stone or heavy timber.

banker mason or **stone dresser** or (Scotland) **hewer** or **blocker and dresser** A *tradesman* working at a bench (*banker*) who marks out, shapes, and finishes by hand or by machine, stones which have been cut to size by the *stone-machine hand*. *See* **machine mason**.

bar A *glazing bar*.

barbed dowel pin [joi.] (USA) A barbed piece of steel wire pointed at one end, used by joiners for fastening the *mortise-and-tenon joints* of doors and windows.

bar door [joi.] (Scotland) A *batten door*.

bare or **tight** A description of a piece which is smaller in one or more dimensions than it should be. For joiners, the term may mean about $0·8$ mm ($\frac{1}{32}$ in.) small, since joiners do not measure closer than about $1·5$ mm ($\frac{1}{16}$ in.). The opposite of *full*. *Scant* is the term preferred by BS 565:1963.

bareface tenon [carp.] A *tenon* with a shoulder cut on one side only. It is used when the piece on which the tenon is cut (a door *rail*, for instance) is narrower than the *stile* in which the *mortise* is cut.

barge board or **verge board** or **gable board** A sloping board (built in pairs) along a *gable*, covering the ends of roof timbers, and protecting the *barge course* from rain. Old barge boards were often beautifully carved.

barge couple [carp.] A name given to the end pair of *rafters* when they oversail the *gable*.

barge course or **verge course** (1) A brick *coping* to a *gable wall*, slightly overhanging the wall.

(2) A course of bricks-on-edge laid across a wall as a *coping*.

(3) The tiles next to the gable, which overhang it slightly.

barium plaster A plaster used on the walls of X-ray rooms to reduce the amount of radiation which passes through them. It contains barytes aggregate with *gypsum plaster* or *Portland cement* as a *binder*.

barn-door hanger (USA) A set of pulleys over a barn door, attached to the door by a steel strap carrying the weight of the door. The pulleys travel on a rail hung from the lintel. This arrangement is often used for garages or other heavy doors.

barrel [plu.] That part of a pipe throughout which the bore and wall thickness remain uniform. Gas pipes were, at first, made from musket barrels, surplus after the battle of Waterloo. *See also C.*

barrel bolt A brass- or chromium-plated or steel cylindrical bolt running in a case. It is pushed home by the finger and does not form part of a lock. *See* **foot bolt, tower bolt.**

barrel light A roof *light* of curved shape, made with curved glass and *glazing bars*.

barrel nipple [plu.] A short length of pipe, with a *taper* thread outside at each end, bare in the middle; a *tubular*.

barrow A *wheelbarrow*.

barrow run A smooth path for wheeling loaded barrows on a building site, usually made by laying *scaffold boards* on the ground.

base (1) The *base course* of a wall. *See also C.*

(2) A widening or *moulding* at the foot of a wall or column.

(3) or **sub-base** (USA) A *skirting board*.

(4) The part of a floor or *stair* tread or *riser* below the finish, usually a granolithic or similar surface laid with a trowel.

(5) [pai.] The *ground*.

(6) [pai.] The main ingredient of a paint, either its main *pigment* (lead base, zinc base) or the main part of the *medium* (oil base).

baseboard or **base** (USA) *Skirting* board.

base course The lowest or lowest visible *course* of a masonry wall, often provided with a *water table*.

base exchange A *water-softening* process in which water is passed through a bed of a mineral reagent called zeolite, which absorbs those salts in the water which make it hard. Periodically, the tank containing the zeolite must be flushed with a salt solution, to remove the absorbed hardness and rejuvenate the zeolite. This process is used in small houses, and gives completely soft water, until the zeolite is spent and needs rejuvenating. *See also C.*

basement A *storey* of which, in USA, less than half is below ground level. In Britain it can be wholly below ground level but is generally living space. *Compare* **cellar, sub-basement.**

base moulding The *moulding* immediately above the *plinth* of a wall, column, etc.

base shoe or **shoe mold** [joi.] (USA) A quarter-round *bead*, *planted* at the junction of *skirting* board and floor boards to cover the joint.

base trim (USA) Any *mouldings* which decorate the foot of a wall, column, pedestal, etc.

basic lead sulphate [pai.] A *white pigment* formed by burning galena (PbS) or zinc minerals containing it. Its properties in painting are like those of *white lead*. It contains some lead monoxide (PbO), but generally not more than 5% zinc oxide. *See* **leaded zinc oxide.**

basil *Bezel.*

bastard ashlar (1) or **bastard masonry** Facing stones of a *rubble wall*, which are dressed and built like *ashlar*.

(2) Stones for ashlar work which are not completely dressed at the quarry.

bastard-sawn [tim.] *Flat-sawn* (plain-sawn or slash-sawn).

bastard tuck pointing or **bastard pointing** *Pointing* with a projection from the face of the brickwork which is of the same mortar as the pointing. The joint is thus shadowed. *Compare* **tuck pointing.**

bat (1) A brick cut across, larger than a quarter brick; *see* **closer.**

(2) A *lead wedge.*

(3) or **bat insulation** (USA) A rectangular, paper-wrapped, flat-surfaced insulating *blanket* inserted between the *studs* of external walls. They are therefore 5 to 8 cm (2 to 3 in.) thick, 38 to 58 cm (15 to 23 in.) wide, and 20 to 120 cm (8 to 48 in.) long.

batch One mixing of concrete, mortar, plaster, etc.

batch box or **measuring frame** or **gauge box** A box of size calculated for the *mix* (C) in which the stones or sand are measured, before mixing with cement and water, to make concrete. The box has usually four sides only and no bottom, and before being filled it is placed on a mixing platform or *banker*. The box thus never needs to be lifted full.

baton or **baton rod** [joi.] Scots for a *guard bead*.

batted surface or **tooled surface** A vertically incised *ashlar* surface made with a *batting tool*, after the surface has been rubbed smooth.

batten [tim.] A piece of *square-sawn* softwood timber 48 to 102 mm (2 to 4 in.) thick and 102 to 203 mm (4 to 8 in.) wide (BS 565).

(2) Horizontal strips 5 × 2 cm (2 × 1 in.) approximately, which are used as fixings for wood lathing or slates or tiles (slating or tiling batten). *See* **cleat, battening, guard bead.**

battenboard [tim.] Built-up board like *blockboard* but with a core of strips each less than 76 mm (3 in.) wide. (Obsolescent, BS 565.)

batten door or **ledged door** [joi.] A door faced with vertical boards

31

fixed on to two or more horizontal *ledges* at the back, which are often *clench-nailed* to the facing and sometimes diagonally braced. It has no frame round the edges.

battening [carp.] *Common grounds* put on to a wall as a base for *lathing*.

batter or **rake** (1) An artificial, uniform, steep slope.

(2) Its inclination, expressed as one unit of length horizontally for so many units vertically.

batter board or **batterboard** (USA), known in Britain as a **profile**. Horizontal boards fixed on edge at a datum level outside the foundation dig for a building. The level at which they are set is usually basement or ground floor or a convenient number of feet above or below it. Nails or saw cuts in the top edges of the boards show the dig lines, footing lines, *building lines*, and any other important lines for setting out the lower part of the building. One batterboard is fixed at the end of each line, that is two for each corner, so that strings can be stretched between the nails to show any required line at any time.

batter brace [carp.] (USA) A *diagonal brace* (*C*), usually in a *truss*.

batter peg or **batter post** Pegs which are driven into the ground to set out the limits of an earth slope.

batting or **broad tooling** or **droving** or **angle dunting** Surfacing a stone with a batting tool in parallel strokes, each traversing the full depth of the stone face. The strokes may be vertical, when it is often referred to as tooling, or oblique at 45° to 60°. The result is a regular pattern of fluted cuts in the stone face. *Compare* **boasting**.

batting tool or **broad tool** A mason's chisel 7 to 11 cm (3 to 4½ in.) wide for surfacing sandstones.

baulk [tim.] or **timber** or **squared log** A square-sawn or *hewn* softwood timber of equal or approximately equal cross dimensions but greater than 102×114 mm (4×4½ in.) (BS 565).

bay (1) One of several uniform divisions of a building, such as the space enclosed between four columns or two beams.

(2) A development of (1) – the volume of concrete poured at any one time, or the area of plaster, asphalt, or other roofing, etc., laid at any one time.

bay window A *window* formed in a projection of the wall beyond its general line. It resembles an *oriel* window, except that it is carried on foundations outside the wall line and an oriel is carried on *corbels*. *See* **bow window**.

bead (1) [joi.] or **beading** Usually a semi-circular moulding that masks a joint, sometimes with a *quirk*. A *guard bead* is nearly rectangular.

(2) [joi.] A *glazing bead*. *See* **patent glazing**.

(3) In zinc or copper roofing, at eaves or flashings, an edge bent round to a tube shape or to 180° for stiffening the edge of the sheet and fixing it.

(4) [pai.] The accumulation of a *paint* or *varnish* at a lower edge due to excessive *flow*.

bead-and-batten work [carp.] Rough partitions made of *battens* with a *bead* along one edge.

bead and quirk [joi.] A bead separated by a narrow groove (*quirk*) from the rest of the surface which it decorates.

bead butt or **bead and butt** [joi.] A thick door-panelling which on one face is flush with the frame and on the other is recessed without *mouldings*. The flush side is usually decorated with mouldings cut on the solid in the direction of the grain only, so that the panel butts against the rails of the frame without a bead. The panelling is not elegant, but such thick doors have high *fire grading*.

bead butt and square [joi.] Door panelling which is *flush* one face and *bead butt* the other.

beading [joi.] A *bead*.

beading tool [joi.] A *bead plane* or a *cutting iron* for a bead plane.

bead plane or **bead router** or **beader** or **beading tool** [joi.] A *plane* for shaping, that is *sticking*, beads on wood or for cutting grooves in which to insert beads. *See* **hollow**.

beam box A *wall box*, or the formwork for a beam.

beam filling or **wind filling** *Brick nogging* or masonry between floor or ceiling *joists* at their supports. This stiffens the joists and provides a *fire stop* in the floor or ceiling.

beam hanger A *stirrup strap*.

bearer [carp.] A horizontal timber, often a *joist*, which supports other timbers.

bearing bar A wrought-iron bar laid on brickwork, instead of a *wall plate*, to provide a level support for floor joists.

bearing plate A plate in a wall which supports a *beam* (*C*) or *column*. The plate spreads the load from the beam over an area of wall large enough to prevent the wall crushing locally.

bearing wall A *loadbearing wall* (*C*).

bed The under-surface of a *brick*, building stone, *slate*, or *tile*, as laid in a building, or the mortar touching this surface. *See also* **back** (*6*).

bedding A layer, or the act of placing a layer on which something rests continuously. For example, *glazier's putty* under glass, *concrete* under a drain, the *mortar* under bricks or stones, the jute serving under the steel wire armour of a cable.

bedding stone A plane marble slab used by bricklayers and masons to check that rubbed surfaces are flat.

bed dowel A *dowel* in the centre of the *bed* of a stone.

bed joint The horizontal *joints* in *masonry* or brickwork and the radiating joints of an arch.

bed putty or **back putty** The *glazier's putty* under glass, on which it is bedded. *Compare* **face putty**.

beech [tim.] (Fagus sylvatica) A *hardwood* of Europe, Asia Minor, and

Japan with typical spindle-shaped markings on *flat-sawn* boards. It is light reddish-brown in colour and slightly more easily worked than English oak.

beeswax [pai.] The wax from honeybees, used in wax *stains*, *stoppings*, *matt varnishes*, and polish for wood.

beetle (1) or **maul** or **mall** A heavy **mallet** used for striking pegs, paving slabs, and material which would be damaged by a sledge hammer.

(2) *Death-watch* beetle, *pinhole* borer, and *powder-post* beetles are examples of those insects which damage *timbers*.

bel A unit of intensity of sound, equal to 10 *decibels*.

Belfast truss [carp.] A wooden *bowstring girder* (*C*) for spans up to about 15 m, built of relatively small boards. The *truss* is curved at the top and has a horizontal tie, called the string, joined by sloping members to the curved top *chord* (*C*).

bell-and-spigot joint [plu.] (USA) A *spigot-and-socket*.

bellcast eaves (Scotland) Eaves with *cocking pieces*.

belt or **belt course** A simple *string course*.

bench holdfast [joi.] A *cramp* for holding wood to the bench while it is being worked.

bench hook or **bench stop** or **side hook** [joi.] A flat piece of wood placed on the bench. It protects the bench top from injury by the tools. It is fitted with a small *cleat* at each end on opposite sides, giving it a Z-shape.

benching *Concrete* cast in a *manhole* (*C*) round a half-round drainage channel to ensure that after a flood the flow falls gradually and no solids are left behind. It is also used by men for standing on when *rodding*.

bench knife [joi.] A knife blade projecting about 4 cm (1½ in.) from a bench surface and used, like a *bench stop*, for steadying the end of a timber being worked.

bench planes [joi.] Planes used more at the bench than elsewhere, for reducing and smoothing flat timber, the *jack*, *smoothing*, and *try planes*. The term is sometimes also used to include the *block plane* and *compass plane*.

bench sander [joi.] A stationary *sanding machine*. It is too heavy to be carried on to a building site and is used in the workshop, the wood being brought up to it. Three types exist, disc sanders or grinders, belt sanders, and reciprocating pad sanders. All of them grind with strong glass paper or glass cloth, unlike the *grinder*.

bench screw [joi.] The thread of a wooden vice.

bench stop [joi.] A wood or metal peg projecting through a hole in the top of a workbench by an adjustable amount. A board can be planed without clamping, if it is merely placed against the stop.

bench trimmer [joi.] A *trimming machine*.

bench work Hand work done at the bench rather than by machines.

bend A curved length of pipe, tubing, or conduit. A 90° bend is called a quarter-bend, a 45° bend is called a one-eighth-bend. *See* **fittings.**

bending iron or **bolt** or **pin** or **bend iron** [plu.] A curved steel bar up to 60 cm (2 ft) long for straightening or widening lead pipe. *See* **plumber's tools** (illus.).

bending spring [plu.] Circular, close-packed, spiral steel springs 60 cm (2 ft) long, used for maintaining the circular cross-section of copper or lead pipes between 12 and 50 mm (0·5 and 2 in.) bore when they are being bent to a curve. This is the most convenient method of *loading pipes*, but for the larger diameters it is probably impossible for one man to bend a pipe alone even if the pipe is dead soft. The spring must be of the right diameter for the pipe, and well oiled before insertion.

bent [tim.] A timber piece curved by lamination or steaming. *See C.*

benzene [pai.] A highly inflammable paint-remover, derived from coal, usually considered to be C_6H_6.

benzine [pai.] A highly inflammable petroleum fraction which boils at a lower temperature than *white spirit*, and is used in quick-drying finishes.

best hand-picked lime or **lump lime** [pla.] Large lumps of *quicklime* chosen for their size. It is the best quality lime since it contains no ash. *Compare* **small lime.**

best reed (Arundo phragmites) The true thatching *reed. Compare* **mixed stuff.**

bevel [joi.] (1) A surface meeting another surface at an angle which is not a right angle. A *chamfer* is a special bevel. *See also* **bezel.**

(2) or **tee-bevel** or **bevel square** or **sliding bevel** A hinged blade passing through a stock (the handle) to which it is fixed by a screw and *wing nut* at any desired angle. It is used for setting out angles. *See* **mason's tools** (illus.).

(3) *See* **mitre square.**

bevel halving or **bevelled halving** [carp.] A *halving* joint in which the contact surfaces are at an angle to the plane of the timbers, to prevent them pulling apart when fixed.

bevelled closer A brick cut longitudinally along a vertical plane, starting at the middle of one end, to the far corner. One-quarter of the brick is cut off. *See* **closer.**

bevel siding [tim.] *Clapboard.*

bezel or **basil** [joi.] The *bevel* at the cutting edge of a chisel, plane iron, or other cutting tool.

bib or **bibcock** [plu.] A water tap which is fed by a horizontal supply pipe, not, as in the usual washbasin, by a pipe from below.

bid (USA) A *tender.*

billet [carp.] A piece of wood with three sides sawn and the fourth left round. *See also C.*

billing [q.s.] Writing out a *bill of quantities* with the description and *quantities* of each *item.*

bill of quantities [q.s.] A list of numbered items, each of which describes the work to be done in a civil engineering or building *contract*. Each item shows the *quantity* of work involved. When the procedure of tendering is adopted (as is usual in Britain), the bill is sent out to *contractors*. Those contractors who wish to do the work return the bill, with an *extended price* opposite each item. This *priced bill* constitutes the contractor's offer (or *tender* or bid) to do the work. It is possible for a bill to include the *specification*. *See* **abstracting**.

binder (1) *Cement*, tar, bitumen, *gypsum plaster*, *lime*, *synthetic resin*, etc. for joining stones, sand, *glassfibre*, etc.

(2) or **binding joist** [carp.] A wooden or steel beam covering the full span of an opening from wall to wall, and supporting *common joists*. *See* **framed floor**.

(3) [pai.] That part of a paint *medium* which holds the *pigment* in a coherent *film* and is therefore not *volatile*. It may be *linseed oil*, another drying oil, a *size*, or *resin*.

binding beam or **binding joist** [carp.] A *binder*.

binding rafter [carp.] A *purlin*.

birch [tim.] (Betula pendula) The silver birch. This and other birches are the hardest European *hardwoods*. The silver birch grows farther north and at higher altitudes even than pine or fir trees. It is white to pale brown without *figure*, more easily workable, but stiffer also, than *oak*. It is used for making furniture and *plywood*. Norway *burr* birch resembles *birdseye* maple, and is used as a *face veneer*.

bird peck [tim.] A small hole or patch of distorted *grain*, sometimes discoloured, attributed to birds.

birdseye or **peacock's eye** [tim.] The *figure* formed by small pointed depressions in the *annual rings* of many consecutive years, one of the few figures which show up well when cut in a lathe (*rotary cutting*), particularly with maple.

birdsmouth joint [carp.] (1) A cut into the end of a timber to fit it over a cross timber, particularly the cut in a *rafter* to fit it over a *wall plate*. This is called a *foot cut* in USA.

(2) A *bridle joint*.

bit [carp., etc.] (1) An interchangeable cutting point used by a carpenter in a *brace*, by a miner in a rock drill, or by an oil-well driller in a *rotary drill* (*C*), etc.

(2) [plu.] The working head of a *soldering iron*, usually made of copper.

bitch A steel spike like a *dog* with the spikes pointing in directions at right angles to each other. Instead of being two-dimensional like a dog, it is three-dimensional. *See* **fasteners**.

bit extension or **brace extension** [carp.] A steel rod of which one end is held by a brace, the other grips a *bit*. It is made for drilling holes deeper than the length of the bit in use.

bit gauge or **bit stop** [joi.] A small metal piece fixed temporarily to a bit by a *wing nut* to prevent the hole being drilled too deeply.

bit stock [carp.] A *brace*.

bit stop [carp.] This piece fixed on a drilling bit makes sure that the hole is not drilled too deep.

bitty [pai.] A paint or varnish is bitty if small *nibs* of broken paint skin or flocculated material stick up above the paint surface.

bitumen felt or **bituminous felt** or **roofing felt** or (USA) **roll roofing** or **rag felt** or **composition roofing** Sheets of asbestos, flax, or other fibres matted into a felt for roofing, generally treated with bitumen or pitch, sold in Britain in rolls 1 m (40 in.) wide, 11 m (36 ft) long, weighing from 0·7 to 3·7 kg/m² (0·15 to 0·75 lb/ft²). Felt can be used in single layers or in multiple layers of *built-up roofing* joined with *compound*. *See* **impregnated flax felt, mineral-surfaced bitumen felt, sarking felt**; and BSCP 144, Roof Coverings.

bituminized fibre pipe North American for *pitch fibre pipe*.

bituminous paint A paint with a high proportion of bitumen, usually dark in colour; for example, *black Japan*. It consists usually of gilsonite dissolved in *drying oil* with sometimes also *resin* and black *pigment*.

bituminous plastics *Plastics* made from natural bitumens, such as that from Trinidad, used for door furniture and electrical fittings.

black Japan [pai.] The best black *varnish*, also called *bituminous paint*.

black mortar *Mortar* containing ashes either for cheapness or for its colour in pointing. *Small lime* is sometimes used in it. It is usually of very low strength.

blade [pla.] The part of a plasterer's *trowel* which touches the plaster.

blanc fixe [pai.] Barium sulphate ($BaSO_4$), an amorphous *extender* made fine by chemical precipitation, not, like barytes, by crushing.

blank [joi.] A piece of timber cut to a specified size and shape, from which the finished article is made (BS 565).

blank door or **window** A walled-up door or window.

blanket or **quilt** *Insulating material* consisting of *eel grass*, *glass silk*, *mineral wool*, etc., with paper lining on each face. Some quilts are backed with *aluminium foil* to reduce heat losses by radiation. Blankets can be bought in strips 30 to 90 cm wide in lengths which vary with the material. *See* **asbestos blanket, bat**.

blank wall A wall without openings.

blast-furnace slag Slag from iron smelting. It is sometimes ground with *Portland cement* to form a water-resisting concrete for dams, etc. *Compare* **foamed slag**.

bleaching [pai.] The removal of *colour* by chemical action, often an oxidation caused by sunlight and air together.

bled timber Timber from trees which have been tapped for *resin*. It is usually inferior to other timber.

bleeder tile (USA) Pipes built into the foundation of a building to drain water from outside the basement *retaining wall* (*C*) into drains provided in the building.

bleeding (1) [tim.] or **bleed-through** The penetration of *glue* through a *veneer* to the surface of the veneer. *See also C.*

(2) [pai.] The passing of a *colour* or *vehicle* through the material above it. This can usually be prevented by a *sealer*.

blemish [tim.] Anything which mars the appearance of timber without affecting its strength and is therefore not a *defect*.

blender [pai.] A round softish *brush* of badger hair with a blunt tip, used for blending colours and removing brush marks left by coarser brushes.

blind area A *dry area*.

blind casing or **sub-casing** or **ground casing** [joi.] *Common grounds* used in USA as fixings for the *outside lining* and the *inside lining*.

blind floor (USA) A *sub-floor* or *rough floor*.

blind header A half-brick or a *header* not seen on one face.

blind hinge [joi.] A concealed *hinge*.

blind hole A drilled or cored hole which does not pass right through the material. The opposite of a *bottomless hole*.

blind mortise [joi.] A *mortise* which does not pass through the timber and which encloses a *stub tenon*.

blind nailing [joi.] *Secret nailing*.

blister figure [tim.] A *quilted figure*.

blistering [pai.] (1) Bubbles in a paint surface, not to be confused with a bitty film. Caused by vaporization of moisture or resin under the surface.

(2) (USA) **blub** Local swellings on finished plaster often caused by badly matured lime, a fault which may cause the finishing coat to fall away from the backing. *Compare* **blowing**.

bloated clay *Expanded clay*.

block (1) [carp.] *See* **angle block.**

(2) A masonry unit larger than a brick, laid usually in mortar. The term is also used for any preformed building unit such as glass blocks, or solid or hollow clay, or hardwood floor blocks. *See* **building blocks.**

(3) A small piece of lead or soft wood about 5 cm long and as wide as the glass thickness, which supports the glass in *glazing*.

(4) (USA) An area of land bounded by streets.

blockboard [tim.] Board built up from wooden *core* strips up to 25 mm (1 in.) wide, glued between outer *veneers* whose grain runs in the opposite direction. *See* **laminboard.**

block bonding Connecting several *courses* of brickwork of one wall into a number of courses of another wall. It is commonly used for *bonding* shallow facing bricks into thicker common bricks in the backing. It is also used for bonding new work into old. *Compare* **toothing.**

block bridging [carp.] (USA) *Solid bridging*.

block flooring *See* **wood-block paving.**

blocking course One or several courses of bricks or *masonry* in the wall over a stone *cornice* to hold it in position by weight.

block plan A small-scale plan showing the broad outlines of existing buildings or a project.

block plane [joi.] A small metal *plane* about 15 cm long for cutting end-grain, originally the end-grain of butcher's blocks. Compared to other *bench planes*, the cutting *bevel* is reversed; also there is no *back iron*.

block tin [plu.] Commercially pure tin.

blockwork Masonry of hollow *precast concrete* (*C*) blocks that now usually have $45 \times 22 \cdot 5$ cm (18×9 in.) face size to course with the metric *brick format*. Another block of face size 40×10 cm (16×4 in.) exists to meet the demand for a 10 cm (4 in.) *module*. These dimensions include 10 mm ($\frac{1}{2}$ in.) for jointing. Thicknesses vary from 5 cm (2 in.) for the thinnest partition blocks up to $22 \cdot 5$ cm (9 in.) for loadbearing walls. Blockwork is usually more thermally insulating and sometimes cheaper than brickwork. *See also* (*C*) *and* BSCP 121 and 122.

bloom [pai.] (1) A thin film, like the bloom on fruit, which forms on old glossy paint or varnish. It veils the colour or reduces the gloss. The cause is not known, but it can sometimes be removed by wiping with a cloth.

(2) *See* **efflorescence.**

blowing or **popping** or **pitting** [pla.] Small pits formed by the expulsion of plaster from the surface by material expanding behind it. The cause may be lime which slakes slowly, or occasionally the oxidation of coal in the lime. *See* **water-burnt lime.**

blow lamp or (USA) **blow torch** [plu.] A paraffin lamp which gives a powerful flame for melting solder or lead or burning off paint. It is being superseded by lamps fuelled by *bottled gas* that are much easier to control, though the fuel is more expensive. *See* **oxy-acetylene.**

blown joint or **cup-and-cone joint** [plu.] A joint in which one lead pipe is opened and fitted over another which is rasped to fit it. The two are then joined with *fine solder* poured in molten. *See* **taft joint.**

blown oil [pai.] *See* **boiled oil.**

blow torch [plu.] A *blow lamp*.

blub [pla.] (1) (USA) *Blistering.*

(2) A hole in a plaster cast, formed by an air bubble.

blue bricks *See* **Staffordshire blues.**

blueing [pai.] Increasing the apparent whiteness of a white *pigment* or paint by adding a trace of blue.

blue stain or **sap stain** [tim.] A blue fungal discolouration of *sapwood* which does not harm the timber.

blushing [pai.] Milky opalescence in a *lacquer*, a fault usually caused by moisture or a lack of *compatibility* in the paint.

board [carp.] (1) Sawn *softwood* 102 mm (4 in.) or more wide, under 48 mm (2 in.) thick. *See* **plank, baulk.**

(2) For *hardwood* the term is usually interchangeable with *plank*.

(3) (USA) *Lumber* 20 cm (8 in.) or more wide, and less than 5 cm (2 in.) thick.

board foot [tim.] The North American unit of *board measure*. The amount of *lumber* in a piece which before sawing, planing, and *shrinkage* measured $30 \times 30 \times 2 \cdot 5$ cm ($12 \times 12 \times 1$ in.). It may well measure 3 mm ($\frac{1}{8}$ in.) short in all three directions. *See* **dressed size.**

boarding or **close boarding** [carp.] Boards closely laid over *rafters* or *studs* to act as a surface for fixing insulation, *cladding*, tiles, slates, *flexible metal* sheet, and so on. *Compare* **weather-boarding.**

boarding joists [carp.] *Common joists.*

board lath [pla.] (USA) (1) Wood *laths.*

(2) *Gypsum plank.*

board measure [tim.] The North American method of measuring timber by the *board foot*, not to be confused with *face measure*.

boasted ashlar or **boasted surface** A rough finish to ashlar made by *boasting*.

boaster or **bolster** or (Scotland) **drove** A mason's *boasting* chisel, 4 to 8 cm ($1\frac{1}{2}$ to 3 in.) wide, which is struck by a *mallet* in dressing the surface of stone, and used after the *claw chisel*. *See* **mason's tools** (illus.).

boasting Surfacing a stone with roughly parallel, oblique, or vertical strokes from a *boaster*, which are not usually uniform nor carried across the face of the stone. There are about nine strokes per inch. *Compare* **batting.**

boat scaffold A *cradle*.

boat spike [carp.] A *ship spike*.

bobbin [plu.] *See* **boxwood bobbin.**

body [pai.] The stiffness of a paint or the solidity of the dried *film*.

bodying in or **bodying up** [pai.] The early stages of French polishing, including staining, filling, and the first polishing before *spiriting off*.

boiled oil [pai.] *Linseed oil* which has been heated (not boiled) for a short time at about 260° C. (500° F.) with soluble lead or manganese *driers*. Air may also be blown through it (blown oil). The drier and the air oxidize the oil, so that paint made with it dries more quickly than with *raw linseed oil*. Boiled oil can be obtained pale or dark in colour. (*See* **pale boiled oil**.) Dark boiled oil contains litharge as drier.

boiler A domestic water heater in which, generally, the water should not boil. It is heated by coal, oil, gas, or electricity. *See* **back boiler,**

boot boiler, electrode, immersion heater, magazine, pot-type boiler. *See also C.*

boiling hole [pla.] A hole in the ground, about 90 cm cube, in which putty is prepared from a magnesian *quicklime*. Excess water should be able to drain away slowly. During slaking, the hole is covered with boarding, and preferably also sacking, to keep heat in. *Compare* **boiling tub, maturing bin.**

boiling tub [pla.] A tub used like a *boiling hole* for making magnesian *lime putty,* and generally preferred to it.

bole [tim.] The tree trunk. *See* **merchantable bole.**

bolster or **corbel piece** or **crown plate** or **head tree** or **saddle** [carp.] (1) A short timber cap over a post to increase the bearing area under a beam.

(2) A bricklayer's *chisel* about 11 cm wide. *See* **boaster.**

bolt (1) [joi.] The tongue of a *lock,* which prevents a door opening. *See* **draw bolt.**

(2) [mech.] *See C.*

bolting iron [joi.] A narrow *chisel* for mortising drawer locks.

bolt shooting *See* **stud shooting.**

bond (1) The laying of bricks or stones regularly in a wall according to a recognized pattern for strength. *English bond* has been used in England since the sixteenth century. Every *cross joint* is about one-quarter of the length of a brick or stone from the nearest cross joint above or below. Bond was essential to brickwork when mortars had little or no adhesion, but now that cement and cement-lime mortars with high strength and *adhesion* are used, it is often unnecessary. Other important bonds are *Flemish, heading, stretching, garden-wall, diagonal, American,* and *English cross.*

(2) The placing of slates or tiles to exclude rain in such a way that the joint between two slates in each course is at or near the centre of the slate or tile in the course below it.

(3) [pla.] *Adhesion* (also called interface strength) resulting from *mechanical bond* (*C*) and *specific adhesion* (*C*).

(4) [tim.] The layer of adhesive in a *plywood* joint.

(5) *See* **lashing,** *also C.*

bond course A course of *headers,* particularly in America.

bonder A *bond stone.*

Bonderizing [mech.] *See* **phosphating.**

bonding *See* **bond.**

bonding brick A specially moulded brick used instead of a *wall tie* for holding together the two leaves of a *cavity wall.* It fits as a *header* into the facing and into one *course* higher in the *backing,* being Z-shaped.

bonding compound In *bitumen-felt* roofing, an oxidized bitumen melted and applied hot to fix layers of felt to the roof and to each other. *Sealing compound* is applied cold.

bonding conductor [elec.] A length of wire or cable which earths cable sheaths or the metal frames of electrical apparatus. It should have very low resistance to earth, so as to ensure that nothing to which it is connected can rise appreciably above earth potential, even if there is a power leakage.

bonding timber Long timbers built into brickwork to strengthen it, an obsolescent practice owing to the danger of the timber rotting. A more effective reinforcement is *hoop iron* or expanded metal built into the *bed joints*.

bonding treatment [pla.] Any pre-treatment for very smooth backgrounds (dense concrete, glazed bricks or tiles or smooth sound paint) that makes plaster stick to them. Bituminous solutions are often used on walls; polyvinyl acetate emulsion or other polymers are used on *soffits* as well as on walls and for reducing suction. *See* **concrete-bonding plaster.**

bond stone or **bonder** or **through stone** A long stone laid in a wall as a header, and seen on one face, or in a thin wall on both faces. In thick walls it is best for bond stones not to be of the full thickness of the wall but about two-thirds of it, and to be laid alternately from opposite faces to cross in the middle. The number required can be estimated from the area of the headers on the face, which should be about one-eighth to one-quarter of the total surface area. When the facing is a *veneer* held by metal cramps, no bonders are needed, nor are they practicable.

bone black [pai.] A black *pigment* consisting of pure carbon. It is extremely fine grained, and, like *lamp black*, after screening only 5% of it is left on a mesh with 0·06 mm square openings. It is made from charred bones, ivory chips, etc., and so is called ivory black or drop black according to the grade. *See also* **carbon black.**

boning pegs Small hardwood cubes placed at the corners of large stones to help in dressing their faces truly plane.

bonnet (1) A roof over a *bay window*.

(2) A wire netting sphere inserted into the top of a vent stack or chimney to prevent birds coming in and sparks going out.

bonnet tile or **bonnet hip** or **cone tile** A *hip tile* with a rounded top. It is bedded in mortar on the next bonnet tile.

bonus system of wages An *incentive system* of payment where direct piece rates are not used. Men receive a day-wage which is increased by an additional payment, the bonus, if a set quantity of work is achieved.

book matching or **herring-bone matching** [tim.] A way of placing successive sheets of *veneer* sliced from the same *flitch*, so that alternate sheets are placed face up and face down as they come from the knife. If the veneers are laid side by side, each sheet will be a mirror image of its neighbour.

booster heater An extra electric water heater which raises the water temperature in one part of the system.

boot A projection from a concrete beam or floor slab to carry the *facing brickwork*.

boot boiler A small hot-water heater at the back of a domestic fire, a modification of the *back boiler*, L-shaped to increase its heating surface.

borrowed light A window in an internal wall or partition (BS 565).

boss (1) A rounded projection down from a *ceiling*, often at an intersection of ribs, and carved if of wood or stone.

(2) A keystone to a dome.

(3) [plu.] A boxwood cone for opening lead pipes.

bossage Roughly dressed *ashlars* projecting from a wall, carved after they are built in.

bossing [plu.] The shaping of sheet lead, zinc, etc., to fit a roof or other shape, with boxwood shapes and a mallet. *Compare* **hammering**.

bossing stick [plu.] A *boxwood* shaper about 30 cm long for shaping a sheet lead lining to a tank, etc.

Boston hip or **Boston ridge** or **shingle ridge finish** (USA) *Weaving*.

bottled gas *Propane* or *butane* (C) compressed and sold liquid in steel cylinders, in districts where piped gas is not available. Since its *calorific value* (C) per cubic metre is more than that of piped gas, the distribution pipes may be much smaller. Rubber pipes should not be used, copper or plastics tubes being preferable.

bottle-nose drip or **bottle-nose curb** [plu.] A rounded edge to a *drip* on a sheet lead roof.

bottle nosing [joi.] (Scotland) A half-round *nosing*.

bottom glazing flashing In corrugated asbestos roofing, an accessory below *patent glazing*, which receives a *flexible-metal flashing*.

bottomless hole A hole which passes through a material and is therefore difficult to plug except with a *screw anchor* or similar fixing. *Compare* **blind hole**.

bottom rail [joi.] The horizontal bottom member of a door, casement, or lower sash.

bottom shore [carp.] In *raking shores* holding up a building, the bottom shore is the shore next to the building.

boule [tim.] A *hardwood* log, *flat-sawn* and reassembled into the shape of the log.

bow or **camber** [tim.] A warping of timber at right angles to its face. *Compare* **cup, spring, twist**.

bow drill or **fiddle drill** [carp.] A primitive boring device in which the *bit* is rotated by the grip of a cord attached to a bow pulled to and fro like a saw.

bowled floor (USA) A floor sloping down towards one end, at about 1 in 25, as in a theatre or church.

bow saw or **turning saw** [carp.] A saw consisting of a removable blade held in a wooden frame which is tightened by twisting a string. It is used for sawing round curves.

bow window A *bay window* which is curved in plan.

box casing [joi.] (USA) *Inside lining*.

box cornice or **closed cornice** [carp.] A hollow *cornice*, built up from wood, and used on American houses. The box shape is enclosed by gutter or shingles above, *fascia* in front, the *planceer* below, and the wall of the building behind. Unlike the *open cornice*, the *soffits* of the *rafters* are hidden.

boxed frame or **boxing** or **box frame** [joi.] A *cased frame*.

boxed heart [tim.] Timber sawn so that the *pith* is cut out, usually within a 10 cm. square. This is done with some Australian and other *hardwoods* which have poor heart.

box gutter or **trough gutter** or **parallel gutter** A wooden gutter of rectangular cross section with *flexible metal*, asphalt, or *roofing felt* lining. It is used in *valleys* or behind *parapets*.

box-head window (USA) A *window* built with a wide slot in the *head*, through which one or both *sashes* can pass, to increase the available ventilation opening.

boxing (1) [joi.] A *cased frame*.

    (2) [joi.] The part of a window recess hollowed out to take a *boxing shutter*.

    (3) [carp.] (USA) *Sheathing* for buildings.

boxing shutters or **folding shutters** [joi.] *Shutters* inside a window which fold away into a recess at the side, the *boxing*.

boxing up [carp.] (USA) Nailing sheathing to studs, or otherwise encasing something with timber.

box stair (USA) *Closed stair*.

box staple or (USA) **box strike** [joi.] The metal box on a door post into which the latch of a *rim lock* passes when the door is shut. *See* **striking plate**.

box up *See* **boxing up**.

boxwood [tim.] A hard hardwood used for making chisel handles, chessmen, *boxwood dressers*, etc.

boxwood bobbin [plu.] An egg-shaped tool drilled through the centre and used for truing bends in lead pipes. It is threaded on a strong cord with a brass follower behind it, and pulled through the pipe until the distortions are removed. The follower is used as a sort of hammer, being pulled back and forth until the bobbin passes the obstacle.

boxwood dresser [plu.] A tool for straightening lead sheet and pipe. It may be of some *hardwood* cheaper than boxwood, such as hornbeam. *See* **bossing stick**.

boxwood tampin or **turning pin** [plu.] A conical *hardwood* tool used by plumbers for opening out the end of a lead pipe.

brace or **bit-stock** or **carpenter's brace** [carp.] A cranked tool used for turning a drilling *bit* and thus making holes in wood. *See* **carpenter's tools** (illus.).

brace blocks [carp.] In built-up beams, wooden *keys* which prevent one part of the beam sliding relatively to the other.

braced frame [carp.] A wood building frame consisting of widely spaced, heavy corner posts into which *binders* or girders are framed. The studs between the posts carry no floor load, as they do in *balloon framing*. The corner posts rise to the roof, being framed into each floor or ceiling as they pass it. Some bracing is needed in the frame, though it is appreciably stiffer than balloon framing. The term is also used for any *frame construction* intermediate between this and balloon framing.

brace extension [carp.] A *bit extension*.

brace jaws or **brace chuck** [carp.] The clamp for a *bit* in a *brace*.

bracket (1) A projecting support, of any material, in *masonry* or brickwork called a *corbel*, in metal or reinforced-concrete construction, a *cantilever* (*C*).

(2) [carp.] A short vertical board fixed to the *carriage* of a *stair* to support the *tread* directly.

bracket baluster A steel *baluster* bent to a right angle at its foot and built into the side of stone or concrete stairs, common in France.

bracketed cornice [pla.] A plaster *cornice*, on brackets carried from the wall or ceiling.

bracketed stairs [joi.] Stairs carried on a *cut string*. The overhanging *nosings* are usually ornamented.

bracketing [pla.] *Cradling* consisting of shaped wooden *brackets* which carry the lathing for the plaster of cornices. *See* **Scotch bracketing**.

bracket scaffold A light *scaffold* for men repairing a wall. A *grappler* is driven firmly into a brick joint and the steel-framed bracket hooked over it.

brad [carp.] or **floor brad** (1) A *cut nail* of constant thickness but tapering width, with a square head projecting on one edge only. Used for fixing floorboards, they are 5 to 6·2 cm (2 to 2½ in.) long.

(2) An *oval-wire brad*.

bradawl [joi.] A short *awl* with a narrow *chisel* point pushed in to make holes for nails and screws.

brad setter or **brad punch** [joi.] (USA) A *nail punch*.

branch [plu.] An inlet or outlet from a *run* (3) of pipe.

branch circuit or **final sub-circuit** [elec.] The conductors installed between the wall *plug* or other *point* and the fuse which protects them.

branched knot [tim.] Two or more knots branching from a common axis (BS 565).

brander or **counter-lath** [pla.] A fillet from 19 × 19 mm (¾ × ¾ in.) to 5 × 2·5 cm (2 × 1 in.) nailed at 30 to 38 cm (12 to 15 in.) centres on to the *soffit* of joists. Ceiling laths are nailed to the branders.

brandering or **counter-lathing** [pla.] When floor joists are over 8 cm (3 in.) wide, it may be difficult to get a good *key* on laths nailed directly to them. Branders are therefore nailed to the joists, and the plastering laths nailed to the branders.

brash or **brashy** A description of timber which breaks with small resistance to shock and little or no splintering. This so-called 'short grain' may be due to fungus.

brazier An iron basket or old oil drum which contains burning coal or coke. It was once used in Britain by the men to make tea and sometimes to fry sausages on a shovel.

brazing [mech.] An ancient method of *capillary jointing* of metals with a film of copper–zinc alloy (hard *solder*) between the red-hot contact surfaces. The method is used for brass, copper, steel, and cast iron in any combination, using a flux such as borax. Unlike *bronze welding* (*C*), any method of heating may be used, though gas is usual. For copper tube, *silver brazing* was introduced to British plumbers in 1959, and seemed to be highly promising for all diameters, at least up to 10 cm (4 in.).

break (1) A change in direction of a wall face.

(2) An overlap at a *purlin* between successive tiers of *patent glazing*.

breaking down [tim.] *Conversion*.

break iron [joi.] (1) A *back iron*.

(2) A *dressing iron*.

break joint or **breaking joint** or **staggering joints** The locating of structural joints, such as *veneer* ends, mortar *cross joints*, or the ends of wooden *lath*, or reinforcement in concrete slabs, out of line with each other, so that each joint is overlapped sufficiently by other laths, bricks, or sheets of veneer.

breast (1) A projection into a room, containing the flue and hearth of a fireplace.

(2) [joi.] (Scotland) A *riser* of a *stair*.

(3) The wall under the *sill* of a *window*, down to floor level.

breast drill [mech.] A drill which, like the *brace*, is held with both hands. Unlike the brace, it is provided with an upward extension, topped by a flat bar on which the user can bear the weight of his chest to increase his drilling speed in metal. Breast drills can usually take *twist drills* of 12 mm dia. The bit is turned by a handle through bevel gears, often with a choice of two speeds. The *hand drill* is a small breast drill without the bearing for the chest.

breast lining [joi.] Wooden panelling inside a window below the *window board*.

breast wall A breast-high, *retaining wall* (*C*) for earth.

breeze Small coke which, before 1939, may have been used in Britain to make *building blocks*, but is no longer used for this purpose.

breeze blocks (1) *Building blocks* of *coke breeze*, no longer made in Britain.

(2) A misnomer commonly used for *clinker blocks*. *See* **pan breeze**.

bressummer or **breastsummer** A long heavy beam, usually timber, carrying a considerable load of brickwork or *masonry*, often placed over a shop window. It is a large *lintel*. Steel or concrete girders are now more usual.

brick In temperate climates, bricks are made of clay with some coarser material such as silt or sand. They are burnt, not baked, in a kiln and slightly fused. This fusion can be seen from the surface glaze of *hard-burnt* bricks. In hot climates bricks are often sun-dried (baked) from *adobe* with straw to bind it (as in the Bible). In Roman Britain, bricks varied from 15×15 to 45×30 cm (6×6 to 18×12 in.). From 1936 until metrication, the standard British brick had an actual size of $8\frac{3}{4} \times 4\frac{3}{16} \times 2\frac{5}{8}$ in. ($222 \times 106 \times 67$ mm) making a *brick format* of $9 \times 4\frac{1}{2} \times 3$ in. ($228 \times 114 \times 76$ mm). Four courses made 12 in. height, 30·5 cm instead of the present 30·0 cm. American sizes are less uniform than the British. The 'standard modular brick' measures $190 \times 89 \times 55$ mm ($7\frac{1}{2} \times 3\frac{1}{2} \times 2\frac{1}{6}$ in.), lays up three courses to 8 in. (203 mm) and has a brick format in plan of $203 \times 101·5$ mm (8×4 in.) including the half-joint each side. The 'standard common brick' is about $190 \times 95 \times 57$ mm ($7\frac{1}{2} \times 3\frac{3}{4} \times 2\frac{1}{4}$ in. high) but many other sizes exist. *See* **concrete, economy, engineered, engineering, moulded, Norman, pressed, Roman, sand-lime, solid masonry unit, wirecut brick, tolerance.**

brick-and-a-half-wall In Britain a brick wall 34 cm ($13\frac{1}{4}$ in.) thick.

brick and brick (USA) *Gauged brickwork*.

brick and stud *See* **brick nogging.**

brick axe A *bricklayer's hammer*.

brick clay *Brick earth*.

brick construction Building with *loadbearing* (*C*) brick walls.

brick core Brick filling under the *soffit* of a *relieving arch*, or hidden behind the lintel of a *flat arch*.

brick definitions BS 3921:1969, part 2, defines four types of brick. 'Cellular' bricks have holes closed at one end (*frogs*) exceeding 20 per cent of the brick volume. 'Solid' bricks have a frog not exceeding 20 per cent of the volume, and any small holes passing right through the brick do not exceed 25 per cent of the volume. 'Perforated' bricks have small through holes that exceed 25 per cent of the volume. 'Hollow' bricks have through holes that exceed 25 per cent of the volume and the holes are not small. 'Small' holes are less than 2 cm across or 5 cm^2 in area. A *block* is larger in any dimension than a brick.

brick earth or **brick clay** Any earth which is good for making bricks.

brick elevator A small endless-chain elevator for raising building materials on to a *scaffold*. Like the *portable belt conveyor* it is easily moved.

brick format The dimensions of a brick plus one half mortar joint each side (10 mm). Thus with bricks of $215 \times 102·5 \times 65$ mm the brick format will be $225 \times 112·5 \times 75$ mm ($9 \times 4\frac{1}{2} \times 3$ in.). Four units long make 90 cm and four units high make 30 cm. Other formats exist, from $300 \times 100 \times 100$ mm ($12 \times 4 \times 4$ in.) to $200 \times 100 \times 75$ ($8 \times 4 \times 3$ in.).

bricking (1) Laying bricks.

(2) Imitation brick on plastered surfaces.

bricklayer A *tradesman* who builds and repairs brickwork, lays and joints salt-glazed *stoneware* drains, sets chimney pots, manhole frames, and fireplaces, and renders brickwork, including the insides of manholes. A sewer and tunnel bricklayer is a specialized bricklayer. In some districts, bricklayers also fix wall and *flooring tiles*, and slating, and lay plaster and granolithic floors, but elsewhere these are plasterer's specialities. *See* **mason**.

bricklayer in firebrick A *bricklayer* specialized in setting firebrick or refractory blocks in fireclay or other *refractory mortar*, in thin joints not more than 3 mm. thick.

bricklayer's hammer or **brick axe** or (USA) **ax** or **axhammer** A small *hammer* with a sharp *cross peen* as well as a striking face, used for dressing bricks. *Compare* **scutch**. *See* **mason's tools** (illus.).

bricklayer's labourer A skilled *labourer* who mixes *mortar* and carries it either on a hod (hod carrier), on a *head board* (tupper), or in a wheelbarrow to the bricklayer. Before bricklaying on the site has begun, he helps digging, concreting, drainlaying, and shoring. He may own his own shovel and hod and eventually become a *builder's labourer*.

bricklayer's scaffold A *scaffold* supported by *putlogs*, of which one end is left in holes in the brickwork, the other end being carried on *ledgers* held by the *standards*. *Compare* **mason's scaffold**.

brick mason (USA) A *bricklayer*.

brick masonry (USA) *Brickwork*.

brick nogging or **bricknogging** Brickwork infilling between the studs of a wooden *framed partition* or building frame. *See* **nogging**.

brick-on-edge coping A *coping* of *headers* laid on edge.

brick-on-edge sill A door or window *sill* of *headers* laid on edge.

brick set (USA) A *bolster* for cutting bricks.

brick trimmer A *trimmer arch* made of brick.

brick trowel A bricklayer's large triangular trowel for spreading mortar, with a blade about 28 cm (11 in.) long.

brick truck A modern barrow with one or two wheels, pushed by one man and used for moving bricks from place to place without re-stacking them each time (a necessity with the ordinary *wheelbarrow*).

brick veneer An outer covering, usually to a timber house, consisting of a half-brick wall, used in North America and Australia where timber is plentiful. *See* **veneered wall**.

brickwork (1) *Bricks* built into a wall or other structure. *See* BSCP 121.

(2) The art of *bonding* bricks effectively; in USA known as brick *masonry*. *See* **bond**.

brickyard A works where bricks are made and burnt, usually near a clay pit.

bridge board [carp.] A *cut string*.

bridge stone A flat stone spanning a gap.

bridging [carp.] (1) The spanning of a gap with *common joists*.

(2) The stiffening of adjacent common joists against each other by a row of *solid* or *herring-bone strutting* against the joists, under the floorboards, at the midspan of the joists. This ensures that the joists deflect together when one is heavily loaded, and increases the effectiveness of the floor, since the load on each joist is shared by its neighbours. Because of the possibility of confusion between these two meanings, it might be better to use only the word strutting for this sense. *See* **double-bridging**.

(3) [pai.] The covering over of a gap in a *ground* by the film. Bridging by a *paint* or *varnish* weakens the *film*.

bridging floor [carp.] A floor carried on *common joists* only.

bridging joist [carp.] A *common joist*.

bridging piece [carp.] A short bearer either between or across *common joists* to carry a *partition*.

bridle [carp.] (Scotland) A *trimmer joist*.

bridle joint [carp.] *See* **birdsmouth**. A development of the *mortise-and-tenon* joint for heavy framing. The tenoned member is usually a post, cut at the sides to leave a central tongue which enters a notch or mortise in the upper member (usually a beam carried by the post.

bright [tim.] A description of freshly sawn timber without discoloration.

brilliance [pai.] (1) The clearness of a *varnish* or *lacquer*, the absence of opalescence and similar defects.

(2) The cleanness and brightness of a *colour*.

brilliant cutting Cutting a design on *glass* by pushing it down on a revolving sandstone wheel.

brindled bricks Bricks which have a striped surface and are therefore unsuitable for *facing* but are otherwise perfect.

brise-soleil A shield from the sun, used in tropical or Mediterranean countries. It often consists of vertical or horizontal precast concrete strips which prevent intensively hot, direct sun from entering a room.

bristle brush [pai.] A *brush* made from hog's hair, now (1973) being replaced by synthetic fibres.

British Standard or **British Standard Specification** or **BS** A numbered publication of the *British Standards Institution* describing the quality of a material, or the dimensions of a manufacture, such as pipes or *bricks*. Frequently the dimensions and the quality are described in two separate standards. The use by architects or engineers of British Standards in their *specifications* can reduce the volume of the description to a reference (for steel frames, for example) to BS 449: 1970.

British Standards Institution or **BSI** The British organization for standardizing, by agreement between maker and user, the methods of test and dimensions of materials as well as *codes of practice* and nomenclature. Corresponding organizations in other countries are AFNOR (Association Française de Normalisation); ASTM (American Society for Testing and Materials); ASA (American Standards Association). A German standard number is prefixed by DIN (Deutsche Industrie Norm).

British thermal unit or **Btu** The amount of heat required to warm one pound of water from 39° F. to 40° F. It is equivalent to 0·252 kilogram *calories* and to one hundred-thousandth part of a *therm*. 1 Btu = 1·055 kilojoule and 1 Btu/hour = 0·293 watt.

brittleheart [tim.] Weak brittle wood at the heart which may need to be *boxed*, common in tropical timbers.

brittleness [pai.] A finish is brittle if it cracks when stretched or when a knife or needle is pulled across it. The opposite of flexibility.

broach (1) A pin inside a *lock*. It locates the barrel of a key passing on to it. This sense is related to the old sense (now little used) of a reamer in mechanical engineering.

(2) The *mason*'s pointed *chisel*.

broached work *Punched work*.

broad axe [carp.] A wide-edged axe with a cutting *bevel* on one side only. It was used for log-cabin building in USA, and is still used for rough-dressing timber. The handle may be offset sideways from the blade.

broad tooling *Batting*.

broken-joint tile A *single-lap tile*.

broken-range ashlar (USA) *Uncoursed rubble*.

broken white [pai.] A white which has been toned down, usually to a creamy colour.

bronze An alloy usually of copper and tin without other elements. Aluminium bronze is mainly copper and aluminium.

bronze powder [pai.] *Gold bronze* powder.

bronzing fluid [pai.] A *varnish* or *lacquer* which is used for applying aluminium or gold bronze powders by mixing them in with it.

brooming [pla.] Scratching a floating coat with a broom to make a *key* for plaster. *See also C.*

brotch A *spar* in thatching.

brown coat or **browning coat** [pla.] A *floating coat*.

brown rot [tim.] A *decay* of timber to a brown, soft mass.

Brunswick black [pai.] A *bituminous paint*.

Brunswick green [pai.] *Lead chrome green*.

brush [pai.] A tuft of animal, or artificial (nylon), or vegetable fibres held on to a handle, usually wooden, and used for putting on *paint* or for wetting or dusting down surfaces. Brushes may be of *bristle*, *sable*, badger, squirrel hair, etc. *See* blender, **camel-hair mop, dabber, distemper brush, dusting brush, fitch, flat paint brush, flat wall brush, ground brush, lining tool, mottler, overgrainer, rigger, sable pencil, sable writer, sash tool, softener, stippler.** *See also* **house painter** (illus.) and BS 2992.

brushability [pai.] The ease with which a liquid can be brushed on. Brushable paints are not *gummy* or *ropy* and can be joined easily with paint put on some minutes earlier.

BS *British Standard.*

BSCP British Standard *Code of Practice.*

BSI *British Standards Institution.*

Btu *British thermal unit.*

bubbling [pai.] A defect of films containing very volatile *solvents*. It consists of bubbles of air or solvent vapours which may disappear before the film dries.

buckle A *spar* in thatching.

budget A pocket for carrying nails, used by *slaters and tilers*, and *bricklayers*.

buff To polish or grind down a floor finish of *terrazzo* or screeded material. It is derived from the high-speed buffing wheels of mechanical engineering which were originally made of buffalo leather and are used for polishing with a slight abrasion.

builder's equipment Plant used by builders, from *scaffold boards* to 100-ton *Goliath cranes* (*C*). Also called builder's plant and machinery, a term which has the advantage that it cannot be confused with *building equipment*.

builder's handyman A *jobber*.

builder's labourer A *semi-skilled man* who can mix *mortar* or *concrete*, dig, clean old bricks, load or unload material, place *concrete*, demolish houses, and so on. Like the *jobber*, who does everything inside a house, the builder's labourer does everything outside a house except those trades which are well paid. He may own his *hod*, *pick*, and shovel. *See* **bricklayer's labourer.**

builder's ladder or **pole ladder** A wooden *ladder* with half-round *stiles*. Its rungs are made of *oak*, *ash*, or *hickory*. The stiles may be a *spruce* or *silver fir* pole sawn in two. *Compare* **standing ladder.**

builder's level (1) A *level tube* (*C*) set in a long wooden or *light-alloy straight edge*.

(2) A *dumpy level* (*C*) used on a building site, and therefore sturdy, but not usually very sensitive.

builder's staging or **scaffold** A *mason's scaffold*.

builder's tape A linen *tape* (*C*) or steel tape, often 30 m (100 ft) long, which rolls into a round leather or plastic pocket case.

building blocks Hollow or solid walling blocks of *burnt clay, gypsum, concrete*, or other material, larger than *bricks*, often $46 \times 23 \times 23$ cm ($18 \times 9 \times 9$ in.) and therefore quicker to lay. (*See* **blockwork.**) They may also be used for building *partitions*, in which case they can be as thin as 6·4 cm (2½ in.) hollow or 5 cm (2 in.) solid, and may also be of *diatomite* brick or other insulating material. *See* **hollow blocks, clinker blocks.**

building board A facing to interior walls or *ceilings* or a background to plaster, made of compressed wood pulp, cane fibre, plaster, plastics, paper, etc., sometimes veneered with *plywood*. It is usually about 1 cm (½ in.) thick or less, and sold in panels more than 60 cm (2 ft) wide. *See* **asbestos cement, asbestos wallboard, chipboard, fibre board, plaster board, wallboard.** *Woodwool slab* and *corkboard* are intermediate between *building blocks* and building board.

building brick (USA) A *common brick*.

building certificate *See* **certificate.**

building code (USA) Local building laws corresponding to *by-laws* in Britain. *Compare* **code of practice.**

building component A part of a building too complex to be called a *building element*; for instance, a wall, door, roof, or window, all of which may be built of several elements. The electrical, plumbing, gas or other services also contain building components.

building element An elementary part of a building, only one stage removed from a *building material*. *Bricks*, and *joists* of metal, wood or precast concrete are building elements.

building equipment *Services*, furniture, and other plant used in a completed building. *Compare* **builder's equipment.**

building in Fixing a wall tie, air brick, bracket, or other building part into a wall or other part of a building by bedding it in mortar and laying bricks or stones over it and round it. Building in may be done while the wall is being built, or later, by leaving a hole, or breaking a hole afterwards. The part can then be grouted into the hole or fixed into it with *dry pack* (*C*).

building inspector An employee in Britain of a *Local Authority*, building society, or insurance company, who tells his employers whether a building is built in accordance with *by-laws* or advises them on its rateable value, fire risk, and mortgage value. He needs wide knowledge of building construction, which may be attested by the Building Inspector's Certificate of the Institution of Municipal Engineers. *See* **inspector** (*C*)**, building official.**

building line (1) The line fixed, usually by the *Local Authority*, as a limit to building near a road.

(2) The outside face of the wall of a building, shown, in plan, as a line on a drawing.

building materials The materials of which *building elements* are made, for example, sand, ballast, clay, cement, round timber.

building module A *module* which may be small, such as 75 or 100 mm (3 or 4 in.), or large, such as 900 mm or 1 m (36 or 40 in.).

building official (USA) An employee of a *Local Authority* whose duty it is to enforce a *building code*. He corresponds to the British *building inspector* or to other employees of the Local Authority.

building owner The owner of the works to be built on a site; he may be an individual, an organization, a government, etc. *See* **client**.

building paper Fibre-reinforced bitumen between layers of kraft paper, laid under road slabs to prevent loss of cement into the earth and damage to the road slab by acid in the soil. It is also used for many other purposes – over the *sheathing* of a wall or roof, for example. In Britain it is made in rolls up to 1·8 m (6 ft) wide and 231 m (250 yd) long. It may also be called roofing paper, sheathing paper, Willesden paper, waterproof paper, concreting paper, etc., some of which may be less strong or less waterproof than that described above.

Building Research Digest An inexpensive leaflet stating the conclusions of the *Building Research Establishment* about a building topic. They are issued at frequent intervals and are both authoritative and useful, covering hundreds of subjects from 'Co-ordination of building colours' to 'Materials for making concrete'. The full title is *Building Research Establishment Digest*.

Building Research Establishment (BRE) The BRE of the British Department of the Environment was created in 1971 by the fusion of four pre-existing government bodies, the Building Research Station (Watford, Herts), the BRE Scottish Laboratory (East Kilbride, Glasgow), the Fire Research Station (Boreham Wood, Herts), and the former Forest Products Research Laboratory (Princes Risborough, Bucks). One example of the international reputation of the BRE is its work on *single-stack* drainage, now beginning to be applied in USA.

Building Regulations British regulations that supersede the *model by-laws* and were first put into operation in February 1966. Metrication and the issue of many amendments made it advisable to issue an edition in 1972. This edition, with its tendency in favour of metrication, is likely to stay in force for some years.

building society Organization for financing building, in Britain usually backed by an insurance company. Most private building is financed by these societies. The society is repaid by the buyer in the form of rent, usually for not less than twenty years, a large part of the rent being interest on the *mortgage*. The buyer must make a down payment of at least 10% of the value of the house, often 25%. *Co-operative housing societies* are very much cheaper in rent and first payment, but demand some physical work or other co-operative effort from the co-operative buyers.

building surveyor A specialist in building construction and repairs and alterations to buildings. He supervises building, decorating, and sanitary work, advises in disputes on party walls, *easements*, light, and other legal matters. He is usually a *quantity surveyor* by training, being an Associate of the Royal Institution of Chartered Surveyors, or of the Incorporated Association of Architects and Surveyors.

building system An arrangement of the *building elements* to form a connected whole. *Box frame construction* (*C*) is one building system, *loadbearing walls* (*C*) with timber floors and roofs another.

building trade One of or all the occupations of *tradesmen* in building.

built in *See* building in.

built-up [carp.] A timber piece built of several smaller ones, usually glued, but sometimes screwed, nailed, bolted, or *fished* (*C*). *See* compound beam.

built-up roofing Two or more layers of *bitumen felt*, laid to break joint and jointed with *bonding* or sealing compound. This roof covering is cheap, lightweight, and popular, but its life is not exactly known, as it has not been in use for long. Guarantees for 3-ply roofing are often given for twenty years and for 2-ply roofing for ten years, therefore they may be anticipated to last generally for thirty and fifteen years each. *See* roof decking.

bulkhead A box-shape built over a *concrete roof* (*C*) slab (usually itself roofed with a concrete slab) to cover a water tank, lift shaft, stair well, etc. *Compare* penthouse.

bulldog clip A *floor clip*.

bulldog plate or **toothed plate** [carp.] A timber *connector*.

bullet catch or **bullet bolt** A *ball catch*.

bull header A *header* with the upper exposed arris rounded, used in brick window *sills*.

bullhead tee [plu.] (USA) A tee in steam and water fitting, in which the branch is longer than the *run*.

bullnose The rounding of an *arris*; in general, any rounded end or edge of a brick, a step, a joiner's plane, etc.

bull's eye A small, circular or oval, window or opening.

bull stretcher A *stretcher* with an exposed *arris* rounded.

bunched cables [elec.] Cables, of which several are in one conduit or groove.

bungalow siding [carp.] (USA) *Clapboard* which is 20 cm (8 in.) or more wide.

burl [tim.] *Burr*.

burlap A term, used most in USA, for *hessian* canvas. As a reinforcement for plaster it is cheaper than *metal lathing*, but also very much weaker.

burning in [plu.] The fixing by a *plumber* of a lead *flashing* into a dovetail groove (*raglet*) in stone by filling the groove with molten lead.

burning off [pai.] Removing old paint by treating it with a *blow lamp* and scraping it off while hot.

burnisher [mech.] A hard-steel tool, often made from a worn-out file smoothed down. It gives a final smoothing and polishing to the cutting edge of a tool.

burnt clay Clay burnt in a kiln to make bricks, tiles, earthenware pipes, etc. For the distinction from baked clay, *see* **adobe, ceramics.**

burnt lime *Quicklime*, calcium oxide (CaO).

burnt sienna or **burnt umber** [pai.] *See* **sienna, umber.**

burr (1) or **burl** [tim.] The curly, much-valued *figure*, got by cutting through the enlarged trunk of certain trees, particularly walnut. It is formed of the dark *pith* centres of many undeveloped buds.

(2) *See below*.

burring reamer [plu.] A tool turned in a brace so as to remove the burr left by a *pipe cutter* inside a pipe.

bus bar [elec.] A bare copper or aluminium *conductor*, usually fixed on a slate slab, or wall, parallel to the other bars of its *circuit*, used for carrying a heavy current.

bushing (1) [plu.] A screwed pipe fitting which connects two others of different diameters. It is threaded inside and outside, and, being short, is convenient to use where there is no space to insert a *taper pipe*.

(2) [elec.] An insulated tube, sometimes of porcelain, which protects cables. *See* **conduit bushing.**

butane *See* **propane blowlamp, bottled gas.**

butt (1) To meet without overlapping.

(2) [tim.] The thick end of a *shingle* or a log, generally the thicker end of anything.

(3) [joi.] A *butt hinge*.

butterfly wall tie A *wall tie* made from galvanized steel wire about 3 mm ($\frac{1}{8}$ in.) dia., bent to the form of a figure 8. These ties are used in house building in Britain, since they transmit less sound than the stronger, stiffer ties of steel strip. *See* **cavity wall.**

buttering The spreading of *mortar* on a vertical face of a brick before laying, usually to form a *cross joint*.

buttering trowel A trowel smaller than the *brick trowel*.

butt gauge [joi.] A *marking gauge*.

butt hinge or **butt** [joi.] The commonest *hinge* for doors. When the door is shut, the two halves are folded together, one half being on the *door frame*, the other on the *hanging stile*. The ordinary steel butt hinge is very cheap and durable, but ball-bearing butts can be obtained which are smoother running, noiseless, longer lasting, and much dearer.

butt joint (1) [carp.] A joint between two pieces of wood (or metal) which meet at their ends without overlapping.

(2) [tim.] In veneering, a joint perpendicular to the grain, as opposed to an *edge joint*.

button [joi.] A small piece of wood or metal held loosely by a screw so that it can turn and be used for holding a cupboard door shut.

button-headed screw or **half-round screw** [joi.] A *screw* with a hemispherical head.

butt stile [joi.] The *hanging stile* of a door.

butt veneer [tim.] *Veneer* having a strong curly figure like *crotch*, caused by roots coming into the trunk at all angles.

buzz-saw [tim.] A *circular saw*.

BX cable [elec.] (USA) An *armoured cable*.

by-laws Regulations governing building in Britain, made by *Local Authorities*, called building codes in USA. Building by-laws are now replaced by the *Building Regulations*.

by-pass [plu.] An arrangement of pipes (or conduits) for directing flow around instead of through a certain pipe or conduit.

C

cabin The hut at a building site where a clerk of works or foreman keeps his drawings.

cabinet file [joi.] A smooth, half-round, single-cut file which enables a joiner to form a smooth finish on a joint. *See* **cut** (*C*).

cabinet finish [joi.] A varnished or polished finish on *hardwood*, like that on good furniture, meaning the same in Britain and USA. *Compare* **carpenter's finish**.

cabinet maker A *joiner* who makes fine furniture and is therefore capable of the finest workmanship.

cabinet scraper or (USA) **scraper plane** [joi.] A flat piece of steel drawn over a wood surface to prepare it for sandpapering by removing plane marks.

cabin hook [joi.] A hooked bar on a cupboard door or window frame which engages in a *screw eye* on the door to hold it shut or open.

cable [elec.] The copper conductors through which an electric lamp or other appliance receives its power. The conductors are separately insulated but laid together within a common insulating sheath nowadays often made of *polyvinyl chloride*. The cable layout is indicated in the wiring diagram.

cabtyre sheathing [elec.] *See* **tough-rubber sheathing (TRS)**.

cadmium plating A *protective finish* for steel articles such as wood *screws*. If *chromated* it takes paint or lacquer better.

caisson or **coffer** A deeply recessed panel in a *soffit*.

caking [pai.] The settling of dense *pigments* into a compact mass which cannot easily be re-dispersed by stirring. A *thixotropic* (*C*) paint rarely suffers from this defect.

calcicosis A disease of the lungs caused by breathing marble dust.

calcium carbonate ($CaCO_3$) The chemical name for chalk, limestone, marble, and so on.

calcium chloride *See* **accelerator**.

calcium hydroxide ($Ca(OH)_2$) *Slaked lime*.

calcium oxide (CaO) *Quicklime*. When water is added, it becomes calcium hydroxide; thus, $CaO + H_2O = Ca(OH)_2$.

calcium silicate The term now preferred to sand-lime in describing bricks made of these materials. *See* **aerated concrete** (*C*).

calcium sulphate ($CaSO_4$) The mineral anhydrite which has the same composition as calcined *gypsum*.

calcium sulphate hemihydrate or **plaster of Paris** or **casting plaster** ($CaSO_4 \cdot \frac{1}{2}H_2O$) The basis of retarded *hemihydrate plasters*.

calk or **caulk** or **tang** A fish-tailed steel bar built into *masonry*. *See* **caulking**.

callus or **rindgall** [tim.] A mass of wood formed by a growing tree over a surface wound.

calorie The quantity of heat required to warm one gram of water from 15° to 16° C. is the gram calorie; that required for one kilogram of water is the kilogram calorie, large calorie, or kilo-calorie. *Compare* **British thermal unit.**

calorifier [plu.] A closed tank, not necessarily for storage, in which water is heated, usually by a submerged coil of pipe with steam or hotter water passing through it. An *indirect cylinder* is one type of calorifier.

cam [mech.] A wheel which is not circular. It bears on a part which is made to move by the variations in the radius of the cam. For example, the *lever cap* fixed to the *back iron* of a steel *plane* holds the back iron in place by the wedging action of a cam. Cams are also used for lifting the valves of internal combustion engines.

camber or **hog** (1) Curvature which is domed to allow water to run off a road, to hide the deflection of a girder, and so on. 'Camber' is used mainly for roads, 'hog' for structures.

(2) [tim.] *See* **bow.**

camber arch or **straight arch** An arch with a level *extrados* and just enough rise in the *intrados*, about 1%, to counteract the appearance of sagging.

camber beam [carp.] (1) A beam cambered on its upper surface.

(2) An old term for the tie-beam of a truss.

camber board A *template* for forming a *camber*.

camber slip or **turning** or **trimming piece** A shaped piece of wood, cambered on its upper surface, used in the *centers* for a flat brick arch to ensure that the arch *soffit* at the midspan is slightly cambered above the springing line.

cambium [tim.] The living material just below the bark, from which new wood grows.

came An H-section strip of lead or of soft copper, shaped to fix each piece of glass to the next one, in *leaded lights* or stained-glass windows.

camel-hair mop [pai.] A *brush* like a *dabber*, made of squirrel's hair.

camp ceiling (Scotland) An attic ceiling with four sloping surfaces.

Canadian latch A *thumb latch*.

Canadian spruce [tim.] (Picea glauca) etc. Several species of *spruce*, *softwoods* exported from Canada, second in quantity after *Douglas fir* alone. It is almost white, with a slight lustre, and is stiff and resilient. It is used for *joinery* and structural *timbers*, including *ladder* stiles.

cant (1) To tilt.

(2) A cant *moulding* is one with flat surfaces and no curves.

(3) Stone built 'on cant' has the natural bedding vertical.

(4) [tim.] To cut *wane* from a log.

cant bay A *bay window* with three straight sides.

cant brick A *splay brick*.

canted Splayed, bevelled, or off-square.

canted wall A wall joining another wall at an angle.

canting strip A *water table*.

canting table [tim.] A *saw bench* with a working surface which can be tilted and thus *rip* at any angle required (bevel cutting).

cant strip (USA) A *tilting fillet*.

canvas Coarse cloth of various sorts, sometimes glued to a plastered wall to make the best backing for paint or paper. *See scrim*.

cap (1) The uppermost, decorative part of a *newel* post.

(2) [plu.] A cover, with internal threads, screwed over the end of a pipe. It thus closes, or caps, the pipe.

cap and lining [plu.] A *union*.

capillary groove or **break** A groove or space left between two surfaces, large enough to prevent capillary movement of water into a building, as in a *water-checked casement*. *See (C)*.

capillary joint or (USA) **sweat joint** [plu.] A strong joint in *light gauge copper tubing* made by inserting one pipe into a fitting so very slightly larger in diameter that the space between the two pipes can accurately be described as a capillary (hair's breadth) space. Molten *solder*, which flows into this space, therefore immediately spreads round the perimeter. Several types of capillary joint exist, of which some have the advantage that they can be made by unskilled, even amateur, labour. They also are neater and smaller than the *compression joint*, but the latter does not need heating. *See copper fittings*.

cap iron [carp.] A *back iron*.

cap lever [carp.] *See cam*.

capping (1) In *flexible metal roofing*, a metal strip covering the wood *roll* either separate from, or welted to the edge of the roofing sheets which are dressed up the sides of the roll.

(2) A metal section held outside some *patent glazing* bars to fix the glass and protect the stem of the bar from the weather.

(3) The crowning part of a screen or panelling which does not reach to the ceiling, such as *dado capping*.

(4) A *coping*.

(5) [plu.] The closing of an end of screwed pipe by a *cap*.

capping plane [joi.] A *plane* which rounds off the top of a handrail.

cap sheet The top layer of *mineral-surfaced bitumen felt* in *built-up roofing*.

capstone A *coping*.

carbonating or **carbonation** The natural, slow, hardening of lime mortars; their conversion into stable *calcium carbonate* by their absorption of carbon dioxide from the air. On the surface of *cast stone*, *crazing* has been caused by the shrinkage which occurs with carbonation, and this can now be eliminated by carbonation during manufacture.

carbon black or **hydrocarbon black** [pai.] A black *pigment*. Chemically, *bone black*, *lamp black*, and vegetable black, being pure carbon, are identical with it, but in Britain carbon black is standardized as an extremely fine powder of which only 0·05% remains on a mesh with 0·06 mm square openings. It is usually made by burning a hydrocarbon gas in insufficient air, or by cracking acetylene gas. Carbon black is very opaque and of high *staining power*, but it absorbs *drier* and therefore slightly retards the drying of *oil paints*.

carcase or **carcass** or **fabric** The loadbearing part of a structure without windows, doors, plaster, or finishes. *See* **carcassing**.

carcassing (1) The work involved in building the *carcase*.

(2) The gas pipes in a building, sometimes also the electrical conduits.

carnauba wax [pai.] A hard wax, obtained from the leaves of the Brazilian carnauba palm, or tree of life. It is used as a polish for wood and in *matt* and stoving varnishes.

carpenter A man who erects wood frames, fits joints, fixes wood floors, *stairs* and window frames, asbestos sheeting and other *wallboard*, and builds or dismantles wood or metal *formwork* (*C*). The two trades of carpenter and joiner were originally the same, and most men can do both, but specialize in one or the other. In USA the term carpenter includes *joiner*. Derived from the French word 'charpente', a wood or metal framework.

carpenter's finish (USA) What in Britain would be called joiner's finish, since it includes all *joinery* but no roughly finished work such as *rafters*.

carpenter's hammer [carp.] A *claw hammer*. Compare **joiner's hammer**.

carpenter's mate *See* **joiner's labourer**.

carpentry The craft of cutting timber to make structural frameworks, the carcassing of wooden buildings. In USA (but not in Britain) carpentry includes *joinery*.

carpet strip [carp.] A strip of wood fixed to the floor beneath a door so that the door shall not have a large gap below when it is closed, and yet shall clear the carpet.

carport (USA) A shelter for a car near a house, usually roofed, but not wholly walled-in, and in this way distinguished from a garage.

carriage [carp.] (1) or **carriage piece** or **rough string** or **bearer** or **stair horse** An inclined timber placed between the two *strings* against the underside of wide *stairs* to support them in the middle.

(2) (USA) A carriage can also be a *string*.

carriage bolt (USA) A *coach bolt*.

carriage piece [carp.] A *carriage*.

carrying capacity [elec.] The current which a *fuse* or cable can carry without being overheated or having too big a voltage drop.

carton pierre [pla.] A material which is used like *fibrous plaster* for making ornamental plasterwork. It consists of paper pulp, boiled

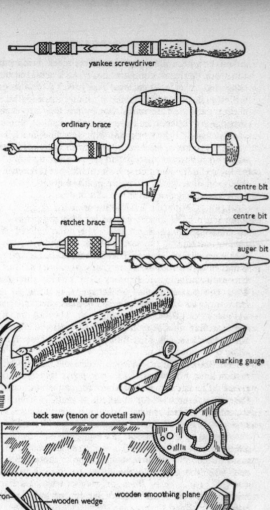

yankee screwdriver

ordinary brace

ratchet brace

centre bit

centre bit

auger bit

claw hammer

marking gauge

back saw (tenon or dovetail saw)

cutting iron — wooden wedge

wooden smoothing plane

cap iron

sole

Carpenter's tools.

and mixed with *whiting* and *size*. This French term means cardboard stone, but the result compares badly with good plaster.

cartridge fuse or (USA) **enclosed fuse** [elec.] A *fuse* sold in a tube of insulating incombustible material which stops any flashing or danger of fire when the fuse melts. Sometimes they are designed so that cartridges of different *carrying capacities* cannot be interchanged.

carved brickwork Brickwork laid with very fine *joints*, about 1·5 mm ($\frac{1}{16}$ in.) like *gauged brickwork*, sometimes much wider than standard *bricks* to allow for material to be carved off. About 1680, much brickwork in the best houses was carved, but this became unfashionable soon afterwards. *Compare* **moulded bricks.**

case (1) The *facing brick* or stone of a building.

(2) [joi.] (Scotland) A *cased frame.*

case bay [carp.] The joists enclosed between two *binders.*

cased frame or **boxed frame** [joi.] The hollow, fixed part of a *sash window*, containing the sash weights and pulleys. It is bounded by visible boards called the *outside lining* and *inside lining.*

case hardening [tim.] During seasoning, the outer skin of timber often dries out, shrinks, and hardens more quickly than the inner part. When this occurs, *checks* or warping may occur as the inner part dries out, and the timber is said to be case-hardened. *See also* C.

casein glues [joi.] *Glues* made from milk. They are much more water-resistant than animal or fish glue. They abrade edge tools, and are not always proof against moulds, but they have fairly good *gap-filling* properties and can be used where the *glue line* may be up to 0·8 mm (0·03 in.) thick. Casein is water-soluble before it hardens, and is sometimes mixed with lime to improve its resistance to bacterial attack. It is often used as a *binder* for paints over cement and lime. Casein has been used for hundreds of years by cabinet makers.

casement The hinged or fixed sash of a *casement window* (BS 565).

casement door A glazed door or pair of doors also called French door or window, or glazed door. *See* **espagnolette.**

casement stay A bar used for fixing open a casement. *See* **peg stay.**

casement window A window in which one or more lights are hinged to open (BS 565). Generally the hinges are vertical, like door hinges.

casing [joi.] (1) A *cased frame* or wooden trim or a finishing board, from which comes the sense of the timber lining round a wooden stair.

(2) A *lining* (1) or window frame of light timber.

(3) A timber or similar enclosure on the face of a wall, floor, or ceiling, made to accommodate pipes or cables in a *chase* or other duct. *See also* C.

cassava glue [tim.] A starch *glue* made from the tapioca plant, little used in Britain.

casting resin or **cast resin** A *synthetic resin* in which shapes can be cast, often a *phenol* or *epoxide* resin.

cast stone or **reconstructed stone** or **patent stone** Imitation stone for facing buildings, now made from a concrete core poured on to a crushed-stone *face-mix* about 2·5 cm thick, laid first in the mould. Examples of cast stone can be seen in the fortifications of Carcassonne, repaired in 1138. The technique was only recently revived. *See* **carbonation**. *Compare* **artificial stone**.

Catalloy *See* **glassfibre reinforced resin**.

catalyst A substance which increases the speed of a chemical reaction, generally without taking part in it itself; for example, an *accelerator* for synthetic resins.

catch bolt [joi.] A spring-loaded door *lock* which is normally extended and locked, but is drawn in when the door is shut or opened.

cat eye [tim.] (USA) A *pin knot*.

cathedral glass Rolled *glass*, slightly obscured by a surface pattern such as hammering or fluting.

cat ladder or **duck board** A ladder, or board with cleats nailed on it, laid over a roof slope to protect it and give access for workmen to repair the roof.

cat walk A gangway round the upper outside walls of high buildings, such as rolling mills, giving access to the roof and eaves for painting or other purposes.

caul [tim.] A sheet of 1·5 mm ($\frac{1}{16}$ in.) thick aluminium in *hot pressing*, or 6 to 9 mm ($\frac{1}{4}$ to $\frac{3}{8}$ in.) thick *plywood* in cold pressing, to protect *veneers* from contact with the presses.

caulking (1) [carp.] *See* **cocking**.

(2) Splitting and twisting the ends of a metal bar to increase its adhesion to mortar when built in.

(3) [plu.] Making a tight spigot-and-socket joint with lead wool or other material driven in by a *caulking iron*. *See also* **C**.

caulking gun or **pressure gun** An injecting tool for sealing joints with *mastic*, for instance in *patent glazing*.

caulking iron or **caulking chisel** [plu.] *See* **plumber's tools** (illus.).

cavil or **kevel** or **jedding axe** An axe with a pointed *peen* and an axe blade for cutting stone. It may weigh up to 8 kg (18 lb), but is usually much smaller.

cavity block Precast concrete blocks, shaped so that if laid over each other they form a *cavity wall*. *See* **V-brick**.

cavity filling Increasing the insulation value of a cavity wall by filling its 5 cm (2 in.) gap with *mineral wool* or granules of expanded plastics or plastics foam that solidifies inside, or other material. The insulation is blown into the gap via holes drilled through the outer 11 cm ($4\frac{1}{4}$ in.) leaf of brickwork. *See* **loose-fill insulation**.

cavity flashing A *damp course* which crosses the gap of a *cavity wall*.

cavity wall or **hollow wall** A very popular wall in Britain, usually built of two 11·2 cm ($4\frac{1}{2}$ in.) thick leaves separated by a 5 cm (2 in.) continuous gap. The two leaves are connected by *wall ties* at a spacing of 2 per m². This wall is very dry and warm, its *U-value* being 1·93 W/m² deg. C. (0·34 Btu/ft² h deg. F.). It is usually 28 cm (11 in.) thick but occasionally 39 cm (15½ in.), in which case the inner leaf is the thicker, 23 cm (9 in.), since it carries the floor load.

cedar [tim.] *See* **western red cedar.**

ceil To cover the *soffit* of a room or other space with plaster, board, or other finishing material.

ceiling (1) A plastered or panelled or boarded upper surface to a space.

(2) [tim.] (USA) Matched boards with beads grooved in them, used as a ceiling covering.

ceiling floor [carp.] (USA) *Ceiling joists* or other supports for a ceiling.

ceiling joist [carp.] A joist which carries the *ceiling* beneath it but not the floor over it.

ceiling strap [carp.] A strip of wood nailed to *rafters* or floor joists for suspending ceiling joists.

cellar A room or rooms, of which more than half is below ground level, usually reserved for storage or for the central-heating boiler, whereas a *basement* is living space.

cellular brick A brick with deep *frogs* or other indentations that amount to more than 20 per cent of its volume. *See* **brick definitions.**

cellular concrete *Aerated concrete.*

cellular or **foamed plastics** There are many of these useful, ultra light-weight insulators, like *expanded polystyrene.*

cellulose acetate A non-flammable material which can be an excellent insulator over ceiling linings when appropriately laminated. It has a *conductivity* of 0·04 W/m deg. C. (0·28 Btu in./ft² h deg. F.). It is not porous and will not rot.

cellulose enamel *See* **lacquer.**

cellulose nitrate *Nitrocellulose.*

cellulose sheet A sheet floor finish made from a mixture of cork dust, sawdust, wood flour, and pigments with gelatinized *nitrocellulose* on a backing of woven jute. It is suitable for restaurants, showrooms, and domestic floors.

cement (1) The bond or matrix between the particles in a rock, particularly that binding the sand grains in a sandstone.

(2) A *binder*, such as *Portland cement*, which binds aggregates into a hard *concrete* or *mortar* within a few days.

(3) [tim.] A term occasionally used for glue.

(4) *See* **Perspex.**

cement-coated nails *Nails* which have been coated with cement are used in USA for fixing *parquet floors* and for nailing green timber because they have high holding power.

cement fillet or **weather fillet** or **mortar fillet** A triangular length of *mortar* which fills in a corner between slates and a wall, or other similar corners, to weather them or keep them clean. *See* **fillet.**

cement mortar Mortar composed of four (or fewer) parts of sand to one of cement, with a suitable amount of water, and either lime or *plasticizer*. *See* **lime mortar.**

cement paint (1) A mixture of cement and water which applied to concrete, masonry, or brickwork makes it waterproof. Ordinary cement is not very pleasantly coloured, but it is cheap. Coloured cement paints are very much dearer.

(2) A paint based on *tung oil* or *casein* or other *alkali-resistant* material which can be used over cement.

cement rendering [pla.] The covering of surfaces with a mix of *Portland cement* and sand. As weather-resistant surfaces, these have failed badly in the past by shrinkage and cracking but a 1:2:9 cement: lime:sand mix is now recommended. Neat cement:sand renderings are still used inside manholes and in permanently wet positions. *See* **stucco, coarse stuff.**

cement-rubber latex A *jointless floor* which is also flexible, thanks to its rubber content. It also contains a setting agent, *Portland cement* or a similar material, and aggregate. The *aggregate* may be marble or other stone chippings or cork or wood chips. A floor made of this can be ground and buffed like a *terrazzo*. It can be as thin as 6 mm (¼ in.). The highly adhesive mixture of cement and rubber is also used as a protective coating over steel reinforcement used in *aerated concrete*. *See* **elastomer, sponge-backed rubber.**

cement screed A *screed* of *cement mortar* laid on a floor, particularly on a concrete slab.

cement slurry A liquid cement-water mix for injection, or used as a wash over a wall.

cement-wood floor A *jointless floor* made of *Portland cement*, sawdust, sand, and pigment. Laid by specialists, it can make a non-dusting quiet floor suitable for offices or living-rooms, though it is unsuitable for bathrooms, kitchens, or workshops.

centers or **centering (centres, centring)** Curved temporary supports, usually wooden, for an arch or dome during casting or laying.

Central American mahogany [tim.] *See* **Honduras mahogany.**

central heating [plu.] Heating of building space by hot water or steam which circulates through the building in pipes, or by warm air which circulates through it in ducts. Warm-air heating began to be widely used in the 1950s (earlier in north America) and the method is quickly becoming popular. It is perhaps more logical than piped systems and is certainly faster in its starting effect because much of the heat from pipes goes to parts of the building, not to the human beings who need it. *See* **small-bore,** *also* BSCP 3006.

centre *Centers.*

centre bit [carp.] A simple bit for drilling with the *brace*. It has a central plain or gimlet point and two side cutters, one of which, the nicker, cuts a circle, while the other, the *router*, ploughs under the wood which has been cut round by the nicker and lifts it out. This simple, cheap bit is, for deep holes, replaced by the *auger bit*.

centre flower [pla.] A *centre piece*.

centre nailing Nailing slates at a point just above their middle, a fixing which holds the slates better in windy weather than *head nailing*. The nails are covered by only one slate, which is a disadvantage in wet places. BS 2717:1956 defines it as 'nailing slates along a line slightly above the head of the slate in the course below'.

centre piece or **centre flower** A central ceiling decoration of plaster or metal.

centre plank [tim.] The plank or planks free of heart, produced from either side of the heart of a log, usually fully *quartered*. Usually applied to hardwoods. *See* **flat-sawn, heart plank.**

centring *See* **centers.**

ceramic mosaic (USA) *Flooring tiles* sold in sheets.

ceramic veneer (USA) *Terra cotta* from 3 to 6·3 cm ($1\frac{1}{8}$ to $2\frac{1}{2}$ in.) thick, of large sizes, hand-moulded or machine-extruded, and of very wide colour range. The thinner slabs are held by *mortar* only to the wall; the 5 cm (2 in.) or thicker slabs are held on by wires round 6 mm ($\frac{1}{4}$ in.) dia. vertical bars grouted into a grout space between wall and slab.

ceramics *Bricks, terra cotta, glazed tiles, stoneware* pipes, burnt clay *blocks, tiles,* and other pottery.

certificate or **building certificate** A statement signed by the *architect* (or engineer) that the builder is entitled to an instalment on work done. Certificates are usually completed and paid monthly during the progress of building.

cesspit or **cesspool** or **cess box** or **cess** (1) A waterproofed box, often lead-lined, at the end of roof gutters. It collects rainwater before it enters the *downpipe*.

(2) A pit in which sewage collects. *See also* C, BSCP 302.200, Cesspools.

chain dogs *See* **dog.** (2)

chain mortiser [tim.] A *mortising machine*.

chain saw [tim.] (1) A power-operated (petrol-engine) *cross-cut saw* with a projecting jib round which the chain travels. The chain carries cutting picks. These saws are used for cross-cutting logs in the forest.

(2) A power-driven saw for cutting building stone. *Hard-faced* (C) metal picks must be used.

chain tongs [plu.] A plumber's heavy pipe grip, which holds the pipe by a chain linked to a bar toothed at the end touching the pipe.

chair rail [joi.] A wooden *moulding* fixed to a wall at *dado* rail height to protect the wall from chair backs. With modern hard *gypsum plasters* it is not needed.

chalking [pai.] The break-up of pigmented *films* on exposure. The binder is so much decomposed by the weather that the *pigment* can be removed by lightly rubbing it. The term is used for all colours, not only for near-white colours, although it originates from these. This ageing fault is preferable to many others because washing may restore the original appearance.

chalk line A length of bricklayer's *line* well rubbed with chalk, held tight and plucked against a wall, floor, or other surface to mark a straight line on it. It is also used by plasterers, surveyors, and mural painters.

chamfer An *arris* cut off symmetrically, that is, at 45°. When cut off unsymmetrically, the surface may be called a *bevel*. *See* **splay, stopped chamfer**.

chamfer stop (1) [joi.] A *stopped chamfer*.

(2) A brick so shaped as to form a stop to a chamfer.

channel (1) *See* **channel pipe**.

(2) *See C*.

channelled quoin An ashlar *quoin* with a rebated upper edge.

channel pipe or **channel** An open pipe, semi-circular or three-quarter round, used in drainage, particularly at *manholes* (*C*).

channel tile The under-tile of *Spanish* or *Italian tiling*.

charge hand A man in charge of work, the next grade below *foreman* or *ganger* (*C*).

chartered quantity surveyor A *quantity surveyor* who has been admitted to membership of the Royal Institution of Chartered Surveyors.

chase A groove cut or built into a wall or floor to receive pipes, conduits, or cables or a flashing (*see* **raglet**). A pipe chase in a floor is either filled with sand and surfaced with a mortar screed, or covered with some other facing. A *duct* can be a large chase.

chase mortising [carp.] Cutting a *mortise* in a timber which is in position. The bottom of the *blind mortise* is arc-shaped so that the *tenon* can be slid into it sideways.

chaser *See* **progress chaser**.

chase wedge [plu.] A wooden wedge with a handle, used in *bossing* lead.

cheapener [pai.] A term sometimes used rather unfairly for *extender*.

check (1) [joi.] (Scotland) **rabbet**.

(2) [tim.] A crack in converted timber along the grain and across the rings, not passing right through the wood, caused by seasoning stresses. In figured *veneers*, fine checks may add character to the pattern and thus be of considerable value. *Compare* **shake**.

(3) [pai.] A crack which penetrates one or more *coats* and usually causes complete failure. *See* **crazing**.

checked back Recessed, *rebated*.

checked ground [carp.] (Scotland) A timber *rebated* (checked) on the edge, to receive lathing or an architrave round an opening. It is flush with the plaster, which may be covered with an *architrave* at its joint with the timber.

checker A storeman or his helper who counts stores or supplies as they arrive on site and checks that they agree with the invoices.

check-fillet An asphalt kerb formed on a roof surface to control rainwater.

check lock [joi.] An arrangement for holding in the locked position the bolt of a door *lock*.

check rail [joi.] A *meeting rail*.

check throat [joi.] A *capillary groove* under a window or door sill.

cheek The side of a *dormer*, *mortise*, *tenon*, etc.

cheek cut [carp.] (USA) The bevelled cut at the foot of a *hip-*, *valley-*, or *jack-rafter*.

cheek nailing (Scotland) Double-nailing of slates through a hole at one side of the slate and a notch at the other side.

cheesiness [pai.] A paint film which is soft and incompletely dry is called cheesy and is the opposite of a flexible, tough film. However, a film which is tough on the surface may be cheesy underneath.

chemical plumber A craftsman who installs and repairs lead tanks and tank linings. He builds large-diameter pipes (30 cm (12 in.) or more) from sheet lead by joining the ends of sheets bent into a circle. *Compare* **plumber**.

chemical test [plu.] A type of *scent test* in which a chemical is put into the drain, giving off a strong recognizable smell when it comes into contact with water.

cheneau (USA) An ornamental upper part of a cornice or gutter.

chequer work *Rubble* walls built of two materials, such as brick and flint or stone, arranged in alternating squares. Sometimes confusingly called *diaper work*.

chestnut or **sweet chestnut** or **Spanish chestnut** [tim.] The castanea species. The timber resembles oak but lacks *silver grain*, is more easily workable, splits easily, and is nailable. It is much used for fencing and for gates and ladder rungs, since it is as durable as oak when cut young.

chilling [pai.] (1) The deterioration of *paints* or *varnishes* which have been stored at a low temperature.

(2) *See* **blushing**.

chimney The brick, masonry or steel shaft containing a vertical flue.

chimney back The 23 cm (9 in.) (or thicker) wall behind a fireplace.

chimney bar or **camber bar** or **turning bar** A wrought iron or steel bar built into the jambs of a fireplace and carrying the brickwork over it.

chimney block (USA) Precast, circular concrete pipe used as *flue lining*.

chimney bond *Stretching bond*, as used for half-brick partitions in chimneys.

chimney breast The chimney wall which projects into the room and contains the fireplace and flues.

chimney can (Scotland) A *chimney pot*.

chimney cap or **chimney hood** An ornamental finish to the *chimney stack*, excluding the chimney pot. It is often designed to improve the draught.

chimney cowl A revolving metal ventilator over a chimney.

chimney cricket *See* **cricket**.

chimney gutter A *back gutter*.

chimney hood A *chimney cap*.

chimney lining (1) A *flue lining*.

(2) *Wall tiles* fixed to chimney *jambs* to protect them from smoke.

chimney pot A burnt-clay pipe at the top of a chimney stack which leads the smoke clear of the brickwork.

chimney shaft The part of a chimney which stands free of other structures, usually restricted to a large chimney containing one or two flues only.

chimney stack The brickwork containing one or more *flues* and projecting above a roof.

chimney throating That part of a flue just above the *flue gathering*.

chipboards or **resin-bonded chipboards** [tim.] Artificial wood compressed from waste wood with *synthetic resins* as *binders* into boards 1·2 m (4 ft) wide and of any length, 3·7 m (12 ft) being the longest boards generally stocked. They are more used for light structural work than for *carpentry*, as their *modulus of elasticity* (C) is only one-third that of wood. Their *moisture movement* (C) is high, about equalling that of wood along the grain (0·02%), but they are more fire-resistant than wood. The correct term is **wood chipboard** (BS 565 and BS 2604).

chipped grain or **torn grain** [tim.] The surface left when small pieces of wood are torn out in planing or machining.

chisel (1) [carp.] A generally wood-handled steel cutting tool, of which the commonest examples are the *firmer*, *socket*, *swan-neck*, and *paring* chisels. The British gas industry now uses all-steel chisels so as to save gasfitters the labour of carrying a mallet as well as a hammer.

(2) Bricklayer's and *mason's* chisels may be *hammer-headed* or *mallet-headed*.

(3) [mech.] *See* **cold chisel** (C).

chisel knife [pai.] A narrow *stripping knife* with a square edge not wider than 4 cm (1·5 in.). *Compare* **stopping knife**.

chlorinated rubber paints Fire-resisting paints that extinguish when the flame is removed. Since, like all paints, they are a blend of several materials, they also contain flame retardants, such as calcium carbonate. They often contain chlorinated paraffin waxes and are blended with alkyd resins. Some chlorinated rubber paints are highly resistant to chemicals and are thus suitable for protecting outside concrete, asbestos-cement sheets or cement renderings; they are fast drying and two coats can be applied in a day by brush or roller or airless spray.

chop [joi.] The movable outer plate of the jaw of a bench *vice*.

chromating [pai.] (1) Priming with lead or zinc chromate to prevent rust forming under the paint.

(2) A protective coating for magnesium alloys (like *anodizing* (C)

but without electrolysis) formed by dipping the article in hot solutions of alkaline dichromate or of chromic and nitric acids.

chrome green [pai.] A *pigment* which may be *lead chrome green,* a mixture, or pure chromium oxide (Cr_2O_3), which is naturally green.

chrome yellow or **Leipzig yellow** [pai.] A yellow *pigment,* lead chromate ($PbCrO_4$), often mixed with *lead chromes.*

chromium plating or **chromium plate** A *protective finish* consisting of an electroplated surface of chromium. When put on to iron or steel, the chromium is best deposited on nickel previously electro-deposited on copper, which is the first coating on the steel. Electro-deposited chromium is almost as hard as diamond. Hard plating is chromium plate without the interposed copper and nickel.

cill The spelling usual in the British building industry for *sill.*

cinder block (USA) *Clinker block.*

circle-on-circle or **double circular** or **of double curvature** [joi.] A description of work which is curved both in plan and elevation.

circuit [elec.] (1) A ring made by conductors, generally containing a power source.

   (2) The conductors which carry the power to a house or other consumer.

circuit vent or **loop vent** [plu.] (USA) That part of a *ventilation pipe* above the last *soil* connection.

circular plane [joi.] A *compass plane.*

circular saw [tim.] A circular steel disc with teeth cut round the rim. It turns at a rim speed of about 50 m/s (170 ft/s). *See* **swage-setting.**

circular stair A *spiral stair.*

circulating water The water contained in the closed circuit of a *central-heating* system.

circulation (1) In *planning,* the proper arrangement and proportioning of rooms and spaces to allow for traffic to be suitably channelled.

   (2) [plu.] A system for *circulating water,* or the flow through the system.

CI/SfB classification An arrangement of the information subjects in the building industry, promoted, administered and developed by an agency of the *RIBA.* It is based on the *international SfB* system and is set out in two manuals. The CI/SfB *Construction Indexing Manual* gives a classification for architects' office libraries and the contents of individual documents. The *Project Manual* advises on the use of drawings, *specifications, bills of quantities,* and the other documents of a contract.

cissing [pai.] *See* **crawling.**

cistern [plu.] (1) or **storage tank** A rectangular, open-topped, cold-water tank usually fixed in the *roof* of a house, *compare* **combination tank.** Nowadays usually made of asbestos cement, galvanized steel sheet, or various plastics. In the past they were made of slate slabs bolted together, or of wood lined with lead or zinc or copper sheet.

(2) or **flushing cistern** or **water-waste preventer** A small tank above a WC fitment which contains enough water to flush the WC once. *Compare* **flushing trough**.

(3) An underground tank for rainwater, often of concrete.

city planning *See* **town planning**.

cladding The non-loadbearing clothing of the walls and roof of a building, the skin used to keep the weather out. An American term which corresponds closely to this is *siding*. *See* BSCP 143, 297, 298.

claire colle *See* **clearcole**.

clamp (1) A *cramp* in *joinery*.

(2) A stack of bricks burnt over flues built up from burnt bricks. This old way of brick making is still used in England for certain *facing bricks*, but most bricks are now burnt in *kilns*.

clamping plate [carp.] A timber *connector*.

clamping time [tim.] The time for which a glued joint must be clamped. It is greater for curved *plywood* than for flat plywood, and varies also with the type of *glue*.

clamp nail [joi.] (USA) A fastener for picture frames and similar mitred (*see* **mitre**) corners. It is a simplified **corrugated fastener**.

clapboard or **bevel siding** or **lap siding** [tim.] (USA) *Weather-boarding*, from 8 to 28 cm (3 to 11 in.) wide, 16 mm ($\frac{5}{8}$ in.) thick on the lower edge, 6 mm ($\frac{1}{4}$ in.) thick along the upper edge (which is covered by the lower edge of the board above). It is *feather-edged* (neither *tongued and grooved* nor *rebated*). *See* **bungalow siding**.

clapboard gauge or **siding gauge** [carp.] A measure to show the amount of board which should be exposed, used by *carpenters* fixing clapboard.

clasp nail A *cut nail* of square section. Its head has two points which sink into the wood.

claw [carp.] A bar with a split end for drawing nails.

claw bar [carp.] A *pinch bar*.

claw chisel A mason's *mallet-headed chisel* with several broad nicks in the cutting edge, used in roughly shaping stone.

claw hammer or **carpenter's hammer** [carp.] A hammer with one split, claw-shaped peen for drawing nails, much used for erecting and dismantling formwork. *See* **adze-eye hammer**, **carpenter's tools** (illus.).

claw hatchet A *shingling hatchet*.

claw plate [carp.] A *connector* for timber.

clay lath A base for plastering, made from copper-plated steel-wire mesh on which clay pellets have been kiln-burnt to the wire intersections. Sold in rolls 5 m (16 ft) by 1 m (3 ft 4 in.) or in sheets half this length.

clay mortar mix Dried clay powder used for making *masonry cement* for American bricklaying.

clay tile (1) A *roofing tile*.

(2) *Quarry tile* for wall-facing or flooring.

clean aggregate Sand or gravel which is free from clay or silt.

clean back The seen face of a header stone. *See* **rough back, skewback.**

cleaning eye An *access eye.*

cleaning hinges Long-leaved metal hinges for outward-opening metal *casements.* The outside of the glass can be cleaned by passing the hand through the gap between the hinges.

cleanout (USA) An *access eye,* soot door, etc.

clean timber Timber which is free from knots (BS 565). *See* **clear timber.**

clean up [joi.] Scraping, smoothing, and sanding of finished *joinery.*

clearcole or **clairecolle** [pai.] Diluted glue *size* containing *whiting,* applied to ceilings and walls before distempering them.

clear timber or **clear stuff** Timber which is free from visible *defects* (BS 565). Since knot-free timber is generally only found in the lower part of the trunks of closely grown trees in virgin forests, clear and clean timber is now scarce and *stress grading* has been adopted. *See* **second growth.**

cleat (1) or **batten** [carp.] A small piece of wood reinforcing another or used to locate positively another timber, or plugged to a wall to carry a shelf.

(2) [elec.] A strip of insulator across a cable, holding it in position.

(3) A *tingle* in *flexible-metal roofing. See also C.*

cleavage [tim.] Wood has cleavage along the grain, particularly *western red cedar* and other woods used for split *shingles. See below.*

cleft-chestnut fencing Fencing made from split-chestnut sticks about 10 cm (4 in.) apart, held with twisted wire to large posts at about 2·1 m (7 ft) apart.

cleft timber Timber split along the grain to nearly the right size.

clenching or **clinching** or **clench nailing** [carp.] Driving a nail right through timber, then bending the point over and on to the back of the timber, a technique used in building *ledged-and-braced doors,* and boats.

clench nail [carp.] A nail used for clenching, such as a *duckbill nail.*

clerk of works The representative on a building site of the *client.* He usually works under the instructions of the *architect* or engineer (who also works for the client). He is responsible for ensuring that the work done is of the quantity and quality specified in the *contract,* and keeps records of such work as foundations which are later covered up. He has a wide knowledge of building construction, which may be confirmed by membership of the Institute of Clerks of Works of Great Britain Incorporated, or he may hold the *Building Inspector*'s certificate of the Institution of Municipal Engineers. He is usually an experienced tradesman who has done considerable evening study.

client The person or organization by whom a builder or consultant is employed, to whom he is responsible, and from whom he draws his fees. In a building *contract* he is usually the building owner's senior consultant, the *architect.*

clinching [carp.] *See* **clenching.**

clink A *seam* between adjacent bays of *flexible-metal roofing. See also C.*

clinker Sintered or fused ash from furnaces. If properly burnt and containing very little unburnt coal, it is an excellent hardcore or concrete aggregate for precast *building blocks*. It is sometimes mistakenly called *breeze*.

clinker block or (USA) **cinder block** A cheap, strong *building block* (sometimes insulating if hollow) of *clinker* concrete. If kept dry they have very little *moisture movement* (*C*), but if laid wet in a wall they are very likely to cause the plaster over them to crack badly while they dry out.

clip or **tack**, etc. A *tingle.*

clipped gable (USA) A *jerkin head roof.*

clocks Clocks built into a building can be electrically driven by one of two methods. Impulse clocks are moved by electrical impulses at one-minute or half-minute intervals from a master clock, and are not connected to the mains. The other sort of clock is connected to the mains and runs in step with the alternators, being controlled by their frequency. If the power station is working they should always be right within a fraction of a second.

close-boarded or **close-sheeted** [carp.] Said of a roof (or wall) that is covered with boards touching each other at the edges, below the slates or tiles. *See* **boarding.**

close-boarded fencing Vertical *feather-edged boarding* nailed to two or three horizontal rails which span between posts about 2·7 m (9 ft) apart. A *gravel board* is often provided along the bottom.

close-contact glue [tim.] A *glue* which will not stick if the surfaces to be joined are further apart than about 0·13 mm (0·005 in.). *See* **gap-filling glue.**

close-couple or **couple-close** [carp.] Term describing a roof of *common rafters* joined at the wall plate level with a *tie-beam*, suitable for spans of about 4 m (13 ft).

close-cut hip or **valley** or **cut-and-mitred hip** or **valley** A *hip* or *valley* in which the slates, shingles, or tiles are cut to meet on the hip or valley line. *Soakers* beneath them (or a metal gutter) keep the water out.

closed cornice [carp.] A *box cornice.*

closed stair or **box stair** (mainly USA) A *stair* walled in on each side and closed by a door at one end. *See* **open stair, fire break.**

closed valley (USA) A *secret gutter.*

close grain or **fine grain** [tim.] A description of *fine-textured wood. Compare* **close-grained.**

close-grained wood [tim.] *Narrow-ringed wood.*

close nipple or **parallel nipple** [plu.] A *nipple* twice the length of the standard pipe thread, with no shoulder in the middle separating the threads.

closer (1) or **bat** A brick or stone cut or moulded to complete the *bond* at the corner of a wall.

(2) The part of a brick which is cut to 57 mm (2¼ in.) on face. *See* **king closer, queen closer, bevelled closer.**

close string [joi.] A *string* having its top and bottom edges parallel (BS 565). Sometimes in USA called a closed stringer. It is also called a housed string because the ends of the *treads* and risers are housed in its face. *Compare* **cut string.**

closet (1) A privy, water closet, etc.

(2) (USA) A small room or a cupboard.

closet lining [joi.] (USA) Thin tongued-and-grooved boards of North American red cedar, used for lining clothes cupboards, because their smell repels moths.

closing stile [joi.] The door *stile* furthest from the hinges.

closure A *closer.*

clothes chute A *laundry chute.*

clouring *Picking.*

clout nail or **felt nail** A short galvanized *nail* from 9·5 to 63 mm (⅜ to 2½ in.) long with a large round flat head, used for fixing *sash cords*, roofing felt, plasterboard, etc.

club hammer or **mash hammer** (Scots) or **lump hammer** A double-face *hammer* with a head weighing from 0·7 to 1·8 kg (1·5 to 4 lb) used by bricklayers and *masons.*

clunch Hard chalk, of which English cottages and barns were built in medieval times. It weathers badly but was protected by a skin of *limewash* which was often renewed. *See* **cob.**

coach bolt or (USA) **carriage bolt** A round-headed bolt enlarged under the head to a square section. This enables it to grip the wood without turning as the nut is tightened up. *See* **fastenings.**

coach screw or (USA) **lag bolt, screw spike** [carp.] A large, *gimlet-pointed* screw for making fixings to wood, it is usually 6 mm (¼ in.) dia. or larger. It is driven into wood by turning the square head with a spanner. Part of the hole must be drilled for it. *See* **fastenings.**

coarse-grained [tim.] *Wide-ringed. Compare* **coarse-textured.**

coarse stuff [pla.] The material for the first and second coats of plaster, usually, in the past, mixed with *hair.* It is made either with *hydrated lime* or with *lime putty.* If made with lime putty, mixes of the following proportions by volume are used: (a) 1 lime, 2 sand; or (b) 1 lime, ⅓ gypsum plaster, 2 to 4 of sand; or (c) 1/10 lime, 1 cement, 4 sand, or 1, 1, 6, or 2, 1, 9 (in decreasing order of strength and speed of set). If made with hydrated lime the following mixes by volume are used: 1 lime, 4 sand; or 2 lime, 1 cement, 8 sand. Generally more sand should be used on brick or stonework than on *lathing, clinker concrete,* or clay *building blocks.* It is generally not desirable to use pure cement-sand mixes since they are very strong, shrink rapidly, and may crack the *finishing coat. See* **cement rendering, floating**

coat, pricking-up coat, rendering coat, three-coat work, two-coat work.

coarse-textured [tim.] A term applied to timber with relatively large wood elements (coarse grain or open grain), or unusually wide *annual rings* for the type of wood. Such timber needs a *filler* before it is varnished. *Oak*, *chestnut*, and *ash* are coarse-textured. *Compare* **fine-textured**.

coat (1) A single layer of asphalt, plaster, etc. applied to a stated thickness.

(2) [pai.] The paint, varnish, lacquer, etc. laid on a surface in one film. Usually each coat is allowed to dry before the next is put on, but when a coat is put on before the previous one has dried there is no clear distinction between one coat and the next. *See* **full coat, glaze coat, ground coat, guide coat, mist coat, priming coat, sharp coat.**

cob (1) An unburnt brick, with straw binder.

(2) (south-west USA) or **pisé de terre** or **rammed-earth construction.** Walling of damp earth sometimes mixed with cement, rammed without reinforcement into *formwork* (*C*). Some mixtures can be laid without formwork. This cheap method of walling has in the past been practised in France, East Anglia, and the west of England, and is now used in Australia and other semi-arid countries. *See* **adobe.**

cobwebbing [pai.] The ejection from a *spray gun* of a series of spiderweb like threads (of chlorinated, natural, or synthetic rubbers). The same difficulty may occur in brushing.

cobwork (1) (USA) Log-house construction.

(2) *See* **cob.**

cock [plu.] A valve for controlling a pipeline of water, gas, or other fluid. *Plug cocks* are usual for gas, *full-way valves* for water supplies with inadequate pressure, and *screw-down valves* for a water supply which has more than enough pressure.

cocked hinge [joi.] *See* **cocking.**

cocking (1) or **cogging** or **corking** [carp.] Where a beam rests on a *wall plate*, a notch is cut out beneath the beam and two notches are cut out of the wall plate so that the beam drops into the notches in the wall plate. *See* **cog.**

(2) [joi.] *Hinges* fixed to make a door rise when it opens are said to be cocked. *Rising butts* do the same work more expensively but more elegantly.

cocking piece or **sprocket piece** or **sprocket** [carp.] A short rafter nailed to each *common rafter* at the *eaves* to give an eaves overhang which is slightly flatter than the slope of the rest of the roof. It is used where the overhang obtained with a *tilting fillet* and ordinary rafter is not enough. *See* **false rafter.**

cockscomb A mason's *drag*.

cockspur fastener A metal fastener for *casement windows*, provided in addition to the usual *casement stay* and pin.

code of practice or **BSCP** A publication issued by the *BSI* describing what is considered to be good practice in the trade described. Codes of practice do not generally have the force of law, but supplement building *by-laws*. American *building codes* generally have the force of law.

coffer A panel in a *ceiling*, strongly recessed to make a decorative pattern.

cog (1) A *nib*, in roofing tiles.

(2) [carp.] In a *cocking* joint, the solid tooth left projecting upwards from the *wall plate* into the beam.

cogging [carp.] *Cocking*.

coil heating (1) Heating of a concrete floor slab by water pipes cast into it.

(2) Heating of a concrete floor slab by *Pyrotenax* or similar electric cables cast into it.

(3) *See* **panel heating**.

coin A *quoin*.

coke breeze Small coke which was unsaleable in Britain before 1939 and was then used for making *breeze blocks*.

cold-cathode lamp or **fluorescent tube** [elec.] An electric lamp consisting of a fluorescent glass tube about 1·2 to 2·7 m (4 to 9 ft) long and 3·8 cm (1½ in.) dia. containing no filament. Alternating current passing through the vacuum in the tube gives no visible light but emits other radiation, which on passing through the glass is changed into visible light by the fluorescence of the glass. These tubes are very costly but give out much more light (and a better diffused light) for less power consumption than a tungsten filament lamp. They are therefore used in drawing offices.

cold-setting resins [pai., tim.] Formaldehyde *synthetic resins* which, when cold, form further, more complicated *polymers* when mixed with an acid accelerator. The action is usually made more rapid if the resin is heated. *See* **thermo-setting resins**.

cold-water service [plu.] The piped cold-water supply into a building, usually brought into the building by a *rising main*.

collapse [tim.] Irregular, sometimes corrugated, *shrinkage* in eucalyptus timbers. *See* **reconditioning**.

collapsible pans *See* **telescopic centering**.

collar (1) A ring of asphalt built up round a vertical pipe passing through an asphalt roof to ensure a watertight joint at the pipe.

(2) [plu.] An enlargement outside a pipe or a reduction within its bore. It is often made to bear on another collar and ensure a tight joint between pipes as in a *union*.

collar beam or **top beam** or **span piece** [carp.] A horizontal tie-beam of a roof as in a *collar-beam roof*.

collar-beam roof [carp.] *Common rafters*, joined half-way up their length by a horizontal tie-beam. This roof gives more headroom in the centre of the room than a *close-couple* roof.

collection line [plu.] (USA) A house drain.

collections [q.s.] Preliminary calculations made by the *taker off* in the *waste* of *dimensions paper* to record minor dimensions and how he arrived at them. After adding them he writes the sum in the dimension column.

colloids [pai.] Gluey, jelly-like substances like glue, gelatin, starch, flour paste, and gums, which stick fast when their solvent evaporates. *See also C.*

colonial siding [tim.] Plain, square-edged *weather boards*. 23 to 30 cm (9 to 12 in.) wide, of which a considerable width is exposed, used in early American buildings.

colour [pai.] Colour in normal speech includes hue, lightness, and saturation. Hues are red, yellow, green, blue, purple, etc. Lightness is the amount of light reflected (regardless of hue) and is also called value or tone. Saturation is the colourfulness or intensity of a hue compared with a neutral grey of similar lightness. The colours of the spectrum are the most intense. Black is considered to be a colour in painting, but white is not. *See* **pigment,** *also* BS 1611.

coloured cement White, rapid-hardening, or ordinary *Portland cement* mixed with mineral *pigments*.

column An upright shaft, generally rectangular or round, of concrete, stone, brick, cast iron, steel, or aluminium, normally designed for carrying axial load (weight) in compression. *Compare* **stanchion** (*C*).

comb (1) or **comb board** A *ridge* of a roof.

(2) A mason's *drag*.

(3) or **scratcher** [pla.] A tool for scratching plaster to give a key for the following coat. It is like a paint brush stock with nails or wire protruding. *Compare* **devil float.**

(4) [pai.] A thin steel, celluloid, rubber, etc. tool used in *graining*.

combed joint or **cornerlocked** or **finger-jointed** or **laminated joint** [carp.] An *angle joint* formed by a series of parallel *tenons* engaging in corresponding slots. *See* **dovetail.**

comb grain [tim.] *Edge grain.*

combination cylinder and tank [plu.] A sheet copper container for hot water (usually a cylinder, either *direct* or *indirect*) which is combined with a feed tank built over it. One advantage of this arrangement is that the feed tank in the roof space is eliminated from house plumbing, and freeze-ups are less likely. The feed tank is at atmospheric pressure and the feed to it from the main is controlled by a ball valve. The cylinder can be heated by *gas circulator*, or *immersion heater*, or if indirect by a water coil connected by flow and return pipes to the boiler.

combination door or **combination window** [joi.] A North-American type of door having an inner removable section consisting of a *storm door* (or window) for winter.

combination plane A *universal plane*.

combination pliers (USA) A tool resembling the British *footprints*.

combination set [mech.] An elaborate *combination square*.

combination square [mech.] A most useful, but costly, adjustable tool which can be used as an inside or outside *try square*, a *mitre square*, *plumb rule*, *marking gauge*, protractor, etc.

combination tank [plu.] A *combination cylinder and tank*.

combined extract and input system A ventilating system which combines the *extract* and the *input systems*. It is more complicated than either, having two fans, and includes automatically controlled heating of the cleaned air.

combing (1) Smoothing the face of soft stone with a *drag* after sawing or chiselling it.

(2) In shingle roofing, a top course of *shingles* which projects above the *ridge* to protect it from the rain blown by the prevailing wind.

(3) [pai.] Partly removing a coat of wet paint with a comb in *graining*.

combustible That which burns. Building materials in Britain are tested by hanging a specimen $5 \times 3 \cdot 8 \times 3 \cdot 8$ cm ($2 \times 1\frac{1}{2} \times 1\frac{1}{2}$ in.) in a furnace at 750° C. for 15 minutes. If the specimen flames, or raises the furnace temperature, it is considered combustible. If neither happens, it is called *non-combustible* (formerly incombustible).

comfort station (USA) WCs and wash basins for public use.

comfort zone A conception used in the design of *air-conditioning* systems. For winter, the most comfortable *equivalent temperature* in England is 16·8° C. (62·3° F.), corresponding to an *effective temperature* of 16° C. (60·8° F.) and a dry bulb temperature of 18·1° C. (64·7° F.). These temperatures are for people doing very light work. In summer the temperature should be about 2·2° C. (4° F.) higher. For heavy work it should be lower. For 70% of people to feel comfortable the temperature should not vary more than 2·2° C. (4° F.) from the above figures. This could be called the comfort zone for 70% of people. *See* degree days.

commode step A *riser* curved in plan, generally at the foot of a stair.

common ashlar Pick-dressed or hammer-dressed *ashlar*.

common bond (USA) *American bond*.

common brick or (USA) **building brick** The locally cheapest *brick*, not necessarily used for *facing*, nor for loadbearing work except in great thicknesses. It may or may not be the *stock brick*.

common dovetail [joi.] A *dovetail joint* in which both members show end grain.

common ground or **rough ground** or **ground** A strip of wood nailed, *plugged*, or otherwise solidly fixed to a wall or sub-frame as a base for plaster, *joinery*, *building board*, and so on. *See* framed ground.

common joist or **floor joist** or **boarding joist** [carp.] Wooden boards laid on edge to span a gap between walls. Floorboards are nailed directly to them. The greatest span of wood joists is usually 5 m (16 ft), but timber can be obtained in greater lengths at greater expense.

common partition [carp.] A wood-*framed partition* consisting of a *head* and *sill* joined by vertical studs at about 45 cm (18 in.) centres, strutted apart by short horizontal wood struts (*nogging* pieces). It carries no load, unlike the *trussed partition*. A common partition 2·4 m (8 ft) high, built of 7·6 × 5 cm (3 × 2 in.) softwood, weighs about 5 kg/m^2 (1 lb/ft^2) without coverings. Nowadays it is uncommon and has been superseded by partitions built of blocks.

common rafter or **rafter spar** or **intermediate rafter** A sloping timber, about 10 × 5 cm (4 × 2 in.) fixed to a *wall plate* at the foot and to a *ridge* at the top, in a *single roof*. On a roof system with *principal rafters* bridged by purlins, the common rafter is a sloping timber carried by the purlins above the principals.

common wall (USA) *Party wall*.

communication pipe In Britain the pipe, between the Water Board's main and the consumer's stop valve or his boundary, whichever is nearer to the main. It is the part of his *service pipe* which belongs to the Board.

comparator [pai.] An instrument for comparing, e.g. *colour* intensities.

compartmentation The division of a building into compartments by *fire-resisting doors*, floors, walls, etc. so as to prevent a fire spreading beyond the compartment of origin.

compass brick or **radial** or **radiating brick** A brick for building circular brickwork. It tapers in at least one direction.

compass plane [joi.] A metal *plane* with an adjustable curved *sole*, to smooth convex or concave surfaces of various radii.

compass roof A roof with curved *rafters* or curved ties.

compass saw or **keyhole saw** or **locksaw** or **fretsaw** [joi.] A handsaw with a blade which tapers to a point. It is used for cutting round sharp curves.

compass window A *bay window* or *oriel window* circular in plan.

compatibility A term more used in painting than elsewhere, but also applicable to plasters. Compatible coats of paint (or materials in the same coat) are those which blend perfectly and look well. An incompatible mix will be cloudy, coagulating, gelling, or precipitating *pigment*. An incompatible film may show *pinholing*, *lining*, *sheariness*, low adhesion, poor *gloss*, slow drying, greasiness or *crawling*.

compo (1) Cement-lime mortar of any composition such as 1 cement: 2 lime:9 sand.

(2) Lead alloy of which gas pipes are made for making a fairly flexible connection to an appliance, such as a meter.

composite board [tim.] (1) *Plywood* containing an insulating or other special-purpose sheet such as asbestos or cork. These boards are used for building cold stores, refrigerators, and other structures which require high insulation. They may be metal-faced both sides as a *vapour barrier*.

(2) A *hardboard* glued to a sheet of other material such as *insulating board* to form a partition *covering*.

composite construction Different materials in conjunction, for example facing and backing bricks in walls, reinforced concrete in-situ *topping* (*C*) over a precast *prestressed concrete* (*C*) floor beam, or brickwork carried on a concrete or steel beam and considered to form its *compression flange* (*C*). It can result in big economies. *See* BSCP 117.

composition floor layer or **granolithic worker** [pla.] A *floor-and-wall tiler* who lays *jointless floors*.

composition nails Brass nails used for slating and tiling.

composition roofing (USA) *Bitumen felt* for roofing.

composition shingles (USA) Imitation *shingles* made of *bitumen felt*.

compound (1) *See* **bonding compound, sealing compound, dressing compound.**

(2) [elec.] An easily melted material like pitch which is poured into a joint box to make a solid, insulating, water-excluding filling round the *conductors*.

compound beam or **built-up** or **keyed beam** [carp.] A wooden rectangular beam *built-up* from several timbers by nailing planks over them, by bolting them together with *connectors*, or by jointing with scarfs, *glue*, or any other method. *See* **compound girder** (*C*).

compound shake [tim.] Several types of *shakes* in combination.

compound walling Walls laid in two or more skins of different materials.

compregnated wood (compressed impregnated wood) [tim.] US term for *high-density plywood*.

compressed cork *See* **corkboard.**

compressed straw slab or **strawboard** Slabs 1·2 m (4 ft) wide, 2·4 to 3·6 m (8 to 12 ft) long, and 5 cm (2 in.) thick, made from compressed straw faced with strong paper each side. It weighs 19·5 kg/m² (4 lb/ft²) and can be used for making 5 cm (2 in.) thick *partitions* or for insulating walls or roofs. Its fire resistance is fairly high and its *k-value* is 0·087 W/m deg. C. (0·6 Btu in./ft²h deg. F.).

compressed wood [tim.] Wood whose density and strength have been increased by pressure, like *high-density plywood*.

compression joint [plu.] A joint in *light-gauge copper tubing*, made by screwing together the ends of the two pipes to be joined, by brass nuts outside them. In the commonest type, the *non-manipulative joint*, the nuts force *glands* into close contact with the copper tube to make a watertight or gastight joint. This joint is less neat than the *capillary joint* and needs more space to make it but no flame. *See* **copper fittings, manipulative joint.**

concave joint A durable *mortar* joint, hollowed out by pushing a 13 mm ($\frac{1}{2}$ in.) dia. bar along it while it is *green*. *See* **jointing**.

concealed gutter (USA) In timber construction, a *box gutter*. There is no *parapet*, but looked at from the ground the gutter is concealed, since it is hidden by, and in, the *cornice*.

concealed heating *See* **panel heating**.

concealed nailing [joi.] *Secret nailing*.

concrete An intimate mixture of water, sand, stone, and a *binder* (nowadays usually *Portland cement*) which hardens to a stone-like mass. Lime and other concretes were used in ancient Rome and in Britain for foundations in the nineteenth century, but the production of strong, cheap, uniform Portland cement has enormously increased its use. *See C*, **aerated**, **lightweight**, BSCP 110.

concrete blocks *See* **blockwork**.

concrete-bonding plaster [pla.] A low-expansion retarded *hemihydrate plaster* to BS 1191 with not more than 5 per cent by weight of shredded wood fibre, *retarder*, etc., used on concrete where other plasters have failed to stick. *See* **bonding treatment**.

concrete bricks Bricks moulded from sand and cement. They can be obtained in many different colours but they represent only one twentieth of the market for clay bricks.

concrete insert A fibre or metal *plug* built into *concrete* or brickwork or fixed into it by drilling. It is used as a fixing for a wood *screw* or *coach screw*.

concrete interlocking tile A standardized British *single-lap tile*, ordinarily measuring $38 \times 23 \times 1$ to $1 \cdot 3$ cm ($15 \times 9 \times \frac{3}{8}$ to $\frac{1}{2}$ in.), provided with a central nail hole at the top end. *Nibs* are provided at both ends, those at the bottom end fitting the water channel of the tiles in the course below. The side lap is $2 \cdot 5$ cm (1 in.) minimum.

concrete nail (USA) A thick, hard steel nail from 1 to 8 cm ($\frac{1}{2}$ to 3 in.) long, for fixing to brick or concrete. *Compare* **masonry nail**.

concrete paint *See* **cement paint**.

concrete tiles Concrete tiles are made to the same shapes and sizes as many of the traditional clay tiles but about eight times more of them are now sold than clay tiles.

concreting paper *Building paper*.

condensation (1) *See* **vapour barrier**. The forming of water on a surface. It can usually be prevented by insulating the inner wall so that its surface is kept warmer. *See* **dewpoint** (*C*).

(2) [tim.] *Cure* of a *synthetic resin*.

condensation groove A part of the *lead sheath* of a lead-clothed glazing bar which projects under the glass to catch water and channel it to the *condensation gutter* at the foot of the glazing.

condensation gutter or **sinking** or **channel** A small gutter at the foot of skylights, *patent glazing*, etc., to carry condensed water through a small hole, pierced to the outside.

condensation washer A shaped fitting which raises the lower end of a *glazing bar* in *patent glazing* above a *purlin*, to allow condensed water to escape.

conditioning [tim.] Bringing timber, hardboard, etc. to the condition in which it will be used, so as to reduce or increase the *moisture content* to the right value. This can be done by wet sponging hardboard or by exposing timber to the room conditions or by kilning it.

conductance The thermal conductance of a building material of a given thickness is the amount of heat which passes through 1 m² (ft²) of it under a temperature difference of 1° C. (F.) between the two faces. It is thus the *conductivity* divided by the thickness in m (in.). For the conductance the temperatures are measured on the face of the material, for the *U-value* they are measured in the air beyond it. For this reason, the *U-value* is always lower than the conductance of the same wall by an amount corresponding to the *surface coefficient*.

conductivity or **K-value** The thermal conductivity of a substance is the number of watts (Btu) passed through an area of 1 m² (ft²), 1 m (in.) thick under a temperature difference of 1° C. ($\frac{1}{2}$° F.) between the two sides. For constructional materials as a whole there is a roughly proportional relationship between conductivity and density. Thus at densities of 720 kg/m³ (45 lb/ft³), K = 0·144 W/m deg. C. (1 Btu in./ft²h deg. F.) while at densities of 2240 kg/m³ (140 lb/ft³), normal for dense concrete, the conductivity is very much higher, K=1·15 W/m deg. C. (8 Btu in./ft²h deg. F.). *See also* U-value, conductance, sound-reduction factor.

conductor (1) or **leader** pronounced 'leeder' (USA) A *downpipe*.

(2) A substance with a high *conductivity* for heat, electricity, or other energy. Generally metals are good conductors.

(3) A *lightning conductor*.

(4) [elec.] or **lead** pronounced 'leed'. A wire or cable of copper, tinned copper, or aluminium, used for leading power from the supply to the consumer.

conductor head (USA) A *rainwater head*.

conduit [elec.] A metal, or plastic, or fibre tube fitted to a wall or ceiling or other part of a building and used as an encasement to cables. For cables laid in the street, earthenware pipe is used, not conduit. *See below*.

conduit box [elec.] A *distribution box*.

conduit bushing [elec.] A short, internally-rounded, threaded, insulating sleeve at an outlet from a *conduit* to prevent the cable insulation being injured.

cone tile or **cone-hip tile** A *bonnet tile*.

congé or (USA) **sanitary shoe** A small concave *moulding* joining the base of a wall to the floor.

conical light A *skylight* or *lantern* built up from straight *glazing bars* and flat panes of glass. It is shaped like a many-sided pyramid.

conical roll or **batten roll** (USA) In *flexible-metal roofing* a roll joint formed over a triangular wood *roll*.

conifers [tim.] Trees of the botanical group gymnosperms, which provide all building *softwoods*. Most of them are fir or pine trees.

connector or **timber connector** [tim.] (1) Joint fasteners for roof *trusses* and similar timber frames which very much increase the shear strength of the timber at the joint. They are steel split rings, shear plates, toothed plates, and corrugated toothed rings, sometimes inserted in precut sinkings made by a *hole saw*, and held to the timbers by a bolt passing through them and the timber. The bolt is screwed up tightly on to large washers at each end and makes a rigid joint.

(2) or **longscrew** [plu.] A piece of pipe with an ordinary *taper thread* at one end and at the other end a long *parallel thread* with a back nut. The length of the parallel thread allows the *coupling* to be screwed back on to it completely so that the connector can be removed and the pipe run unmade at will. The backnut forms a watertight joint with the coupling if a hemp *grommet* soaked in jointing liquid is held between them while they are screwed tight. *See* **double connector, fittings** (illus.), **plug-in connector.**

console French word for a *cantilever* (*C*), generally reserved in Britain for an elaborate *bracket*, usually with an S-shaped scroll.

consultant A registered architect or professional engineer or surveyor who acts on behalf of a *client*. His functions often go very much further than consultation when he, with his staff, provides the complete design of a building.

consumer's supply control [elec.] The electricity undertaking's meter, together with the consumer's main switch and *cutout box*.

consumer's terminals [elec.] The point where the electricity board's cable ends, and the consumer's wiring begins. It is usually at or near this point that the main switch and the distribution switches and fuses are placed.

contingency sum [q.s.] A *provisional sum* for unforeseeable work, such as pumping after storms.

continuity [elec.] The continuous effective contact of all parts of an electrical circuit to give it high conductance (low resistance). *See also C*.

continuous handrail [carp.] A handrail to a *geometrical stair*. *See* **continuous string.**

continuous string [joi.] An *outer string* continued without interruption round a stairwell, usually under a *continuous handrail*.

continuous vent [plu.] *See* **ventilation pipe.**

contour The cross-section of a decorative *moulding* or of the surface of the ground.

contract An agreement between a *client* and a building or civil engineering *contractor* to do certain definite types of work at certain rates. Contracts in Britain are usually based on a *bill of quantities* which consists mainly of a list of numbered *items*, each of which describes a certain type of work and the quantity of that work. The price for supplying the material and labour and doing the work is set against each item by the contractor. If the bill is accepted in this form by the *client* and signed by him, it becomes the principal document of the contract. *See also* **cost-reimbursement contract, etc., fixed-price contract, tendering, variation order.**

contract documents These form the legal *contract*, and consist of the drawings, the *specification*, the *bill of quantities* or *schedule of prices*, the general conditions of contract, and finally a legal *deed* making these all binding on both parties, *contractor* and *client*.

contractor A person who signs a *contract* to do certain specified work at certain rates of payment, generally within a stated time. *See* **main contractor.**

contractor's agent *See* **agent.**

contracts manager A senior architect, surveyor, civil engineer, or tradesman, employed by a building or civil engineering contractor, who has worked for many years in building construction, and takes full responsibility for the completion of the contracts under his authority. He deals with the architect or engineer (or other client) of each site, as well as with his own company directors.

convection When a fluid is warmed it expands, its density decreases, and it rises, its place being taken by denser, cooler fluid. This is the principle of convection of heat. It is the reason why air moves about in rooms where there is no draught and why water circulates in *central heating* systems without circulating pumps. *See* **radiator.**

convector heater [plu.] *A fan convector.*

conversion or **breaking down** [tim.] Sawing logs parallel to their length so as to reduce them to rectangular cross-sections which are of a convenient size to use. *Compare* **cross-cut.**

convert [tim.] To carry out *conversion.*

converted timber [tim.] *Square-sawn timber.*

cooking vat [tim.] A concrete water tank containing steam-heated water pipes. Every *flitch* is stewed in it for some hours before *slicing* or *rotary cutting.*

cooling tower [mech.] A wood or concrete tower, used for cooling condenser water by evaporation. These towers lose considerable quantities of heat to the air and for this reason in closely populated districts *district heating* is now being used to cool the condenser water.

cooperative apartment house (USA) A number of *flats* contained in a building owned by a *cooperative housing society*, of which only the tenants of the flats are members.

cooperative housing society In Britain, a group, approved by a Local Authority, consisting of individuals each of whom requires a house or flat. The group buys the site, commissions the *architect*, orders him to send for *tenders*, and obtains from its members by subscription one tenth of the cost of the site and building. The remaining nine tenths are lent by the *Local Authority* borrowing at a low rate of interest from the Treasury. The loan is paid back by rent from the tenants, all of whom are members of the society. *See* **cooperative apartment house, mortgage, self-build housing society.**

coordination *See* **dimensional coordination.**

copal [pai.] Natural hard *resins* such as Congo, *kauri*, *dammar*, and *amber*, some of which are fossils. They provide the hard dark shiny coating of *varnishes* or *oil paints* and are also used for making *linoleum*.

cope (1) To cover a wall with stones, bricks, or precast slabs which usually overhang as a protection to the wall from rain.

(2) A *coping*.

(3) To fit one *moulding* over another without *mitring* by cutting the first to the profile of the second. A wood moulding is cut with a *coping saw*.

coped joint or **scribed joint** A joint between *mouldings* in which a part of one moulding is cut out but not *mitred* to receive the other. *See* **cope** (3).

coping (1) or **coping stone** or **cope** A brick, stone, or concrete protection, for weathering the top of a wall.

(2) Making a coped joint in a *moulding*.

(3) Splitting stones by drilling them and driving in steel wedges along a line.

coping saw [joi.] A *bow saw* with a narrow blade about 15 cm long and 3 mm. wide, used for cutting sharp curves and *coping* out *mouldings*.

copper A red metal much used for roofing monumental buildings, on which its green *patina* is prized. Copper is also used for *damp courses* and water pipe. Copper pipe was used by the Egyptians about 3000 B.C. *See* **mole plough** (*C*).

copper bit [plu.] A *soldering iron*.

copper fittings [plu.] *Fittings* for *light-gauge copper tube*, many of them not made of copper, but of brass or gunmetal. The two main types are capillary and compression joints. Capillary joints, often of copper, must be completed by heat. *Compression fittings* do not need heat, the joint being completed by tightening *screw threads*. A more recent jointing method, *silver brazing*, can eliminate the use of fittings because all joints can be formed in the tube itself after softening it, by the use of special tools. In the large diameters (10 cm, 4 in.), silver brazing is cheaper because fittings need not be bought, but it requires much heat to reach the brazing temperature. *See over*.

copper glazing *Glazing* with copper *cames* welded to each other. *See* **electro-copper glazing.**

copper nailing Nailing with copper nails, a method of fixing lead sheet on a *dormer cheek*. 'Open' nailing is at 7 to 15 cm spacing (3 to 6 in.), 'close' nailing is at 2 to 5 cm (1 to 2 in.). In contact with

Keene's cement also, copper nails must be used to prevent corrosion. *See* **silicon bronze.**

copper pipe Pipe from 3 to 100 mm bore ($\frac{1}{8}$ to 4 in.) and from 0·6 mm to 3·3 mm (0·024 to 0·128 in.) in wall thickness, used for every plumbing service in building. *See* **light-gauge copper tube.**

copper plating Electro-plating with copper, a *protective finish* to steel nails, wood *screws*, wire, and so on. Used also under *chromium plate.*

copper roofing Flexible-metal roofing, of copper sheet generally from 0·4 to 0·5 mm thick, but in the best work 0·6 mm and in good conditions as little as 0·3 mm. It can be laid in large areas but usually no single sheet should be larger than 1·3 m² (14 ft²). The sheet alone weighs 5 kg/m² (1 lb/ft²).

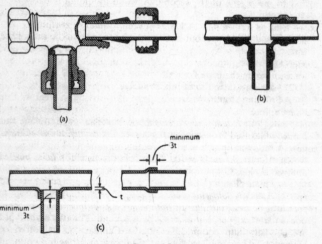

Copper fittings

(*a*) *Compression fitting, joining copper tubes. The left-hand joint is shown in an external view only; the other two joints are sectional, the lower one being completed, the right-hand one being exploded.* (*b*) *Cross-section of capillary fitting for joining copper tube. This tee is provided with three solder rings. When melted, the solder will flow round the thin space between the tubes and fill it.* (*c*) *Joints made by silver brazing; although they are made by capillary action they are not usually called capillary joints. When the pipes have been heated to a dull red, the plumber touches the joint with a strip of silver solder which melts and is drawn in between the two pipes by capillary flow.*

copper slate *See* **lead slate**.

coppersmith's hammer A hammer with a long, bent *ball peen*, used for copper beating.

corbel (1) Brick, masonry, or concrete projecting from a wall face, usually as a support for a beam or roof truss.

(2) A *corbelling iron*.

corbelling Brickwork projecting successively more in each course to support a chimney stack, *oriel* window, etc.

corbelling iron or **corbel pin** A metal support built into brickwork to carry (with other corbelling irons) a *wall plate* instead of corbelling the brickwork out to carry the plate.

corbel piece [carp.] A *bolster*.

corbel pin A *corbelling iron*.

corded way A path on a steep slope, protected from erosion by steps formed with wooden or stone *risers*.

core (1) [joi.] The chips cut from a *mortise*, or the steel bar beneath a *handrail*, or the innermost layer of *three-ply*, *blockboard*, etc.

(2) The *brick core* below a *relieving arch*. The metal, usually steel, within *protected metal sheeting*, the *woodwool* within a paper-covered woodwool slab, etc.

(3) [pla.] A base to a cornice or other complicated plaster work. It may be *bracketing* (hollow core) or a mass of *coarse stuff* reinforced with hair (solid core).

coreboard or (USA) **lumber core** [tim.] (1) Resin-bonded *chipboard* (properly wood chipboard) intended to form the core of sandwich construction (BS 565). (2) Formerly a generic term for *battenboard*, *blockboard*, and *laminboard*.

core driver [carp.] A *hardwood* cylinder the exact size of a hole, pushed through it to clear out chips.

cork The bark of the cork oak, grown in Mediterranean countries and North America. Granulated cork is used as a *loose-fill insulation* weighing 80 to 100 kg/m³. *See below.*

corkboard or **compressed cork** or **baked cork** Granulated cork which has been compressed and baked to form slabs for flooring or insulation. It weighs 120 to 144 kg/m³ (7·5 to 9 lb/ft³) and has a K-value of 0·04 W/m deg. C. (0.29 Btu in./ft²h deg. F.). Loose cork has the same K-value. The slabs are made in thicknesses from 2·5 to 50 cm (1 to 20 in.) and in sizes up to 1·8 by 1·2 m (6 by 4 ft).

cork carpet A floor finish from 3·2 to 6·7 mm thick made like *linoleum* in rolls 1·8 m (6 ft) wide. It is insulating, quiet, and suitable for heavy domestic use but not for hard traffic. It is obtainable in about four colours.

corking [carp.] *Cocking*.

corkscrew stair A *spiral stair*.

cork tile *Flooring tile* of *corkboard*.

corkwood [tim.] *Balsa* wood.

corner bead An *angle staff* protection to a salient corner.

corner chisel [carp.] A *chisel* used for cutting out *mortises*. Its blade is L-shaped, with no wood handle so that both ends are sharpened and used for cutting.

corner cramp [joi.] A *cramp* for gluing *mitres*.

cornerlocked joint [joi.] A *combed joint*.

corner trowel [pla.] (USA) *Angle trowels* of two sorts, used either for internal or for external corners.

cornice (1) A moulding at the top of an outside wall which overhangs it and throws the drips away from the wall.

(2) A moulding at a junction between an inside wall and the ceiling.

corrosion inhibitor Certain chemicals such as sodium nitrite or chromate which protect metals, when they are used in paints or otherwise.

corrugated aluminium A wall and roof *cladding* which is lighter and far more durable, though dearer, than *corrugated iron*.

corrugated asbestos *Asbestos-cement* sheeting for roof or wall *cladding*. The pitch of the corrugations is from 7 to 15 cm (3 to 6 in.). Its appearance is good but it is brittle and unlikely to last much longer than 30 years. *See* BSCP 143.

corrugated fastener or **joint fastener** or **wiggle nail** or **dog** or **mitre brad** [carp.] A corrugated piece of metal, generally steel, driven into the end grain of two boards to join them at a place where fine appearance is not needed, in boxes, etc.

corrugated iron Steel sheet, corrugated and (usually) galvanized on both sides. It will rust badly in damp climates such as Britain, and therefore needs painting. The paint will flake off new galvanizing, but some months after galvanizing the paint will bite better. Corrugations are at 7·5 or 13 cm (3 or 5 in.); sheet thicknesses are from 1·6 to 0·5 mm (0·06 to 0·02 in.). *See* BSCP 143.

corrugated sheet *Cladding* of PVC, steel, aluminium, asbestos, protected metal, or opaque or translucent *glassfibre reinforced resin* corrugated so as to stiffen it and enable it to span the gaps between *purlins*. *See* BSCP 143.

corrugated toothed ring [carp.] A timber **connector**.

Coslettizing *See* **Phosphating**.

costing The process of calculating the cost of doing a unit of work, from the amount of work done by a certain amount of labour with a certain quantity of materials. The contractor's *agent* or general foreman calculates his costs per *item* at frequent intervals so as to be sure that they are below the rate which has been written in the *priced bill* by his firm.

cost-plus-fixed-fee contract Like a *cost-plus-percentage contract*, this form is only used for the most urgent work which cannot await preparation of final drawings. The *client* pays to the *contractor* the full cost of all labour, materials, plant, and services used on the site. Contractor's *overheads* and profits are paid for by a fixed fee.

cost-plus-percentage contract A *contract* used only for very urgent work which cannot await completion of the drawings. The *contractor* is paid the full cost of all labour, materials, plant, and services used on the site plus a percentage of these for his *overheads* and profit. Since the contract is in this form, the contractor has a direct incentive to enlarge the scope of the work and to delay completion. For this reason it is undesirable, except under the closest and strictest site supervision.

cost-reimbursement contract A *contract* based on *cost plus percentage*, *cost plus fixed fee*, or *value cost*. Whenever sufficient time is available to complete the drawings before handing over the site to the contractor, a *fixed-price contract* is used, and this type of contract avoided.

cottage roof A roof without *principals*, having only *common rafters* resting on *wall plates* and joined at the top by a *ridge*.

cotter A steel wedge driven into a *cottered* or similar joint to tighten it.

cottered joint [carp.] A joint used between the *king post* of a *truss* and the *tie-beam* below it. The king post is held to the tie-beam by a metal U-strap passing below the tie-beam. The strap holds the tie-beam tight to the king post by wedges and *gibs* through king post and strap.

coulisse or **cullis** [carp.] A timber grooved to enable a sluice or similar frame to slide in it.

coumarone resins or **cumarone** or **cumarone-indene resins** [pai.] *Synthetic resins* used as a *medium* for painting on metal, as electrical insulating varnish, or as an *alkali-resistant* finish.

counter battens [carp.] (1) *Battens* fixed across the back of several boards to stiffen them, often held by screws in slots to allow moisture movements. They are used on a drawing board.

(2) Battens parallel to the *rafters* and nailed over them on a boarded and felted roof. The *slating* or tiling battens are nailed over them. Counter battens are used to ensure that any water passing on to the roofing felt flows straight down. Otherwise it might be held by the slating battens.

counter ceiling A *false ceiling*.

counter cramp [joi.] A *batten* with small pieces of wood fixed to it, against which several boards can be cramped with *folding wedges* when they are being glued at the edges (*counter wedging*).

counter-flap hinge [joi.] *Hinges* for fitting to the lifting flap of a shop counter, so made that the flap can lie back on the counter. They are usually of double *dovetailed* shape.

counter flashing (USA) A sheet metal *flashing* built into a joint of a chimney or parapet wall and left projecting, to be turned down later when the roof is covered.

counter floor or **sub-floor** [carp.] The lower of two sets of floor boards, often laid diagonally to carry *parquet* or other finished flooring. It is almost universal in USA, where it is also called a blind floor or rough floor.

counter gauge [joi.] A *mortise gauge*.

counter-lathing *Brandering*.

countersink (1) or **countersink bit** [mech.] A drill-bit for metal, with cone-shaped cutting edge used for making a conical sinking round the upper end of a hole drilled for a screw.

(2) The conical sinking made by a countersink bit.

counter wedging [joi.] The use of *counter cramps* for fixing boards together as if for a shop counter.

couple-close roof [carp.] A *close-couple roof*.

couple roof [carp.] A pitched roof with *common rafters* and no *tie-beam*, used for short spans of up to 3 m (10 ft).

coupling or **coupler** (1) [plu.] or **socket** A *fitting* to join two *screwed pipes*, consisting of a short tube threaded at each end, the opposite of a nipple. *See also* **copper fittings, fittings.**

(2) In *tubular scaffolding*, a piece which clamps two or more tubes together, usually by a screwed clamp.

courses Parallel layers of bricks, stones, blocks, slates, tiles, shingles, etc. usually horizontal, including any mortar laid with them. The term is also used for setts or wood blocks laid in rows or for any horizontal layer of material.

coursed ashlar (USA) *Regular-coursed rubble*.

coursed random rubble Random rubble built to courses every 30 cm (12 in.) or so.

coursed snecked rubble *Snecked rubble* built to occasional courses.

coursed squared rubble or (USA) **random ashlar** *Squared rubble* built to occasional courses.

coursing joint (1) A *bed joint*.

(2) A *joint* in an arch, concentric with, and separating two *string courses*.

cove or **coving** A concave moulding joining a wall to ceiling or floor.

coved ceiling A ceiling curved at its junction with the walls.

cove lighting *Indirect lighting* from above a *cove* or *cornice*. The light is thrown up to the ceiling which reflects the light in a pleasant, diffuse way downwards with no glare.

cover The covered width of a slate, tile, or shingle. *See also* **C.**

cover fillet or **cover strip** [joi.] A thin narrow strip used to cover joints in wall or ceiling board.

cover flap [joi.] A panelled flap which covers *boxing shutters*.

cover flashing A vertical *flashing* which overlaps the vertical parts of *soakers*, *lead slates*, etc. *See* **raking flashing, stepped flashing.**

covering power [pai.] A term which should not be used because of its ambiguity. It is being replaced by two other terms, *hiding power* and *spreading rate*.

coverings for framed partitions These are used where the stiffness and sound insulation of rigid *infillings* (*bricknogging*) are not needed or where for other reasons (e.g. weight) they would be unsatisfactory. Apart from timber panelling, *plywood*, and lath and plaster, there are many other such materials, e.g. *plasterboard*, *hardboard*, *insulating board*, *composite board*, *corkboard*, sheet metal, *asbestos-cement*, *asbestos wallboard*, *woodwool*, and *strawboard*. The last two, being 5 cm (2 in.) thick or more can be built up to form a partition without framing. *See* **framed partitions.**

cover moulding [joi.] A moulding *planted* to cover a joint on a flush surface.

coving A *cove*.

cowl A metal cover, often louvred or rotating, fixed on a *chimney* to improve the draught.

crab A portable windlass or hand winch for lifting stone. *See also C.*

cracking in paint The breakdown of a paint film on exposure, with cracks through at least one *coat*. Some of the various sorts of cracking are *hair cracking, crazing, crocodiling, crawling*.

cracking in plaster Plaster cracks when it shrinks, when laths twist or are too thinly plastered, or when the structure settles differentially. *Gypsum plasters* do not shrink on hardening; they expand slightly (Portland cement plasters shrink). All cracks except those due to building settlement and shrinking can be avoided. *See* **fire cracks.**

cradle or **boat scaffold** A removable *scaffold* hanging on ropes, used by painters for painting tall buildings. *See also C.*

cradling [pla.] Rough timber fixed round steel beams as a ground for *lathing*. *See* **bracketing.**

cradling piece [carp.] A short *joist* from the wall each side of the *chimney breast* to the *trimmer joist*, carrying floor boards.

craft or **trade** An occupation which needs intelligence and manual skill, a man who does this work being called a craftsman or *tradesman*.

craftsman A *tradesman* in a building or other *craft*.

cramp (1) or **clamp** [joi.] A tool for squeezing together wood parts during gluing, for example a *counter cramp, floor cramp, hand screw*.

(2) A metal U-shaped bar, from 15 to 30 cm (6 to 12 in.) long and 9 to 25 mm ($\frac{3}{8}$ to 1 in.) wide, which holds ashlars to each other or to a steel or concrete beam behind. It is often called a cramp iron although for monumental work it is often made of non-rusting metal like copper or bronze. *See* **lead plug, slate cramp.**

(3) A metal strap fixed to a door frame or lining and built into a wall to fix the frame.

crampon or **crampoon** *Nippers*.

crank [mech.] A bar with a right-angle bend in it which gives a leverage in turning, for example the starting handle of a car, a carpenter's *brace*, or a crankshaft of an internal combustion engine.

crank brace [carp.] A carpenter's *brace*.

cranked sheet An asbestos-cement corrugated sheet bent to fit the junction of two roof slopes.

crawling [pai.] A *cracking* in glass-finish topcoats containing *drying oils*, which shrink and reveal a *ground* (3) or (4). It may be caused by grease on the *ground*. Cissing is mild crawling.

crawlway A *duct* at least 1·05 m (3½ ft) deep, but not high enough to walk through.

crazing (1) Hair cracks on the surface of concrete or *cement rendering*, caused generally by excessive water or by steel trowelling of a too rich mix. Map crazing is random crazing over the whole surface.

(2) or **shelling** [pla.] A fault in a *finishing coat* which forms many intersecting cracks on its surface, and leaves its bed, owing to a weak *floating coat*. The cure is to cut out bad patches down to the brick, and relay.

(3) [pai.] *Cracking*, consisting of broad deep *checks*. See **crocodiling**.

creasing or **tile creasing** Under a brick-on-edge coping, one or two courses of plain tiles are laid, projecting about 3·8 cm (1½ in.) from the face of the wall. The projection is covered with a *cement fillet* to throw off the water. A similar arrangement is used at window sills.

creep trench A *duct* below floor level less than 1·05 m (3½ ft) high. *Compare* **crawlway**.

cremorne bolt [joi.] An *espagnolette bolt*.

cricket (USA) In a chimney *back gutter*, a small *saddle* built to throw off the water in both directions.

cripple (1) A bracket anchored at the ridge, carrying a scaffold for slaters.

(2) Any framed member shortened at an opening. A *jack-rafter*, for example, is a cripple rafter.

critical path scheduling A development of the *progress chart*, which on urgent and complicated civil engineering contracts is particularly valuable. As on the progress chart, the normal time required for all the operations on a contract is worked out. From these data can be calculated the longest time required to reach the end of the contract. This is the critical path. By putting extra labour or machinery on it this path could be shortened and cease to be critical. Another path would then become the critical one. If a punched card machine is used, with each card carrying the duration and sequence of one operation, the critical path can be changed in only a minute, even with about 1,000 cards.

crocodiling or **alligatoring** [pai.] Bad *crazing* showing a pattern like alligator skin.

crook [tim.] (1) A *knee*.

(2) (USA) The *warping* called *spring* in Britain.

cross [plu.] A *fitting* consisting of two branches meeting at right angles on a *run*.

cross band or **cross banding** or **crossing** [tim.] In *plywood* or *coreboard*, the layers of *veneer* perpendicular to the *core*. They prevent it from shrinking and cracking or expanding like ordinary wood. In five-ply the cross bands are the layers between face and core and between back and core (face crossing, back crossing). In *three-ply* the cross bands are the two outer veneers.

cross bridging [carp.] (USA) *Herring-bone strutting*.

cross-cut [carp.] (1) To cut with a saw at right angles to the grain of wood.

(2) A saw-cut so made.

cross-cut saw [carp.] A saw with its teeth set and sharpened to cut across the grain of wood. The larger the saw the coarser are its teeth; thus a 35 cm (14 in.) cross-cut has 3 to 4 points per cm (8 to 10 per in.), while a 66 cm (26 in.) cross-cut has 2 to 3 points per cm (6 to 8 per in.). *See* **panel saw, pendulum, rip saw, setting.**

cross furring [pla.] (USA) *Brandering*.

cross-garnet [joi.] A *hinge* built of a long mild-steel strap fixed on to the face of a door. A cross bar hinged to the strap is screwed to the *door frame*. This hinge is often used for hanging gates, or *ledged and braced doors*, but is not nowadays used in the best joinery, since the hinge is so conspicuous. The decorative wrought-iron hinges of entrance doors of old cathedrals are of this type.

cross grain [tim.] Fibres which do not run parallel with the length of the piece. They may be partly *end grain, diagonal grain, spiral grain, interlocked grain,* or alternating. All of these are the contrary of straight grain and make the wood hard to work.

cross-grained float [pla.] A wooden float about $30 \times 10 \times 2.5$ cm ($12 \times 4 \times 1$ in.), like the *hand float* but thicker, with the grain parallel to the short side, and used for *scouring*. *See* **plasterer** (illus.).

crossing (1) [tim.] A *cross band*.

(2) [pai.] Laying on a *coat* of *paint* with a *brush* by a series of strokes each at right angles to the previous series. With *oil paints* each series becomes progressively lighter, helping to give a uniform coat.

cross joint or (USA) **head joint** Vertical mortar *joints* perpendicular to the face of a wall. They are less easy to fill than *bed joints* and are usually weaker, being made by *buttering*.

cross-lap joint [carp.] (USA) A *halved joint* between two pieces of wood which cross each other.

cross nogging [carp.] *Herring-bone strutting* of *common joists*.

crossover [plu.] A pipe bent to a U-shape to pass another pipe.

cross peen In the description of hand *hammers*, a wedge shape opposite the face of the hammerhead. This wedge is horizontal when the handle is vertical. The opposite of *straight peen*.

cross tongue or **loose tongue** [joi.] A piece of *plywood* or a slip of wood with *diagonal grain*, glued into a saw cut between two members to stiffen an angle joint. *See* **feather, key, straight tongue.**

cross welt A *seam* between adjacent sheets of *flexible-metal roofing*, usually parallel to the *ridge* or *gutter. See* **staggering.**

crotch or **crutch** [tim.] The join between a large branch and the tree trunk. The *grain* is curly, since the fibres of the branch must be locked into the fibres of the trunk. The *veneer* is often highly figured and therefore valuable. *See* **curl, plume.**

crown course A *cranked* or curved corrugated asbestos sheet used as an alternative to *ridge capping*.

crown cover [tim.] A protective adjustable cover over a *circular saw*.

crown moulding [carp.] (USA) A *moulding* above the cornice, immediately below the roof.

crown plate [carp.] A *bolster*.

crown post [carp.] A *king post* or the short vertical posts near the middle of a *hammer-beam* roof.

crowsfooting [pai.] Minor wrinkling like the imprint of bird's feet, with the wrinkles not joined up. It can be regarded as an incompletely *crystallized finish*.

cruiser or **timber cruiser** [tim.] (USA) A timber estimator who measures and counts some of the trees in a stand and so arrives at a figure for the amount of *standing timber*.

crystallized finish [pai.] The wrinkles formed in paints which contain *tung* or similar oils, or the true crystallizing of certain *lacquers. See* gas-checking.

Cuban mahogany [tim.] (Swietenia mahogoni). The first exported mahogany, now very scarce, also called West Indian, Spanish, or San Domingo mahogany. A close-grained red timber with a fine silky texture, hard, with little moisture movement, obtainable in large sizes, stronger than *oak* but slightly easier to work, used for the very best *joinery*.

cubing [q.s.] (1) Determining volumes in m³ (ft³), etc.

(2) From the volume of a proposed building, in the earliest stage of its design, determining its probable cost by multiplying the cost per m³ (ft³) of recently completed similar work by the volume of the building in m³ (ft³).

cullis [carp.] *See* **coulisse.**

culls American term for material which has been rejected as of too low a quality to be used, particularly bricks or timber.

cup (1) [tim.] A sort of *warp* which is common in *flat-sawn timber* and is sometimes seen in floor boards when they bend up at the edges (or in the middle) so that they are either channel-shaped or cambered across.

(2) [joi.] A hollow metal cone fitted into a countersunk hole in the best joinery, to take the thrust of a countersunk screw.

cupboard latch [joi.] A ball catch or other simple catch for securing a cupboard door. Some are operated by push-button or by lever, and very neat ball catches are now made entirely of white nylon.

cup joint [plu.] A *blown joint*.

cup shake [tim.] *Ring shake*.

cup-square bolt A *coach bolt*.

curb [carp.] or **kerb** A timber upstand, sometimes used as a roll. *See below*.

curb joint (Britain, **curb roll**) or **knuckle joint** [carp.] (USA) The horizontal joint between the two surfaces of a *mansard roof*; there called a *gambrel roof*.

curb rafters [carp.] The rafters of the flatter, upper slope of a *mansard roof*.

curb stringer or **curb string** [joi.] (USA) A three-member *outer string*, consisting of one *close string* carrying the stair, surmounted by a *moulding*, called the shoe rail, from which the *balusters* rise, and faced with a facing string.

cure or **curing** [tim.] The chemical change (polymerization or condensation) which occurs when *thermo-setting* resins are heated, or when an *accelerator* is added to a *cold-setting resin*. They become insoluble in water, strong and hard, a result of additional linkages between the molecules. *See* **polymer** *and* C.

curl [tim.] The fine *figure* obtained by skilful *conversion* of *crotch*.

current-carrying capacity [elec.] *Carrying capacity*.

curtaining or **sagging** [pai.] Excessive *flow* in paint films, particularly on vertical surfaces, producing bow-shaped ridges, which look like hanging curtains and may collapse and tear down. *See* **run**.

curtain wall (1) (USA) A non-loadbearing external wall between columns or piers, not always carried on the floors which it passes.

(2) A wall which acts as a screen merely to hide something.

(3) An enclosing wall round a property.

curtain walling Modern wall *cladding*, often framed in *light alloy* and consisting of two or more layers of opaque glass or other lightweight fire-resistant sheet material. It can be erected quickly after the floors, and clamped to the columns or floor slabs, and is therefore convenient for multi-storey buildings.

curtilage The total land area occupied by a dwelling house and its garden.

cushion (1) A *padstone*.

(2) A seating for glass along the full length of a *patent glazing* bar, usually of asbestos, plastic cord, or lead.

cut-and-mitred string [carp.] A *cut string* in which the end grain of the *risers* is hidden by mitring with the vertical part of the notch in the string.

cut-and-mitred hip or **valley** A *close-cut hip or valley*.

cut brick A brick cut to shape with an *axe* or *bolster*. It is much rougher in shape than a *gauged brick*.

cut nail [carp.] A heavy *nail* of rectangular cross section made by cutting (shearing) it from a piece of steel plate, as opposed to the *wire nail* made by forging pieces of round or oval wire. It cannot be bent over and clenched like a wire nail since it breaks easily.

cutout (1) [elec.] A circuit-breaking device such as an electric *fuse* or circuit breaker.

(2) The upper end of a *patent glazing* bar which is cut away to enable the glazing to be flashed.

cutout box or **fuse box** or **lighting panel** [elec.] The cast-iron box on an electricity consumer's premises, in which the *fuses* or circuit breakers are fixed, close to the main power switches. *See also* **house service cutout.**

cut rubble *Rubble walling* of which the face is coursed or squared.

cut stone or **natural stone** A stone cut to shape with *chisel* and *mallet*. *Compare* **cast stone.**

cut string or **open string** or **bridge board** [carp.] An *outer string* with its upper edge cut in steps so that the treads overhang it, used for the dignified stairs of the eighteenth century. *See* **bracketed stairs, string.**

cutter or **cutter and rubber** A *rubber* brick for *gauged brickwork*.

cutter block [tim.] A steel block fixed to the spindle of a *spindle moulder*, *surface planer*, or similar wood-working machine. It carries two or more knives which shape or smooth the timber. Cutting is done by an adzing action. All cutter blocks turn at high speeds, at least 4,000 rpm. *See* **solid moulding-cutter.**

cutting gauge [joi.] A tool like a *marking gauge* with a thin blade in place of the marking pin. It is used for cutting laths or rebating timber.

cutting in [pai.] Painting a clean edge, usually a straight line, at an edge of a painted area.

cutting iron [joi.] The sharpened steel blade of a *plane*, opposed to the *back iron*. *See* **carpenter's tools** (illus.).

cutting list [carp.] A list showing the sizes and sorts of timber needed for a job. *See also* **C.**

cutting pliers [elec.] A pair of pliers with wire cutters on one side as well as the usual flat jaws.

cyclone cellar (USA) *A storm cellar.*

cylinder [plu.] or **storage calorifier** A closed circular tank for storing hot water to be drawn off at the taps, often now made of copper instead of the galvanized steel sheet that was once the only material. If it is horizontal it is domed at both ends, if it is vertical its top is domed and its bottom is concave. *See* **direct, indirect cylinder, calorifier, cistern.**

cylinder lock [joi.] A *lock* for an entrance door (which may be of rim or *mortise* type) which is opened by key from outside and by knob from inside. It can usually be set with the *latch* permanently in the shut or open position by a catch inside the door. In this lock, the latch has the same metal tongue as the lock ('Yale' lock).

D

dabber [pai.] A dome-shaped *brush* of soft hair for applying *spirit varnish* or for polishing and finishing gilding.

dabbing or **daubing** *Picking* a stone face.

dado A border or panelling over the lower half of the walls of a room above the *skirting*.

dado capping or **surbase** or **dado rail** or **dado moulding** [joi.] The highest part of a framed *dado*.

dado joint [joi.] (USA) A *housed joint*.

damages or **liquidated damages** When a *contract* contains clauses relating to sums payable for delays amounting to a breach of contract, these sums can generally only be recovered from the *contractor* if it can be proved that they are related to the loss caused by the delay.

dammar [pai.] A natural *resin*, soluble in many organic *solvents*, of a pale yellow colour or colourless, used in *varnishes*.

damp course or **damp-proof course** or **dpc** A horizontal layer of impervious material laid in a wall to exclude water, usually at 15 cm above ground level, as well as above the junctions of *parapet* walls with a roof and above door or window openings. Vertical damp courses (*tanking*) of asphaltic material are also provided to keep basements dry. Damp courses may be of asphalt, bitumen sheet, copper, lead, zinc, aluminium, vitreous brick (blue brick) or tile, plastics sheet or other impervious material laid in cement mortar even, occasionally, slate. *See* BSCP 102.

damper or **register** A metal plate across a *flue*, used for blocking or regulating the draught.

damp proofing Putting a horizontal or vertical *damp course* in a building.

dap or **dapping** [carp.] (mainly USA) A sinking such as those made for timber *connectors*, or a *housing* for a *ribbon board* in a *stud*.

dancing step A *balanced step*.

Darby float or **Derby float** [pla.] A two-handled wooden or light alloy float 1 to 1·5 m ($3\frac{1}{2}$ to 5 ft) long, about 13 cm (5 in.) wide and 1·6 cm ($\frac{5}{8}$ in.) thick used for levelling surfaces, particularly ceilings.

daubing (1) *Dabbing*.

(2) [pla.] Rough plastering.

day The opening in a window, usually now called the *daylight width*.

daylight factor or **sky factor** At a given point in a room the daylight factor is the percentage illumination of a horizontal surface at that point compared with the illumination which it would have from the whole hemisphere of sky, if the sky were uniformly bright.

daylight-factor protractor An instrument which simplifies the calculation of *daylight factors* and the solid angles and slopes involved.

daylight prediction, daylight forecasting An architect designing buildings and consequently deciding their heights and spacings needs to know, for each living room, whether enough light will reach all parts of it at all seasons. He can forecast the amount of daylight in any part of the room at any season by using the methods of the *Building Research Station*, including the Waldram diagram or the *daylight-factor protractor*.

daylight width or **sight width** The width of the opening which admits light through a window. *See* **sight size**.

daywork A method of payment for building work, involving agreement between the *clerk of works* and the *contractor* on the hours of work done by each man, and the materials used. Proof of this agreement is shown by the clerk of works's signature of the contractor's day-sheets. Payment to the contractor consists of his expenses in labour and materials, plus an agreed percentage for overheads and profit. Daywork is *cost-plus-percentage* payment on a small scale.

dead [elec.] The contrary of *live wire*, said of any conductor which is disconnected from its source of power.

dead bolt [joi.] A square-section *bolt* which is driven home by turning a key in a *lock*, as opposed to the ordinary bevelled *latch* worked by a door knob, or the *barrel bolt*.

dead door or **flue** or **window** A bricked-in door, flue, or window.

dead end [plu.] That part of a pipe between a blocked-off end and the first branch. If it contains air it may be useful in preventing the noise of *water hammer* (*C*).

deadening or **deafening** or **dead sounding** The *pugging* of floors or walls.

dead knot or **encased knot** [tim.] A knot whose fibres are not inter-grown with surrounding wood. It is a knot which can be easily knocked out and is thus a worse defect than a *live knot*.

dead leg [plu.] A hot-water connection in which the water is stationary (not circulating) except when it is being drawn off. The water cools down between draw-offs and the dead leg therefore wastes both water and heat. Some local authorities specify the maximum length of dead leg which they allow. *See* **indirect cylinder**.

deadlight [joi.] A light in which the glass is fixed direct to the framing, in other words a window, or part of one, that does not open, also called a fixed sash, fast sheet, or stand sheet.

dead lock [joi.] A *lock* which is worked by key only from both sides and therefore has no door knobs.

dead shore [carp.] A heavy upright timber (one of two) carrying the weight of a wall below a *needle*.

dead-soft temper The softness of *copper sheet* required for roofing. The copper may become work-*hardened* (*C*) during placing but can easily be *annealed* (*C*) again by heating it to a dull red heat with a *blow lamp* and allowing it to cool in air or water.

deadwood [tim.] Timber from dead standing trees (BS 565).

deal [tim.] (**1**) A piece of *square-sawn* softwood timber 48 to 102 mm

(1⅞ to 4 in.) thick and 229 to under 279 mm (9 to 11 in.) wide (BS 565). It is usually about 23 by 75 cm (9 by 3 in.). *See* **plank.**

(2) *See* **redwood.**

deal frame [tim.] A *frame-saw* used for cutting *deals*, having vertical feed rollers to keep the deal close to the fence.

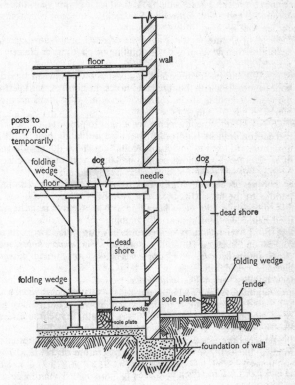

Dead shores.

death-watch beetle [tim.] A *beetle* which burrows deeply into structural timber, in particular the *sapwood* of *English oak*, and is therefore difficult to kill. The adult makes a ticking noise.

decay or **fungal decay** or **rot** [tim.] Decomposition of *timbers* by fungi and other micro-organisms, for example, brown, *dry*, *pocket*, and *wet rot*. *See also* **dote, moisture content, weathering.**

decibel The unit of sound reduction, defined as one tenth of the logarithm to the base 10 of the ratio of the two sound intensities. It is much used in calculating average *sound-reduction factors* of *partitions* and floors. The unit of loudness of sound is the *phon*. *See* **reverberation period.**

deciduous trees [tim.] Trees which lose their leaves every year, that is all *hardwoods* and a few *softwoods*. *See* **larch.**

deck *See* **roof decking,** *also C.*

deducts [q.s.] Quantities such as areas of door or window openings deducted from areas of wall or pointing or painting or plaster; *but see* **adds.**

deed (1) A document forming part of a *contract* which, when signed by both parties, is legally binding, and holds the contractor to perform the work according to the *contract documents*. The *client* is equally bound to pay for the work.

(2) A legal document which gives a right to property.

deep bead or **draught bead** or **sill bead** or **ventilating bead** [joi.] An upright board fixed to and rising about 8 cm above the *window board* of a *sash window*. It allows some ventilation without draught at the *meeting rail* while the bottom *sash* is nearly closed.

deep cutting or **deeping** [tim.] The *re-sawing* of timber lengthwise parallel to the faces (BS 565). *Compare* **flat cutting.**

deep-seal trap [plu.] An *anti-siphon trap* with a 7·6 cm (3 in.) deep *seal*, used in the *one-pipe* system of plumbing.

defect [tim.] An irregularity or weakness in wood which lowers its usefulness or structural suitability, such as *wane*, *shake*, *knots*, *warp*, *decay*, insect damage, and poor *conversion* or machining. *Compare* **blemish, open defect.**

defects liability period The *maintenance period*.

degree-day value A figure which describes the relative coldness of a site. It is based on the number of days yearly by which the average temperature falls below 60° F. (15·5° C.) in Britain, where it is assumed that buildings do not need heating at 60° F. (15·5°C.). The days when the temperature is lower than 60° F. (15·5° C.) are *weighted* (*C*) as follows. Each day with an average temperature of 59° F. (15° C.) counts as 1; 58° F. (14·5° C.) counts as 2; 57° F. (13·9° C.) counts as 3, and so on. In USA, comfort is more valued than in Britain, and the temperature chosen for degree-day charts is 65° F. (18·3° C.). London has a degree-day value about 3500, Aberdeen, also at sea level but 644 km (400 miles) farther north, 5500. The effect of altitude is generally to add one degree-day for each foot of height above the sea level. The degree-day value is used by heating and ventilation engineers to calculate the annual fuel consumption in heating a building. No UK metric definition existed in 1973 but French measurements are in °C. below 18° C.

dehumidifier An air-conditioning unit which cools the air below the

dewpoint (*C*), and thus reduces its humidity. It usually cools the
air by a spray of chilled water.

Delabole slates Grey, green, or red, *sized* or *random slates* from Delabole, Cornwall, an English quarry worked since the sixteenth century.

deliquescence The liquefying of certain salts by their absorption of
water, generally from the air. When patches of chlorides occur in
plaster or brickwork, they may deliquesce and be seen as damp dark
areas on the plaster. The opposite of *efflorescence*.

demolisher or **mattock man** or **topman** or **housebreaker** A *skilled man*
who pulls down a wall by standing on top of it and breaking pieces
off below him, or by pulling a loose wall with a winch and rope, or by
means of a *concrete breaker* (*C*).

demolition contract A *contract* for building on a city site is usually preceded by demolition of the old building. Although the demolition
may be included in the building contract, work is sometimes started
more quickly by separating the two, since very few drawings need to
be prepared for demolition work. *See* BSCP 94 Demolition.

densified impregnated wood [tim.] *Improved wood.*

depth gauge A tool consisting of a graduated rule or rod held in a crosspiece. The rod is inserted into a hole and the crosspiece can be moved
to touch the edge of the hole.

Derby float [pla.] A *Darby float.*

desiccation [tim.] The drying e.g. of *timbers* in a *kiln*.

detector or **automatic call point** A device such as a thermostatically-
controlled circuit which gives a fire alarm.

devil float or **devil** [pla.] A *hand float* with a nail head projecting from
each corner about 3 mm. The nail heads scratch the surface of fresh
plaster to make a *key* for the next coat. *Compare* **comb.** *See* **plasterer.**

devilling [pla.] Scratching plaster to prepare a rough surface for the
next coat.

dextrin or **starch gum** [pai.] A water-soluble gum made from starch,
used as a binder in *water paints* and distempers, and for hanging
heavy wallpaper.

diagonal bond or **herring-bone bond** or **raking bond** (1) For very thick
brickwork about one course in six is a header course with the bricks
at 45° to the face.

(2) A *bond* for facing bricks laid in a decorative herring-bone
pattern on a floor or in a wall.

diagonal grain or **oblique grain** [tim.] A *defect* in which the fibres of
wood are at an angle to the length of the piece owing to faulty
conversion of straight grained timber.

diagonal slating or **drop-point slating** The laying of asbestos-cement
diamond slates with one diagonal horizontal. The corners on this
diagonal are cut off and the head is nailed. *See* **honeycomb slating,**
diamond slating.

diamond matching [tim.] *See* **four-piece butt matching.**

diamond saw A *circular saw* about 1·8 m (6 ft) dia., and 9 mm ($\frac{3}{8}$ in.) thick which cuts stone with *black diamonds* (*C*) set in its perimeter.

diamond slates *Asbestos-cement* slates 30 or 40 cm (11$\frac{3}{4}$ or 15$\frac{3}{4}$ in.) square with two corners cut off, laid in *diagonal slating*.

diamond washer A curved washer used over roof sheeting and fitting its corrugations like the *limpet washer*.

diaper or **diaper work** (1) *Facing bricks* arranged with darker or lighter *headers* showing in a diamond pattern.

(2) *See* **chequer work**.

diaphragm tank A closed tank with one connection to the hot circulating water of a *sealed* heating system, and with its inner space divided in two by a skin or diaphragm which allows more space for the hot water as it gets hotter. The space the other side of the skin contains air which is compressible and the pressure in the system therefore does not rise excessively.

diatomite or **diatomaceous earth** or **moler earth** or **kieselguhr** A soil composed of the hollow siliceous skeletons of tiny marine or freshwater organisms (diatoms) found in Denmark and elsewhere. It is an *extender* in paints, an absorbent for the *nitroglycerin* (*C*) in *dynamite* (*C*) and an *aggregate* for lightweight *building blocks* or flue bricks. It is a very light material and bricks made of it float on water. When laid in mortar, the density of moler brickwork is about 720 kg/m^3 (45 lb/ft^3). Its *coefficient of expansion* (*C*) and thermal *conductivity* are both low. It is therefore an efficient walling material for flue lining, though not very strong and inclined to lose its strength when wet. It can be used as a *refractory* (*C*) at temperatures up to 1300° C.

die (1) [joi.] At the upper and lower ends of a *baluster*, the enlarged square part which meets the rail or plinth.

(2) An internally threaded metal block for cutting male threads on bars or pipes, held in a *die stock*. *Compare* **tap**. *See also C*.

diesquare [tim.] A squared timber generally 10 × 10 cm (4 × 4in.) or larger. *See* **baulk**.

die stock [plu.] A holder for *dies* with which *screw threads* can be cut by hand.

diffuse porous wood [tim.] *Hardwood* in which the pores are uniform in size throughout the *annual ring*, without any abrupt transition in size between *springwood* and *summerwood*. *Compare* **ring-porous**.

diffuser A *register* (2).

diffuse reflection Reflection of light from a surface as smooth, matt, white paper reflects light equally in all directions.

diffuse-reflection factor The ratio of the light which is diffusely reflected from a surface, to that which falls on it.

diluent [pai.] A *thinner*.

dimensional coordination Agreement between *building-component* makers and *architects* about the dimensions of components so that they can fit into a *modular system*.

dimensional stability A property of building materials. A material is dimensionally stable if it has no *moisture movement* (C), little temperature movement, and does not shrink or expand for any other reason. Cement and wood products are generally unstable, but *plywood* is much more stable than wood, owing to the different directions of the grain in alternate *veneers*.

dimension lumber [tim.] (USA) *Lumber* stocked in a lumber yard, from 10 to 30 cm (4 to 12 in.) wide, and from 5 to 13 cm (2 to 5 in.) thick, not planed.

dimension shingles *Shingles* cut to uniform instead of to random widths. They are 13 and 15 cm (5 and 6 in.) wide by 40, 46, or 61 cm (16, 18 or 24 in.) long, and usually of *western red cedar*.

dimensions paper [q.s.] Foolscap paper used for *taking off*, having the eight columns of vertical ruling shown below. Columns *a* and *e* are for *timesing*. In columns *b* and *f* the dimensions are written. The squaring columns *c* and *g* are for writing the areas, volumes, etc. calculated from *b* and *f*, while columns *d* and *h* are for the descriptions of the items. The right-hand part of *d* and *h* is called the waste.

a	b	c	d		e	f	g	h	
timesing	dimensions	squaring	description	waste	timesing	dimensions	squaring	description	waste

dimension stock [tim.] Timber sawn to exact dimensions for a special purpose (BS 565).

dimension stone *Ashlar*.

diminishing courses or **graduated courses** Courses of slating so laid that the *gauge* between them diminishes from eaves to ridge. The slate sizes also diminish. Extremely careful work is needed but the appearance is very pleasant. *See* **random slates**.

diminishing piece or **diminishing pipe** [plu.] A *taper pipe*.

diminishing stile or **gunstock stile** [joi.] A door *stile* which is narrowed from the *lock rail* upwards and is glazed above the lock rail. The stile is made narrower to give more space for glass.

DIN Deutsche Industrie Norm (German industry Standard).

dinging [pla.] Rough, single-coat *stucco* on walls, consisting of cement and sand, sometimes marked with a *jointer* to imitate masonry joints.

direct cylinder [plu.] All the earliest hot-water systems (and some modern ones) use a direct cylinder; with this, the water passes directly from the boiler to the hot-water cylinder, and then out through the hot tap. *Compare* **indirect cylinder.**

direct heating The heating of a room by a heat source within it, such as an electric or gas or coal fire. *See* **indirect heating.**

direct labour The employment of building *tradesmen* and *labourers* by the *client* or his agent (engineer or *architect*) directly, without the mediation of a *contractor*. About one in every three *Local Authorities* in Britain makes use of direct labour. With good management, direct labour saves money in building and civil engineering.

dirty money Additional pay to a building worker for working in difficult or unusual conditions. *Compare* **boot man** (*C*).

disappearing stair A *loft ladder*.

discharging arch A *relieving arch*.

disconnecting trap [plu.] An *intercepting trap*.

discontinuous construction Sound insulation, with minimum *insulating materials*, of one quiet or noisy room from its neighbours in *skeleton construction* by making breaks in the construction round it, in walls, floors, and ceiling. The ideal is to build the room to be isolated in a separate box resting on cork or other sound-absorbent pads on the loadbearing concrete floor, but this is extremely expensive and a common practice now is *floating floor* construction. *See* **double partition, hollow partition, maisonette.**

disc sander A small *sanding machine* with a rotating stiff rubber disc which acts as a backing for *glasspaper*, used for *sanding* awkward corners of wood floors which the ordinary floor sander cannot approach. It can be fitted to a hand electric drill.

dispersion [pai.] Finely divided drops of liquid in an *emulsion paint* or other *emulsion* (*C*).

distemper [pai.] A heavily pigmented matt paint which can be thinned with water. The *binder* is *size* made from *casein* or other *glues*. In the washable distempers there are some *drying oils*. By this definition an *emulsion paint* cannot be a distemper because it is not heavily pigmented and sometimes has a slight gloss, yet *oil-bound distempers* are emulsions and the confusion seems well established. The distinction between 'hard' and 'soft' distempers is that washable distempers are 'hard' while 'soft' distempers can be removed by water. *See* **water paint.**

distemper brush [pai.] A flat *brush* from 12 to 25 cm (5 to 10 in.) wide, well packed with long bristle. The older two-knot brush is wire bound with two knots fixed on to one stock. *See* **house painter** (illus.).

distribution box [elec.] A small metal box joined to *conduit* to give access for connecting branch circuits.

distribution line [elec.] The main feed of an electrical circuit, to which branch circuits are connected.

distribution panel or **distribution fuseboard** [elec.] An insulated board, at which connections are made from the distribution line to the *branch circuits*. It includes *fuses* or circuit breakers for each branch.

distribution pipe [plu.] A pipe leading water from a storage *cistern*.

district heating [t.p.] Any method of heating flats or houses in one part of a town from a central store of heat. This generally means using waste heat from electric power stations or other industrial plant. Battersea power station, London, sends hot water through a tunnel under the Thames to the Pimlico (Churchill Gardens) flats. The *cooling tower* requirements of the power station are correspondingly reduced, and its thermal efficiency increased.

district surveyor A *civil* or *structural engineer* (*C*), an official peculiar to London, whose responsibility is the approval of building design and construction from the point of view of safety (fire, stability, etc.) in his district, which is usually a former London borough. He is now an employee of the Greater London Council.

diversity [elec.] It is unlikely that all the appliances in a building will be switched on simultaneously, if only for the reason that one appliance may have to be disconnected from a socket outlet in order to connect another to it. For various other reasons also the mains do not need to carry the full load of all the appliances in the building, but only a proportion of it. This proportion, the diversity factor, is the actual maximum demand divided by the sum of the full power demands of all the appliances in the building.

division wall or (USA) **fire wall** A *fire-resisting* wall from the lowest floor up to the roof. It continues 45 or 90 cm (18 or 36 in.) above the roof, if this is not fire resisting. In London its purpose is to limit each building to 7000 m^3 (250 000 ft^3) max. capacity between division walls. *Compare* **fire division wall**.

doat [tim.] *Dote*.

dog (1) A *dressing iron*.

(2) Dogs and chains are a pair of hooks hung from a chain, used for lifting building stone. The hooks grip the stone when the chain is tightened.

(3) A steel U-shaped spike used for fixing together heavy timbers, for instance a *dead shore* to a *needle*. *See* **fasteners**, *also* (*C*).

dog ear or (Scotland) **pig lug** A box-like external corner formed by folding a *flexible-metal roofing* sheet without cutting. *See* **gusset piece**.

dog-leg chisel [joi.] A chisel bent for cleaning out groves. *See* **swan-neck chisel, corner chisel**.

dog-legged stair A *stair* with two *flights* between storeys, a rectangular *half-landing*, and no stair *well*, so that the *outer string* of each flight is housed in the same *newel post*.

dog-tooth course or **dog's-tooth course** *Corbelling* out from a brick wall or making a *string course* by laying a course of *headers* diagonally with one corner projecting.

dolomitic lime *Lime* which is roughly half CaO, half MgO. *Compare* **magnesian lime.**

dome A hemispherical *vault*, circular in plan. The horizontal thrust from the dome must generally be carried by *reinforcement* (*C*) in the ring at the foot of the dome. Domes are often crowned with a *lantern*.

dome light A roof light glazed with glass curved in one or two directions. Glass domes are cast in Britain up to 1·8 m (6 ft) dia. with 25 cm (10 in.) depth, spherically curved.

door buck [joi.] (USA) A door *sub-frame* of wood or pressed metal, to which the *door case* is fixed.

door case [joi.] *See* **lining** (1).

door casing [joi.] (USA) The *architrave* or other *trim* round a door opening.

door check and spring or **door closer** [joi.] A device fixed to the top of a door, which closes it automatically and prevents it slamming.

door cheeks or **door posts** [joi.] The vertical members (stiles) of a door frame.

door frame [joi.] The surround to a door opening, usually *rabbeted*, which carries the door. It is much stronger than a *lining*, being made of timber about 10×8 cm (4×3 in.). It consists of two *door posts* (the *jambs*) and a horizontal (the *head*).

door furniture [joi.] Handles, *escutcheons*, *finger plates*, *locks*, *bolts*, *latches*, *hinges*, and so on.

door head [joi.] The horizontal wood member forming the top of a *door frame*.

door jack [joi.] A wood frame made by a *joiner* to hold a door vertical with one edge on the ground while the other is being planed.

door jamb [joi.] A *door post*.

door lining [joi.] A *door case*.

door post [joi.] A vertical in a *door frame*.

door screen [joi.] (USA) A wire screen fixed in a door frame in summertime to keep the flies out but let the air in.

door sill [joi.] A horizontal timber at the foot of the frame of an outside door. It is connected to the *door posts* and is specially designed to keep out rain.

door stop [joi.] (1) A stop for the door, cut out of the solid, or *planted*, on the *door jamb*.

   (2) A catch set in the floor to hold a door open or to prevent it opening too far (floor stop).

door switch [elec.] A switch operated by the opening or closing of a door.

dope [pai.] Quick-drying cellulose *lacquer* which was much used for tightening and protecting the fabric of aeroplanes and is used for coating textiles or leather. *See* **acetone.**

dormer or **dormer window** A vertical window or opening, coming through a sloping roof and usually provided with its own pitched roof, often with a *gablet* or *hips*. An *internal dormer* is one which has no roof other than the general roof.

dormer cheek The upright side to a *dormer*.

dosy timber [tim.] Wood which is beginning to decay. *See* **dote.**

dote or **doat** [tim.] Early *decay*, indicated by dots or speckles, described as dosy, foxy, etc.

dots (1) A wiped soldered dot is a fixing for sheet lead to timber on a *dormer cheek* or other steep surface. The lead is fixed to the close boarding by a wood screw through it, which is then covered with solder. (The term is also used for a lead peg or dowel used when *burning in*, preferably called a poured-lead dot.)

(2) [pla.] Dabs of plaster or short laths fixed at intervals in a *floating coat*, plumbed or levelled and used in fixing *screeds*. Nails are also used but are removed after floating.

dotting on [q.s.] In the *timesing column* of the sheet on which the *quantities* are taken off, a dot added to the timesing figure indicates that the quantity occurs once more. Some surveyors regard it as a source of error and condemn the practice. *See* **taking off.**

double-acting hinge [joi.] A *hinge* which allows a door to swing through 180°, and usually makes the door self-closing by a spring contained in it.

double bead [joi.] Two parallel *beads* separated by a *quirk*.

double brick A brick 23 cm (9 in.) square in plan, with two *frogs*, made in the UK since 1959.

double-bridging [carp.] Two rows of *herring-bone strutting* to each span, dividing the floor area into three equal parts.

double connector [plu.] A short piece of pipe with, at each end, a long *parallel thread* fitted with a back nut and a socket. It enables a gas supply to be interrupted in an emergency since it is a convenient method of connecting and disconnecting pipes in any position, whoever awkward. *See* **connector.**

double door [joi.] *See* **folding door.**

double-door bolt [joi.] An *espagnolette bolt*.

double-dovetail key or **dovetail feather** or **hammerhead key** [joi.] A hardwood key shaped like two *dovetail* pins joined at their narrow ends, driven into a butt joint between two timbers to hold them together. It is of the same shape as a *slate cramp*.

double dwelling *See* **double house.**

double eaves course or **doubling course** A double row of *shingles*, *plain tiles*, or *slates* laid at the foot of a roof slope or vertical section of slating, tiling, or shingling. *See* **eaves course, tilting fillet, under-cloak.**

double faced [joi.] A description of *architraves*, *skirting boards*, etc. with *mouldings* facing in two different directions.

double-face hammer A *hammer* having two striking faces, one at each end of the head.

double Flemish bond Brickwork which shows *Flemish bond* on both faces of the wall.

double floor (1) or **double-framed floor** [carp.] A floor consisting of *joists* spanning between *binders*. *See also* **single floor, framed floor**.

(2) A *single floor* consisting of a counterfloor carrying a second, finished timber floor.

double glazing *Glazing* in which two layers of glass are separated by an air space for thermal or acoustic insulation. Generally the two layers are mounted on a metal window frame and the air space is sealed, an impossibility with two wooden frames. With a 13 mm ($\frac{1}{2}$ in.) thick air space the *U-value* is 3·06 W/m² deg. C. (0·54 Btu/ft²h deg. F.), that for a single glazed window being 5·85 W/m² deg. C. (1·03 Btu/ft²h deg. F.). Heat losses are therefore nearly halved and condensation usually prevented, but this type only slightly reduces the volume of sound passing through the window. The *storm window*, or any other type with the two panes 5 cm (2 in.) apart or more (preferably 10 cm (4 in.)), will give much better sound insulation, particularly if the two panes of glass are set in different frames. In a very severe climate the inner window frame should be insulated from the building frame.

double-handed saw [carp.] A long *cross-cut* saw pulled by one man at each end.

double header [carp.] (USA) A *trimmer joist* near an opening in a floor or wall, made by nailing together two ordinary joists.

double house (USA) A pair of *semi-detached houses*. *See* **duplex**.

double-hung sash window [joi.] The term recommended by BS 565 for a *sash window*.

double jack-rafter [carp.] (USA) A *rafter* which joins a *valley* to a *hip*.

double-lock welt A *cross welt*.

double-margined door or **double-margin door** [joi.] A door which is hinged at one side only but looks like a pair of double doors.

double measure [joi.] A description of *joinery* moulded on both faces.

double partition [carp.] A *partition* built from two separate rows of studding either for sound reduction or to form a cavity for a sliding door. *See* **discontinuous construction**.

double-pitch roof (1) A *mansard roof*.

(2) A *pitched roof*.

double-quirk bead or **return bead** [joi.] A *bead* recessed into a surface or corner by a *quirk* each side.

double rebated [joi.] A description of a wide *door post* or jamb lining which is rebated on both edges. The door may thus be hung to open inwards or outwards.

double-return stairs or **side flights** A *stair* with one wide *flight* up from the lower floor to the *landing* and two flights from the landing to the next floor.

double-roll verge tile A *single-lap tile* with a roll on each edge so that the *verges* both right and left are similar.

double Roman tile A *single-lap* standardized British clay *roofing tile* measuring ordinarily $42 \times 36 \times 1 \cdot 3$ to $1 \cdot 6$ cm ($16\frac{1}{2} \times 14 \times \frac{1}{2}$ to $\frac{5}{8}$ in.). It has $7 \cdot 6$ cm (3 in.) side lap. There are no *nibs* but two nail holes are provided, of about 5 mm ($\frac{3}{16}$ in.) dia. It differs from the single *Roman tile* in having a roll up the centre. The central roll resembles the edge roll which overlaps the next tile at one side. *Compare* **Poole's tile.**

double roof [carp.] A roof in which the *common rafters* are carried on *purlins* which may rest on a roof *truss* or on other intermediate supports. *See* **single roof.**

double-skin roof An *asbestos-cement* corrugated roof covering which consists of an upper layer, the weathering, and a lower layer, the underlay, which is flat and forms the ceiling.

double skirting [joi.] A *skirting* made higher than normally by rebating a second, upper board into the ordinary skirting board.

double step [carp.] In heavy timber framing, such as the support of a *rafter* on a tie-beam at *eaves* level, a W-shaped notch designed to reduce the likelihood of horizontal shear in the tie-beam. *See* **step joint.**

double-tier partition A *framed partition* two *storeys* high.

double time Double payment customary in the British building industry for men working after 4 p.m. on Saturday until Monday morning, or later than 4 hours after normal finishing time on a weekday. *See* **time-and-a-half.**

double window (1) A *storm window*.

(2) *Double glazing.*

doubling course A *double eaves course*.

doubling piece [carp.] A *tilting fillet*.

Douglas fir (Pseudotsuga taxifolia) or **Oregon pine** or **fir** or **British Columbian pine** [tim.] The most important building *softwood* in the British Commonwealth, also very important in USA. It has a remarkably high *modulus of elasticity* (*C*) of 16 000 N/mm^2 ($2 \cdot 3 \times 10^6$ psi) at 12% *moisture content*, at which its density is 590 kg/m^3 (37 lb/ft^3). Since this value is nearly twice that of other softwoods it deflects only half as much. Otherwise it resembles European *redwood*, though it is less dense. *See* **fir.**

dovetail [joi.] A joint often used in the corners of boxes or fine joinery. It differs from the *combed joint* in that the interlocking tenons (called pins) are fan-shaped, like a pigeon's tail. They are thicker at the end than at the root, and can therefore not easily be pulled out.

dovetail cramp A *cramp* for stone which is of double-dovetail shape and may be of slate or metal.

dovetailed lathing or **dovetail sheeting** [pla.] Steel or plastics sheet bent into dovetailed shape with corrugations of about 19 mm ($\frac{3}{4}$ in.) depth and greatest width. It is used as *metal lathing* since it can be plastered on both faces, or used as permanent *formwork* (*C*) for a concrete floor.

dovetail feather [joi.] A *double-dovetail key*.

dovetail halving or **dovetail halved joint** [carp.] A joint made by *halving*, in which the halved pieces are dovetailed.

dovetail joint [joi.] A *dovetail*.

dovetail margin [joi.] A *banding* which is dovetailed.

dovetail saw [joi.] A *back saw* about 20 cm (8 in.) long, with about 7 points per cm (18 per in.).

dovetail sheeting *Dovetailed lathing*.

dowel (1) A short, round *hardwood* rod used instead of or together with a *tenon* for holding two wood parts together, by inserting it into a hole drilled in each. It should be grooved (keyed) to enable air and excess *glue* to escape. *See* **treenail**.

(2) A short steel rod cast into a concrete floor, over which a door post can be fixed by dropping the bored post over the dowel.

(3) A piece of slate or a *cramp* used for locking adjacent stones to each other in a wall.

dowel bit or **spoon bit** [carp.] A drilling bit of half-cylindrical cross section, similar in shape to a *pod auger*.

dowel pin [joi.] (1) A short round *wire nail* pointed at both ends.

(2) (USA) A headless nail with one point and a barbed shank. It is driven into a *mortise-and-tenon joint* to fasten it permanently.

(3) A wood *dowel*.

dowel plate [carp.] A steel plate in which holes are drilled to the diameters of dowels. It can be used for verifying the diameter of a dowel or for making a dowel by driving the peg through it to remove surplus wood.

dowel screw or **handrail screw** [joi.] A wood *screw* threaded at each end.

downcomer [plu.] A pipe leading water from the cistern to the WC, water heater, wash basin, and bath. Some of these are now directly connected to the main supply. *See also* **downpipe**.

downpipe or **downcomer** or **rainwater pipe** A vertical or very steep pipe which brings rainwater to the ground from roof *gutters*, the successor to the gargoyle. The earliest pipes were of lead, which began to be replaced in Britain by cast iron in the nineteenth century. This in turn is being superseded by asbestos-cement, sheet steel, copper, aluminium and plastics. Corresponding American terms are downspout, conductor, and leader.

dozy [tim.] *See* **dote**.

draft A smooth margin or strip worked on a stone face to the width of a *draft chisel*, generally for levelling or squaring the stone. It may be straight or curved.

draft chisel or **drafting chisel** A *chisel* struck with the *mallet* and used for making a *draft* on the face of a stone.

drafted margin A smooth, uniform margin 1·9 to 5 cm (¾ to 2 in.) wide, worked round the edges of the face of a stone.

draft stop (USA) A *fire stop*.

drag (1) A steel plate about 15 cm long, 10 cm wide (6 × 4 in.), with toothed edges, used for levelling plaster surfaces, for producing a key for the next coat of plaster, or for smoothing *ashlars*. *See* **plasterer, mason's tools** (illus.), **comb**, *also C*.

(2) [pai.] or **brush drag** or **pulling** A resistance to the *brush* from a *paint* or *varnish* while it is being put on, sometimes a serious defect. *See* **gummy**.

dragged work Stone which has been smoothed with a *drag*.

dragon beam or **dragon piece** [carp.] A horizontal timber into which the end of the *hip rafter* is framed. The outer end of it is carried on the corner of the building where the *wall plates* meet, the inner end at the *angle tie*.

dragon's blood [pai.] Red *resins* obtained from palm trees, used for tinting *varnishes*.

dragon tie [carp.] An *angle tie*.

drain chute A *special* drain pipe tapered in its upper half at the point where a drain pipe enters or leaves a manhole. It is shaped in such a way that *rodding* (C) shall be easy.

drain cock A cock placed at the lowest point of a water system, through which the system can be drained when required.

drain ferret A thin glass bottle containing strongly smelling smoke which is broken inside a drain to reveal leaks.

drain pipes Pipes, as well as their joints, may be either rigid or flexible. Flexibility is desirable in areas where the ground may move, e.g. near mines. Flexible pipes, of *pitch fibre*, plastics, ductile cast iron, or steel, deform appreciably before collapse. Rigid pipes, of asbestos cement, vitrified clay, concrete, or grey cast iron, break before their deformation becomes noticeable. Pipe joints made flexible by a plastics or rubber ring have the advantage that they resist most types of corrosion. Rigid pipes are not softened by standing in the hot sun, as pitch or pvc pipes may be. *See* BSCP 301 and 304.

drain plug or **drain stopper** A *bag plug* or *screw plug*.

drain tests [plu.] Drains must be tested for leakage, after installation and before they are covered up with earth, so that the local authority can have alterations made if it does not approve the work. Prospective house buyers or building societies also may insist on a drain test. *See* **air test, chemical test, rocket tester, scent test, water test**.

draught (1) The pressure difference at the foot of a *chimney* between the air outside and that inside the chimney. This pressure difference (caused by the air inside being hotter and lighter than that outside) draws air up through the fuel bed into the chimney. The draught is sometimes measured in inches of water.

(2) [carp.] In *drawboring*, the amount by which the holes are out of line so as to ensure a tight joint.

draught bead [joi.] A *deep bead*.

draught fillet or **windguard** In *patent glazing*, a strip to fill the space between the underside of the glass and its support.

draught stop A *fire stop*, *see also* **sash stop**.

draw bolt [joi.] A *barrel bolt* or other simple bolt pushed by hand, not driven by a key.

drawbore [carp.] To drill holes through a *tenon* and the mortised piece about 3 mm ($\frac{1}{8}$ in.) out of line so that a tapered steel pin driven through both pieces will draw them more closely together. This method of cramping during gluing is more effective than *cramps* since the pressure is never relaxed. The method is such a solid fixing that gluing is not always considered necessary. *See* **draught, drawbore pin.**

drawbore pin or **drawpin** [carp.] A tapered steel pin used in drawboring to bring the holes into line, and then withdrawn before the *treenail* is inserted.

draw-in system [elec.] A carefully planned wiring system, in conduits or ducts; the cables can be pulled in and replaced when required.

drawknife [carp.] A knife blade bent into a U-shape with *tangs* at each end, each tang carrying a handle so that the knife can be pulled by both hands towards the user. The handles and blade are in the same plane.

drawn sections *Architectural sections* made by drawing through a die.

drencher system A system of water sprays to protect the outside of a building from fire. They are now, usually, automatically operated like *sprinklers*, but they have been hand operated as at the safety curtain of a theatre.

dress (1) [carp.] To plane and sandpaper timber.

(2) To cut stones to their final shape.

dressed and matched boards [tim.] Planed *matchboards*.

dressed size [joi.] The dressed size of timber may be 9 mm ($\frac{3}{8}$ in.) less than the *nominal size* in both directions.

dressed stone Stone which has been squared all round and smoothed on the face.

dressed timber (Scotland) *Surfaced timber* (planed timber).

dresser [plu.] A *hardwood* tool shaped for beating lead; *see* **bossing.**

dressing (1) Finishing building stones. *See also* **dressings.**

(2) [plu.] *Bossing*.

dressing compound Bituminous liquid used hot or cold for dressing the exposed surface of *roofing felt*. *Compare* **bonding compound, sealing.**

dressing iron or **break iron** or **dog** or **traverse** A 46 cm (18 in.) long steel straight edge with spikes at each end for fixing into a work bench. A *slate* is laid on the dressing iron and the *zax* brought down on to the slate to cut it neatly. *See* **plasterer's tools** (illus.).

dressings *Masonry* or *mouldings* round openings, or at the corners of buildings, of better quality than the remainder of the facing brick or stone. Rubble walls or brick walls were often built with dressings of stone or of *gauged brick*.

drier [pai.] Any compound of lead, cobalt, manganese, etc. which encourages the *oxidation* of *drying oil* in a paint or varnish. *See* **lead drier, paste drier, soluble drier.**

drift bolt [carp.] A steel pin usually not less than 2·2 cm ($\frac{7}{8}$ in.) dia. driven into holes bored 1·5 mm ($\frac{1}{16}$ in.) smaller as a fixing between heavy timbers. *See also* **drift** (*C*).

drift plate [plu.] A steel plate for dressing one lead sheet over another.

drift plug [plu.] A wooden plug driven through a lead pipe to straighten a kink.

drill bow [joi.] The bow of a *bow drill*.

drip (1) or **throat** A groove under an overhanging edge of a *cornice, coping, moulding*, or roof, designed to stop water flowing back to the building and to throw it off at the outer edge of the overhang.

(2) In *flexible-metal roofing*, a step formed in a flat roof at right angles to the direction of fall. The lower sheet is turned up and over the riser, under the *overcloak*, which is bent down the riser and sometimes on to the flat.

(3) A *drop apron*.

drip cap (mainly USA) A small weathered projection over a door or window opening, with a *drip* under it.

drip channel A *drip* or throat.

drip edge The free lower edge of a *flexible-metal* roof which drips into a gutter or into the open. It is often stiffened with a *bead*.

drip-free paint *See* **thixotropic** (*C*).

dripping eave An *eave* with no gutter.

drip sink A shallow sink near floor level to take drips from a tap. A *lead safe* is a drip sink of lead sheet.

drive screw or **screw nail** [carp.] A galvanized nail 2 mm (0·08 in.) or more thick, with a steep screw *thread* formed round it. It can be driven in by a hammer but is very difficult to withdraw except by turning. It is used for fixing roof sheeting. The spring-head roofing nail is similar, with an even steeper thread.

drop annunciator An *annunciator* in which a signal drops to show the number of the room from which the signal originated.

drop apron or **drip** In *flexible-metal roofing*, a strip of metal fixed vertically down at *eaves, verges*, and *gutters*, held by a *lining plate*.

drop ceiling A *false ceiling*.

drop connection or **drop manhole** A *back drop*.

drop elbow or **tee** [plu.] A small *elbow* or tee with *ears* for screwing it to a wall.

drop escutcheon or **drop key plate** [joi.] A small metal plate pivoted above a keyhole to cover it when the key is not in the *lock*. It matches the *escutcheon*.

drop moulding [joi.] *Moulding* in a panel below the surface of the framing.

drop-point slating *Diagonal slating*.

drop siding or **rustic siding** [tim.] Rebated and overlapping, or tongued and grooved *weather-boarding*.

drop system [plu.] A heating circuit in which the *flow* pipe rises directly to its highest level, from which it feeds downward branches which drop as nearly vertically as possible to a return main.

drop window [joi.] A *sash window* which descends completely into a pocket beneath the *sill*, leaving the whole space for ventilation.

drop wire [elec.] A cable from the nearest pole of an overhead supply line, connecting a house with the supply.

drove (Scotland) A *boaster*.

drowning pipe An inlet pipe to a *cistern*, which has its outlet below the water level so that its noise is much reduced.

drummer or **drummer up** The boy or man who makes the tea on a building site.

drunken saw or **wobble saw** [tim.] A *circular saw* which is deliberately set slightly off the perpendicular to its own shaft so that it makes a wide cut. It is used in *joinery* for cutting *open mortises*.

dry area or **blind area** A narrow roofed *area* between a basement wall and the retaining wall outside, designed to keep the basement wall dry.

dry construction Building without plaster or mortar as in methods using much wood, *plywood*, *precasting* (*C*), or other *prefabricated-building* methods. The building is ready for occupation very quickly but may be expensive.

dry hydrate *Hydrated lime* powder bought in bags.

drying [pai.] The hardening of a coat of paint or varnish by *evaporation* of the *vehicle* or chemical change (usually *oxidation*) or the two together. Air drying is drying at air temperature; forced drying or *stoving* is drying at higher temperatures. *See* **dry to handle, drying oil, hard dry, dust dry, surface dry, touch dry.**

drying oil [pai.] A vegetable or animal oil which forms a tough film by *oxidation* when exposed to air in a thin layer. Drying is often made more rapid by a *drier*. Most drying oils, properly treated, should dry within 48 hours. *Linseed oil* is the commonest drying oil.

dry masonry Walling laid without *mortar*.

dry partition A modern partition wall, prefabricated often in areas as large as 4 m × 1 m (13 × 3 ft) that can therefore be erected without plastering, and very quickly. Some are made of plasterboard with honeycomb paper core, others are of expanded polystyrene faced both sides with hardboard, or of hollow precast plaster. All can be sawn to fit. Precast plaster *coves* fitted to the ceiling joint can improve the looks. Before the coves are nailed, the position of the ceiling joists must be found and marked.

dry press Making *cast stone* with a very dry *mix* (*C*) so that the stone can be demoulded quickly.

dry-press brick (USA) A *brick* of good quality made from nearly dry clay (5% to 7% moisture) pressed in moulds at pressures from 3·4 to 10 N/mm² (500 to 1500 psi).

dry rot [tim.] Timber *decay* due to dampness (the fungus usually responsible being merulius lacrymans) but not such intense dampness as that which causes *wet rot*. Dry rot has a noticeable smell, spreads rapidly even through brickwork, and is difficult to eradicate from a house without burning all decaying timber, disinfecting the remainder, and removing the infection from the brickwork by heating it with a *blow lamp* or disinfecting it thoroughly.

dry sprinkler *See* **sprinkler.**

dry stone wall *Dry walling.*

dry to handle [pai.] The last stage in the *drying* of a paint film when it can be freely handled without damage.

dry walling (1) *Rubble walls* built without mortar.

(2) *See* **dry construction.**

dry wood or **dry stock** [tim.] *Timber* after seasoning, having, in Britain, from 15% to 23% *moisture content*.

dual fuel system A heating system in which oil or coal may be replaced by gas fuel or the reverse.

dual system [plu.] The *two-pipe system.*

dubbing out [pla.] Filling hollows in a wall surface with *coarse stuff* or roughly forming a *cornice* before running the *finishing coat*.

dub off [joi.] To remove arrises from, for example, a *tenon*, to enable it to enter a *mortise*.

duckbill bit [carp.] A *dowel bit*.

duckbill nail [joi.] A chisel-pointed nail, very easy to *clench*.

duckboard A *cat ladder*.

duck-foot bend or **rest bend** A right-angle bend often used at the foot of a column of vertical cast-iron pipes. It is provided with a flat seating to carry the weight of the pipes and water and the thrust due to the change in direction of the water.

duct (1) A subway, *crawlway*, *creep trench*, *chase*, or *casing* which accommodates pipes or cables in a building.

(2) Metal, wood, or *asbestos-cement* or plastics tubes, round or rectangular, for distributing conditioned air to rooms or for withdrawing stale air or fumes from them. *See* **cable duct** (*C*) and BSCP 413, Service ducts.

dumb waiter (1) (USA) An *elevator* which raises or lowers food or crockery from one *storey* to another.

(2) (Britain) A piece of table furniture. It stands on the centre of a table and consists of a rotatable circular tray carrying condiments.

dummy (1) (*See* **plumber**) A lump of lead or iron fixed on the end of a long cane or iron rod, used as an internal mallet for straightening large lead pipes.

(2) A round lump of zinc or lead weighing 1 to 2 kg (2 to 4 lb), with a short wooden handle, used as a hammer by *masons* who are working soft stone, such as Bath stone, with wood-handled chisels or gouges.

dunnage [tim.] Waste timber.

dunter (1) A *monumental mason* who prepares large faces of granite for polishing with a pneumatic surfacing machine.

(2) The pneumatic surfacing machine itself.

duodecimal system The feet and inches system or other systems in which 12 small units make up a large unit. Calculations can be made in this system, but British *quantity surveyors* are the only people who use duodecimal calculation.

duplex apartment (USA) A *maisonette*.

duplex burner A gas burner with two sections which can burn together at full load, one being used alone for reduced heating.

duplex dwelling (USA) A two-family dwelling in which the living units are one above the other.

duramen [tim.] The *heartwood* of a tree.

dust dry or (USA) **dust free** [pai.] A stage in the *drying* of a finish, after which dust will not stick to it.

dusting [pla.] Wear of a concrete floor surface, usually caused by excess water in the mix, careless laying or poor *curing* (*C*).

dusting brush [pai.] Any *brush* used for removing dust before painting.

Dutch arch or **French arch** A brick arch, flat at top and bottom, of which only the central bricks are wedge shaped.

Dutch barn A steel-framed building without walls, with a curved roof.

Dutch bond A confusing term which may mean two different things:

(1) *English cross bond*.

(2) *Flemish bond*.

Dutch door (USA) A *stable door*.

Dutchman A piece of material used to cover up a mistake, in carpentry or other trades.

dwang (1) [carp.] (Scotland) *Strutting* between floor joists.

(2) A crowbar.

dwarf partition A *partition* which does not rise to the ceiling.

dwarf wall A low wall such as the wall supporting the ground floor *joists* of a dwelling house.

dye [pai.] A colouring material which, unlike a *pigment*, is soluble and colours materials by penetration. When a dye colours an insoluble base, such as aluminium hydroxide which is used as a pigment, the resulting pigment is called a *lake*.

dyke or **dike** (Scotland) A *dry wall* in stone.

dyker or **stone ditcher** or **stone hedger** A *walling mason* who builds boundary walls of stone.

E

ear [plu.] or **lug** A projection from a pipe or other *fitting*, to fix it to a wall.

earth (1) Excavated material; strictly speaking only *topsoil* (*C*) is earth.
(2) [elec.] or (USA) **ground** An electrical connection to earth through an *earth electrode* and suitable conductors to it. *See* BSCP 1013, Earthing.

earthed concentric wiring [elec.] A cable in which one conductor, a metal tube, is earthed and contains the other conductor insulated within it. *See* **Pyrotenax**.

earth electrode [elec.] A metal plate, water pipe, or other conductor electrically connected to earth, preferably in a position where the earth is always damp, to ensure low electrical resistance. *See* **earth plate**.

earthenware Pottery from brick earth as opposed to *stoneware* which is harder and non-porous. For building drainage, *glazed ware* pipes are widely used.

earthing lead [elec.] The conductor which makes the final connection to an *earth electrode*.

earth plate [elec.] An *earth electrode* consisting of a large metal plate sunk in damp ground or water.

earth table or **ground table** or **grass table** The lowest course of *masonry* projecting near ground level.

earth termination network [elec.] Those parts of a lightning protective system which distribute the discharge to earth.

easement In law, a right which a person may have over another man's land, such as the right to walk over it or to run a pipe through it. It does not include any rights of tenants.

easing the wedges Slackening the *folding wedges* of *shoring* after completion of an *underpinning* (*C*) job, or of arch *centers* after building an arch.

easy-clean hinge [joi.] A *cleaning hinge*.

eave or **eaves** The lowest, overhanging part of a sloping roof.

eaves board or **eaves catch** A *tilting fillet*.

eaves course A first course of *plain tiles*, *slates*, or *shingles* on a roof, including the course of plain tiles at *eaves* on which the first course of *single-lap tiles* is bedded. *See* **double eaves course, eaves tile**.

eaves fascia [carp.] A board on edge nailed along the feet of the rafters. It often carries the *eaves gutter* and may also act as a *tilting fillet*.

eaves flashing A *drop apron* from an asphalt roof dressed into an eaves gutter.

eaves gutter A rainwater *gutter* along the *eaves*.

eaves plate [carp.] A wall plate spanning between posts or piers at *eaves*. It carries the feet of the rafters when no wall is there to carry them.

eaves pole [carp.] A *tilting fillet*.

eaves tile A short tile about 18 cm (7 in.) long used in the *eaves course* (or *under-eaves course*) in *plain tiling*. *Compare* **under-ridge tile**.

eaves trough or **eaves trow** An *eaves gutter*.

echo Repetition of sound by reflection from walls such as the rear wall of an auditorium; the bad effects of echo are reduced by *absorption*.

economy brick (USA) *Brick* to fit the 10 cm (4 in.) height *module*, therefore measuring about $19 \times 9 \times 9$ cm ($7\frac{1}{2} \times 3\frac{1}{2} \times 3\frac{1}{2}$ in.) to make, with its mortar joints $20 \times 10 \times 10$ cm ($8 \times 4 \times 4$ in.). *Compare* **engineered brick**.

economy wall (USA) A 10 cm (4 in.) thick brick wall plastered or rendered, stiffened at intervals with 20 cm (8 in.) *piers* carrying the roof *trusses*, and projecting outwards at each side of doors and windows.

edge bedding *See* **face-bedded**.

edge bend or (USA) **crook** [tim.] *Spring*.

edge grain or **vertical grain** or **comb grain** [tim.] A grain seen in wood which is converted so that the *growth rings* are at 45° or more to the face of the piece. Such wood is called *quarter-sawn*. The best quarter-sawn oak shows *silver grain*.

edge isolation *See* **expansion strip**.

edge joint [tim.] A joint in the direction of the *grain*, between two *veneers*. *Compare* **butt joint**.

edge nailing [carp.] *Secret nailing* of floor boards, etc.

edge runner A grinding mill consisting of circular rolls driven round in a circular steel bowl containing the material to be ground (mortar, putties, etc.).

edge-shot board [carp.] A board with a planed edge.

edge tools [carp.] Tools with a cutting edge, particularly the *hatchet*, *chisel*, *plane*, *gouge*, knife.

edge trimmer [joi.] A *plane* with a perpendicularly recessed sole for making an edge of small timber square to its face.

edging strip [joi.] A *banding* over the edge of a flush door.

edging trowel A rectangular trowel with one edge turned down so as to trim the edges of kerbs, etc.

eel grass A sea plant (Zostera marina) whose dried leaves when loosely packed are very sound absorbent. It weighs 65 to 80 kg/m³ (4 to 5 lb/ft³) in the form of matting stretched between two sheets of *kraft paper*, is sold in rolls or *blankets*, and has a thermal conductivity of 0·039 W/m deg. C. (0·27 Btu in./ft²h deg. F.). *See* **insulating materials**.

effective temperature An American conception which corresponds to the British *equivalent temperature* and was developed by the *ASHVE* from 1923 onwards. It takes into consideration the temperature, humidity, and speed of movement of the air but does not consider radiation. *See* **comfort zone.**

efflorescence Powdery white salts left on a wall surface as it dries out. It is unsightly though usually harmless except that it can lift any paint which covers the brick or mortar. *See* **deliquescence.**

eggshell (1) [pai.] *See* **gloss.**

(2) A smooth, matt face to building stone.

ejector grille A ventilating grille with slots shaped to force the air out in divergent streams.

elastomer Any synthetic or natural rubber, sometimes in the rubber industry defined as a material that is resilient enough to be stretched to twice its length and, on release, to snap back to the original length. *See* **polymer, fleximer.**

elbow [plu.] A sharp corner in a pipe, as opposed to a bend which is a smooth corner. A right-angled pipe *fitting.*

elbow board [joi.] A *window board.*

elbow lining [joi.] The panelling over a window *jamb.*

electric-blanket heating [tim.] A method of accelerating *assembly gluing* by covering the glued joint with an electrically heated blanket.

electrician A *tradesman* who instals or repairs electric circuits (wiring), machines, or plant. *See* BSCP 1017, Use of Electricity on Building Sites.

electric-panel heater A panel heated by electrical resistances. It may be surfaced with marble, plastics, etc., and can work at a high temperature of about 290° C. (550° F.), at a medium temperature below 120° C. (250° F.), or at a low temperature below 81° C. (180° F.). *See* **panel heating.**

electric screwdriver A tool like a hand-held *electric drill* (*C*). It is used on mass-production work because of its speed of operation and the exactness to which all screws are driven with the same force. This is obtained by a clutch which slips at the final *torque* (*C*).

electro-copper glazing Pieces of glass are accurately cut and assembled between flat copper strips (cames) and pressed tightly together. (The copper strips are brought into electrical contact by soldering.) The assembly is then placed in an electrolyte containing copper salts and the copper strips are made the cathode so that more copper is deposited on them. The glass is thus held tightly. In USA called copper glazing or copperlite glazing. *See* **fire-resisting glazing.**

electrode boiler [elec.] A *boiler,* generally larger than domestic size, which differs from the *immersion heater* in that the water is heated by alternating current passing through it. It has the advantage over a fuel-fired boiler that no surfaces are touched or damaged by flame. No overheating or short circuiting can occur if the boiler is empty when the power is turned on. As in fuel-heated boilers, the safety device against high pressure is a safety valve.

electrolier (mainly USA) A hanging electric light fitting.

elemi [pai.] An *oleo-resin* obtained from the tropics, particularly the Philippines, used for making spirit lacquers and *nitrocellulose* products.

elephant spindle [tim.] A special *spindle moulder*.

elevator (USA) A *lift* for passengers or goods in USA.

ell [plu.] A pipe shaped like an L; *see* elbow.

elliptical stair A *stair* with a *well* which is ellipse shaped in plan.

elm or **common elm** (Ulmus) [tim.] A dull brown *hardwood* which *warps* badly if not carefully seasoned. It should be kept either wet or dry but not allowed to alternate between the two. It has twisted grain and is even harder to split than *oak*, but in other respects is slightly weaker. It is cut as *burr* for veneers, or used in the solid for *piles* (*C*) *weather-boarding*, and panelling.

emery cloth *See* glasspaper.

eminently hydraulic lime *Hydraulic* lime burnt from a limestone containing more than about 25% of aluminium silicates. Such limes cannot be obtained as *dry hydrate* without loss of hydraulic strength.

emissivity of a surface *See* absorptive power.

emulsifier system A *sprinkler system* which works at a minimum pressure of 345 kN/m² (50 psi) and is installed in places where oil may catch fire, for example, near *transformers* (*C*). A powerful spray is directed on to the oil, which is emulsified by the jet, and thus prevented from burning, since each drop is thus surrounded by water. After some hours the oil may separate from the water and be re-used, but this does not occur with all oils.

emulsion paint [pai.] A *dispersion* of a liquid, occasionally of a solid, called the dispersed phase, in another liquid, the continuous phase. Water is usually the continuous phase in emulsion paints, and such oil-in-water emulsions, like milk, will take more water. They may be stabilized by colloids like *casein*, or other glues. *Oil-bound distempers*, coloured bituminous emulsions, and *latex emulsions* are all emulsion paints. Because they harden by *evaporation* of water, not by *oxidation*, they offer less *fire hazard* than *oil paints* or *lacquers*.

enamel (1) [pai.] A *hard gloss paint* whose high *gloss* is obtained by a high proportion of *varnish* with reduced *pigment* content. Enamels *flow* well but need good *undercoats* since they are not opaque.

(2) Vitreous enamel is a glass surface, often white, attached by firing to cast-iron or pressed-steel articles like baths. It is very much more resistant to wear than enamel paint but it chips when struck a hard blow. When sawing it, the chipping can be prevented by first covering the surface with adhesive tape.

enamelled brick A *glazed brick*.

encase [joi.] To cover with a case or lining.

encased knot [tim.] A *dead knot*, surrounded with bark or resin.

encaustic decoration Decoration burnt on to ceramics or glass.

enclosed fuse [elec.] (USA) A *cartridge fuse*.

121

enclosed knot [tim.] A knot which does not appear on the surface of timber.

enclosed stair A *closed stair*.

enclosure wall (USA) Any exterior non-loadbearing wall in *skeleton construction*, which may be not carried by the frame at every level, but is generally anchored to it horizontally.

end grain [tim.] The surface of timber exposed when a tree is felled or when timber is *cross-cut* in any other way.

end joint [tim.] A *butt joint*.

end-lap joint [carp.] An *angle joint* formed between two timbers by *halving* each for a length equal to the width of the other.

endless saw [tim.] A *band saw*.

engineered brick (USA) A *brick* which measures, with its mortar joint when laid up, $20 \times 10 \times 8$ cm ($8 \times 4 \times 3 \cdot 2$ in.), that is 5 courses for 40 cm (16 in.) height. *Compare* **economy brick**.

engineering brick Bricks of uniform size with a high crushing strength. Class A British engineering bricks crush at stresses above 69 N/mm² (10 000 psi), class B at stresses above 48 N/mm² (7000 psi). (*See* **Staffordshire blues**.) The maximum water *absorption* for these bricks is limited to 7 % for class B and $4\frac{1}{2}$ % for class A; in fact the latter are sometimes used as a *damp course*.

engineer's hammer or **fitter's hammer** [mech.] A *hammer* with a head weighing from $0 \cdot 1$ to $1 \cdot 4$ kg (4 oz to 3 lb) on which are a striking face, and a *ball peen*, or a *cross peen*, or a *straight peen*.

English bond A brick *bond* in which alternate *courses* are composed entirely of *stretchers* or entirely of *headers*.

English cross bond or **Saint Andrew's cross bond** A *bond* like *English bond*, except that in alternate stretcher courses a *header* is placed next to the *quoin* stretchers. Therefore the vertical joints of alternate stretcher courses are displaced a half-brick from each other and are not in the same vertical line as in English bond. For this reason, this bond can be much more decorative than English bond, since the *cross joints* in the stretcher courses line up, with a quarter-brick displacement, with alternate cross joints in the header courses. In this way a pattern of diagonal lines crossing each other shows on the face of the brickwork. With the judicious use of *flared headers*, diagonal patterns can be picked out in it. *See* **Flemish diagonal**.

English garden-wall bond *American bond*.

English roofing tile (USA) A *single-lap* clay roofing tile, rarely if ever seen in England. The sides of the tile overlap within its thickness so that both the visible and the hidden surfaces of the tile are smooth.

epoxide resin or **ethoxylene resin** A *synthetic resin* used for gluing metal parts of aeroplanes as well as concretes, and useful for house repairs in the same way as the *glassfibre-reinforced* resins.

equilibrium moisture content [tim.] The *moisture content* at which timber neither gains nor loses moisture, when subjected to a given constant condition of humidity and temperature. *See also C*.

equivalent temperature An idea evolved at the *Building Research Station* to make use of the cooling effect of air currents and radiation (due to the temperature of the walls of the room). It is defined as the temperature of an enclosure in which, in still air, the *eupatheoscope* would lose heat at the same rate as in the environment under consideration. Unlike the American *effective temperature*, equivalent temperature does not take humidity into account and is not therefore used in heavy industry where men sweat. *See* **comfort zone.**

erection The positioning and fixing of the parts of a metal, or timber, or precast concrete frame.

ergonomics The interactions between work and people, particularly the design of machines, chairs, tables, etc. to suit the body, and to permit work with the least fatigue.

erosion [pai.] The wearing away of a painted surface by *chalking* or abrasion.

escalator or **moving stair** [mech.] A moving endless belt with steps on it, which allows crowds to move more quickly into or out of shops, underground stations, etc. than by stairs or lifts. It does however need more space than lifts or stairs, particularly since one upward and one downward moving stair may be needed simultaneously. *Compare* **inclinator.**

escape stair A *stair* which is required by law to be provided as an escape in the event of fire. It may be inside or outside a building and is particularly needed for tall buildings, in which the only stairs provided may be escape stairs.

escutcheon or **scutcheon** or **key plate** [joi.] A metal plate round a key hole.

escutcheon pin [joi.] A brass nail 1 cm (0·5 in.) long for fixing an *escutcheon.*

espagnolette or **cremorne bolt** [joi.] A vertical bolt in *casement doors* which runs the full height of the window, being split at the middle, where each half joins the handle. The bolts can be driven home at *sill* and *door head* simultaneously, by turning the handle. The handle is usually designed so that the door can be fixed ajar.

Essex board measure [tim.] (USA) *Board measure.*

establishment charges *Overheads.*

estate agent One who manages, buys, or sells property. He may belong to the Royal Institution of Chartered Surveyors, to the Auctioneers and Estate Agents Institute, to the Incorporated Society of Auctioneers and Landed Property Agents or he may hold a university degree in estate management. Any of these is a qualification for the work.

estimating [q.s.] Determining the probable cost of work by multiplying the volume of different operations by costs per unit of measurement known from recent work. *See* **cubing.**

etching (1) Cutting (usually a decorative pattern) on the surface of glass or metal with an acid. Metal engravings are cut away (etched) by acid where they are not covered by a 'resist' such as a wax, which protects them from the acid.

(2) Removal of the surface of *cast stone* with acid to expose the aggregate.

eupatheoscope A black, electrically heated cylinder 56 cm (22 in.) high, 19 cm (7½ in.) dia., developed at the *Building Research Station* from 1932 onwards for the estimation of *equivalent temperature*.

evaporation (1) The loss of moisture in vapour form from a liquid.

(2) [pai.] The drying of *varnishes*, *emulsions* and *lacquers* may be solely by the loss of vapour, as opposed to the hardening of *drying oils* which always occurs by *oxidation*, never by evaporation alone.

even-textured or **even grain** [tim.] A description of timber with little variation in the size of the wood elements, for example, timber in which there is little contrast between *springwood* and *summerwood*.

excelsior [tim.] Wood shavings turned or otherwise cut from waste wood to make packing material or *woodwool*.

exfoliated vermiculite *Vermiculite* which has been heated and has thus expanded to ten or fifteen times its original volume.

exfoliation The scaling of stone, caused by the weather.

exhaust shaft A ventilating passage to remove air from a room.

exhaust system of ventilation An *extract system*.

expanded clay or **bloated clay** or **(USA) haydite** Vitreous cellular clay pellets which have been burnt in a cement kiln in such a way as to form hard, air-filled cells. The loose aggregate weighs from 320 to 670 kg/m³ (20 to 42 lb/ft³) and has a *conductivity* (K-value) of about 0·144 W/m degree C. (1 Btu in./ft²h deg. F.). The concrete weighs from 640 kg/m³ (40 lb/ft³) upwards and has a K-value from 0·16 to 0·32 W/m deg. C. (1·1 to 2·2 Btu in./ft²h deg. F.). *See* **lightweight aggregate, LECA.**

expanded polystyrene An *insulating material* obtainable as *loose fill*, or in blocks or sheets, weighing less than any known insulator, 16 kg/m³ (1 lb/ft³). At 32 kg/m³ (2 lb/ft³) its thermal *conductivity* is 0·032 W/m deg. C. (0·22 Btu in./ft²h. deg. F.). Its maximum recommended temperature of use, 70° C., is low, but it is strong for so light a material, having a compressive strength of 173 kN/m² (25 psi) and a tensile strength of 207 kN/m² (30 psi). *See* **sandwich construction, polyurethane.**

expanding bit or **expansion bit** [carp.] A drilling bit with a cutter which can be adjusted to varying radii. One bit can thus cut holes from 1 to 4 cm (½ to 1½ in.) or from 2·2 to 7·6 cm (⅞ to 3 in.) dia. Expanding bits have no twist.

expanding plug A *bag plug* or *screw plug*.

expansion *See* **coefficient of expansion** (*C*), **dimensional stability.**

expansion bay A recess in the side of a pipe *duct* to make space for *expansion bends* (*C*) in the pipes which it contains.

expansion pipe In *open vented* domestic hot-water systems, a pipe leading from the *cylinder* to a point over the *expansion tank* so that, if the water boils, steam (or water) will discharge harmlessly into it.

expansion sleeve or **pipe sleeve** A metal, cardboard, asbestos, or plastics pipe built into a wall or floor, through which another pipe passes. This allows the inner pipe to expand or contract without cracking the wall.

expansion strip or **edge isolation** or **insulating strip** or **expansion tape** or **isolation strip** Material used in *discontinuous construction* to fill the joint between a *partition* and a structural wall or column, or to separate a *glass block* wall from any structural material so as to prevent the glass cracking. It is of resilient *insulating material*.

expansion tank (1) In *open vented* domestic hot-water systems a tank above an *indirect* cylinder to allow for the expansion of the water on heating. *Compare* **expansion pipe**.

(2) *See* **diaphragm tank**.

expediter (USA) A *progress chaser*.

extended prices or **extended rates** [q.s.] The rates or prices for the items in a *bill of quantities* which have been multiplied by their appropriate *quantities* to give a sum of money, written in the bill on the same line as the rate. When all the items have been so multiplied, the bill becomes a *priced bill*.

extender (1) [pai.] or **inert pigment** A white powder, always inorganic, often a crystalline mineral with low *hiding power*. It is added to paint to adjust its film-forming and working properties such as *thixotropy* (*C*). Most extenders are crushed as finely as *pigments*. The commonest are *asbestine*, barytes, *blanc fixe*, *diatomite*, kaolin, *mica*, silica, *whiting*. *See* **filler**.

(2) [tim.] A substance such as wood flour added to expensive *glue* to dilute it and thus increase its spreading capacity. *See* **filler**.

extending ladder A telescopic *ladder* which, if wooden, consists of two or three *standing ladders*. They can normally be obtained up to an extended length of 15 m (50 ft).

extension bolt or **monkey-tail bolt** [joi.] A bolt, usually a *barrel bolt*, with a handle which is longer than usual so that it can be pushed home or released without stooping or standing on a chair.

extension rule A wooden *rule* in two parts which slide relatively to each other. Interior measurements for example the width of a door or window opening, can be taken with it.

extensions [q.s.] *Extended prices*.

exterior trim [carp.] (mainly USA) Wooden *mouldings* for *cornices*, *barge boards*, *eaves gutters*, *water tables*, etc.

exterior-type plywood [tim.] *Plywood* in which the *glue* (not the wood) is moisture resistant. *See* **interior-type plywood**.

external glazing *Glazing* on the outside wall of a building. It may be *outside* or *inside glazing*.

external hazard *See* **fire hazard.**

external wall A wall of which at least one face is exposed to the weather or to the earth.

extra or **extra work** Work which was not included in the original *contract* and has to be ordered by the *architect* or engineer in writing, generally by a *variation order*.

extract system A ventilation system consisting of an electrically-driven fan connected to *air ducts*. The fan sucks the air from the room and blows it into a duct which leads it to the open air through louvres or a cowl. *Compare* **input system.**

extrados The upper surface of an arch, the upper surface of the *arch-stones*. *See* **intrados.**

extra-rapid-hardening cement A Portland cement to which an *accelerator* has been added by the manufacturer to enable concrete to be placed in frosty weather.

extruded section The commonest *light alloy* structural sections, formed by **extrusion** (*C*).

eye (1) *See* **access eye.**

(2) [mech.] An opening formed in a metal member, for example, the eye of a hammer head into which the handle fits, or the ring formed at the end of an *eye bolt* (*C*).

eyebrow or **eyebrow dormer** A window or ventilator opening in a roof surface. The roof to the window forms no sharp angles with the general roof, unlike a *dormer*. Instead, the general horizontal line of the roof is continuous, with a slight bump at the eyebrow. It may therefore be nearly an *internal dormer* but it has no flat roof in front of it.

F

fabric (1) The *carcase* of a building.

(2) *See* **wire-mesh reinforcement** (*C*).

face (1) [tim.] A broad surface of *square-sawn timber* (BS 565).

(2) or **face side** The wide surface of *plywood*, blockboard, timber, etc., with the best appearance.

(3) [carp.] The front (cutting) surface of a *saw-tooth*.

(4) The surface of *gypsum wallboard* which can be painted or distempered without plastering. *See* **back**.

(5) The exposed surface of *ashlars* or of other wall *cladding*.

(6) [mech.] A working surface such as the part of a *hammer* which drives the nail in, the opposite end of a hammer head from the *peen*.

face-bedded or **edge-bedded** Stone laid so that the *natural bed* is vertical. The stone is liable to flake away, as at the Houses of Parliament, London. Only arch-stones are correctly laid like this.

face brick (USA) *Facing brick*.

faced plywood [tim.] Plywood faced with metal, plastics, or any sheet other than wood *veneer*.

faced wall A wall in which the *facing* and *backing* bricks are bonded to act together under load. *Compare* **veneered wall**.

face edge or **working edge** or (USA) **work edge** [joi.] The first edge of *joinery* to be prepared, from which the other edges are measured.

face hammer A *mason*'s hammer with a striking *face* and a cutting *peen*.

face joint The part of a *cross joint* which is seen on the face of a wall.

face mark or **X-mark** [carp.] A mark pencilled on the *face edge* to show that other edges are to be trued from this face. One of the two corners of the face edge is also trued on its rough surface to form a straight line from which setting out is done.

face measure or **surface** or **superficial measure** [tim.] The area of one face of a board. It is not the same as board measure except when the board is 25 mm (1 in.) thick.

face mix A mixture of cement and crushed stone, placed on the surface of a mould for *cast stone* and backed with a cheaper, stronger concrete mix which is poured immediately afterwards and therefore bonded to it. *See* **slip-form** (*C*), **granitic finish**.

face mould A *templet* or full size cut-out drawing which is applied to the face of stone (or to moulded *joinery* such as a handrail) to verify its shape in plan. *See* **falling mould**.

face plate [joi.] That part of a *marking gauge* which is pressed against the *face* of timber while the timber is being marked.

face putty The triangular fillet of *glazier's putty* on the exposed surface of glass. *Compare* **bed putty**.

face side or **working face** or (USA) **work face** [joi.] The surface which carries the *face mark*. *See* **face** (2).

face string [joi.] (USA) An *outer string*.

face veneer [tim.] A *veneer* used for decoration rather than strength.

facia *See* **fascia**.

facing (1) or **lining** [joi.] Fixed non-structural joinery, such as an *architrave*.

(2) Preparing a surface of any material.

(3) *See below*.

facing bond Any *bond* showing mainly *stretchers*.

facing bricks or (USA) **face bricks** Bricks of pleasing but not necessarily uniform colour and satisfying texture, used to cover common brickwork. They are generally not so strong as the *common bricks*.

facing hammer (USA) A *hammer* with a notched rectangular head for dressing stone or precast concrete.

factory-chimney builder A *steeplejack*.

fadding [pai.] Applying *shellac* lacquer with a pad called a fad.

fading [pai.] Bleaching of a colour by ageing or weathering. *Chalking* looks like fading, but the colour can be restored by a coat of varnish.

faggot A South African term for a facing brick made about 2 in. wide to cover a *boot*.

faience (1) Glazed *terra cotta*. It is fired twice, once without and once with the glaze.

(2) Very large, glazed wall tiles.

faience mosaic (USA) Glazed plastic floor mosaic.

fair cutting [q.s.] Cutting of facing brickwork, always assumed 11·5 cm thick, therefore measured by length, and not as an area. Like *rough cutting*, the work is done with *bolster* or *trowel* or *scutch*, less often with the saw.

fair-faced brickwork A brickwork surface which is built neatly and smoothly. It is generally impossible to build both faces fair, one brick thick, in *English bond* but this can be done in *Flemish bond*.

fall bar [joi.] (1) A wooden bar pivoted on a primitive door and controlled by the finger through a hole in the door.

(2) A steel bar which serves the same purpose as (1) but is smaller and neater and is used in the *thumb latch*. Both types fasten the door by dropping into a catch fixed on the *door post*.

falling mould [joi.] The developed elevation of a handrail centre line. It does in elevation what the *face mould* does in plan.

falling stile The *shutting stile* of a gate, particularly of one with a *cocked hinge*.

fall pipe A *downpipe*.

false body [pai.] The high *viscosity* (*C*) of a *thixotropic* (*C*) paint which is always reduced when the paint is stirred.

false ceiling or **drop ceiling** or **counter ceiling** A *ceiling* which is built with a gap between it and the floor above, to provide space for cables and pipes.

false header A half-brick, not a *header*.

false heartwood [tim.] Wood which looks like *heartwood* because of *fungus*, frost, or unusual growth.

false rafter [carp.] A *cocking piece* made of better timber than the ordinary *rafter* because it can be seen in the overhang.

false tenon or **inserted tenon** [carp.] A *hardwood* tenon inserted where the *tenon* of the jointed timber would be too weak.

fan During demolition or building of a wall beside a street, a floor of scaffold boards projecting out and sloping slightly upwards over the street, so that any falling objects which hit the boards are deflected back to the wall, and not out into the street.

fan convector A heat exchanger that usually receives pumped hot water from a boiler and is provided with an electric fan that blows or sucks air over its heating tubes and so acts as a warm air heater. Any fan convector, because of its fan, has a much higher output of heat than a conventional radiator of the same size, although the electrical power demand of the fan is usually well below 100 watts. The *unit heater* is hung at ceiling height but the fan convector is a domestic appliance and is usually hung on a wall at low level, often under a window.

fang The *tang* of a tool, hence the fishtailed end of a metal railing built into a wall.

fanlight [joi.] A *light* over a door, originally semicircular, now of any shape within the main door frame.

fascia board (1) A wide board set vertically on edge, fixed to the *rafter* ends or *wall plate* or wall. It carries the gutter round the *eaves*.

(2) The wide board over a shop front.

fasteners or **fastenings** Metal pieces for fixing wooden members together, such as *connectors*, *nails*, *screws*, *bolts* (*C*), *dowel pins*, *bitches*, *staples*, *ship spikes*. In carpentry, flat steel plates with holes drilled in them are used for fishing joints. They may be rectangular, T-shaped, L-shaped, or 3-dimensional *angle sections* (*C*) which replace the *mortise-and-tenon joint*. *See* **three-way strap.**

fast sheet A *dead light*.

fast to light [pai.] A description of a *colour* which is unaffected by light of a certain sort.

fat board A bricklayer's board for carrying *mortar* when pointing.

fat edge [pai.] A ridge of wet *paint* which collects at the lower edge of a painted area because too much has been put on, or because the paint *flows* too well.

fat lime Lime with high volume yield and good workability, usually *high calcium lime*.

fat mix A rich mix with more cement, lime, or other *binder* than usual.

fat mortar [pla.] A *mortar* which sticks to the trowel. *Compare* **lean mortar.**

fattening or **thickening** [pai.] Increase in the *viscosity* (*C*) of paint during

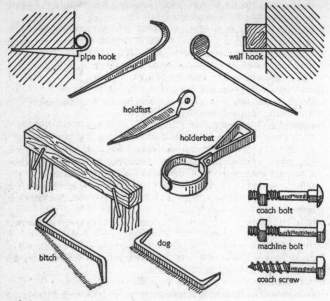

Heavy fasteners.

storage but not sufficient to make it unusable. *Feeding* is very bad fattening.

fattening up or **maturing** [pla.] Increasing the plasticity of *lime putty* by leaving lime in excess water in a *maturing bin* for about a month after slaking.

faucet [plu.] (1) (USA) A small tap, for example at a household sink.

(2) The socket end of a pipe which is joined to a *spigot* (Scots).

faucet ear [plu.] A projection from a pipe socket, for nailing the pipe to the wall.

feather (1) or **slip tongue** or **spline** [carp.] A *cross tongue* joining *match-boards*, not always glued.

(2) or **pendulum slip** [joi.] The wood slip separating the *sash* weights in a sash window.

feather edge [pla.] A *feather-edge rule*.

feather-edge brick A *compass brick*.

feather-edged board [carp.] Tapered boards used as *weather-boarding* or close-boarded fencing, tapering from about 16 to 6 mm from edge to edge.

feather-edged coping or **splayed coping** A coping stone with one edge thicker than the other, thus with its upper surface sloping one way only. (The British Standard term is wedge coping.)

feather-edge rule [pla.] A *rule* from 0·45 to 2 m (18 in. to 6 ft) long with one edge tapered to 3 mm or 1·5 mm ($\frac{1}{8}$ or $\frac{1}{16}$ in.) thick. It is used after a *floating rule* or for working angles.

feather joint or **ploughed-and-tongued joint** [joi.] A joint made with a *cross tongue* between *ploughed* edges.

feather tongue [joi.] A *cross tongue*.

feebly hydraulic lime Lime burnt from a limestone containing 6% to 12% clay. *See* **hydraulic.**

feed (1) [pai.] *See* **feeding.**

(2) *See* **C.**

feed cistern A cold-water storage *cistern*, supplied from the main, usually by a *ball cock*. It supplies a water-heating boiler, and sometimes also the rest of the house except for drinking water.

feeding or **livering** [pai.] A thickening-up of liquid *paint* or *varnish* in the container to a rubbery jelly which cannot be used. *See* **fattening, setting-up.**

feint In *flexible-metal roofing* a slightly bent edge of *cappings* or *flashings* to form a *capillary break*.

felloe [carp.] A segment of the rim of a wooden wheel or of the rib of curved *centers*. Both have roughly the same shape. Sometimes the term is applied to the whole rib or rim.

felt *See* **bitumen felt.**

felt-and-gravel roof (USA) A roof covered with *bitumen felt*, protected by gravel to reduce its *U-value*.

felting down [pai.] *Flatting* a dry *varnish* or *paint* film by a felt pad charged with *abrasive* powder and lubricated with water or other suitable liquid.

felt nail A *clout nail*.

felt paper *Building paper*.

fence [mech.] (1) A guide for timber on a saw bench or other timber cutting machine.

(2) In the joiner's *plough*, a piece of wood or metal parallel to the cut. It holds the blade at a constant distance from the edge of the wood so that the groove is ploughed parallel to the edge of the wood.

(3) A *palisade*.

fender A *baulk* laid on the ground in a street to protect scaffold *standards* from the wheels of traffic. *See* **dead shore** (illus.).

fender wall A *dwarf wall* carrying the hearth slab to a ground floor fireplace and sometimes also *joists*.

fenestration The architectural arrangement of the windows and other openings in the walls of a building, mainly in the façade.

ferro-concrete Obsolete term for *reinforced concrete* (*C*).

ferrule (1) [carp.] A metal band round the handle of a *chisel* to prevent it splitting.

(2) [plu.] Generally a short length of tube such as a *sleeve piece*.

fettle (1) [mech.] To remove the roughness from a casting and to verify that it is free from flaws by hanging it in chains and striking it with a hammer.

(2) An extension of the first sense, the finishing-off work in any trade.

fibre board or **fibre building board** Boards or sheets, commonly sold in sizes 1·2 by 2·4 m and larger, built up by felting from wood or other vegetable fibre. They fall into two main classes, *insulating boards* that are not compressed during manufacture, and *hardboards* that are. All have in common the fact that their main bond is by the felting of woody fibres and not by added cement or glue. All types burn and have in the past been the cause of rapid *flame spread*, but suitable surface treatment can improve this property up to class 1 (very low flame spread) of BS 476. The density of hardboard is always above 480 kg/m³, sometimes as high as 960 kg/m³, and its commonest thickness is 3 mm, though it can be made 19 mm thick. Insulating board is always less dense than 400 kg/m³ and it is not made thinner than 11 mm.

fibre conduit [elec.] Moulded-fibre, insulating *conduit*.

fibreglass *See* glassfibre.

fibre saturation point [tim.] The *moisture content* above which the strength and dimension of timber remain roughly constant, and below which the strength increases and the wood shrinks progressively as the moisture content falls. It cannot be exactly determined, but is about 30% moisture content.

fibrous concrete *Concrete* containing fibrous *aggregate* such as asbestos or sawdust, used for its lightness or nailability.

fibrous plaster or **stick and rag work** [pla.] *Plaster of Paris* shapes made in the workshop by casting in gelatin or plaster moulds. They are reinforced with coarse, open canvas and wood laths, and sometimes with wire netting and tow.

fibrous plasterer [pla.] A *plasterer* who makes or fixes *fibrous plaster* work. He is specialized either as a 'shop hand', or as a 'fixer' who sets the cast plaster work in place and joins the pieces to each other with wet plaster.

fiddleback [tim.] A *mottle* figure, a *ripple* in sycamore or maple, used for veneering violin backs.

fiddle drill [carp.] A *bow drill*.

fielded panel [joi.] A *raised panel*.

figure [tim.] The natural markings of timber, including both *grain* and colour. Figure usually adds to the beauty and value of a timber but is not always a sign of strength.

figuring (USA) Calculating or taking off *quantities* or estimating costs from a drawing.

filler (1) A material added to plastics to vary their mechanical properties and sometimes also to lower their cost. It has a similar function to *extenders* in *glue*. Some fillers are wood flour, cotton, carbon black, barytes, bentonite, asbestos, paper, slate dust, mica, quartz, and, for translucent plastics, glass fibre.

(2) [pai.] A paste containing *extenders* often with white lead, mixed with *gold size* and *turpentine* or *white spirit*. It is put on a surface after priming to fill up indentations. *Glazier's putty* is also used for painted surfaces, beeswax for pale polished wood, litharge and glue for varnished pitchpine. *See* **hard stopping, surfacer**.

fillet [joi.] (1) A narrow strip of wood fixed to the angle between two surfaces, for example, a shelf *cleat*.

(2) A small square wooden *moulding*.

(3) A waterproofing which replaces *flashings* at *abutments* or under *verges* and may consist of a triangular mortar strip (cement fillet), or an asphalt seal formed separately from the rest of the roof asphalt, or of slates cut to shape.

fillet chisel, fillet rasp, fillet saw. Tools used by masons for working stone to fine limits.

filletster [joi.] A *fillister* plane.

filling [pai.] *See* **filler** (2).

filling knife [pai.] A knife like a *stopping knife* but having a thinner, springier blade, used for laying on paste *filler*.

filling-in piece [carp.] A timber such as a *jack-rafter*, shorter than its neighbours.

filling piece [joi.] A piece of wood planted on another to make a plane surface.

fillister (1) or **sash fillister** [joi.] A *rabbet* cut in a *glazing bar* to receive glass and putty.

(2) or **filletster** An adjustable *plane* for rebating *glazing bars*.

film [pai.] The dried *paint* or *varnish* of one or several coats.

film building or **film forming** [pai.] The property of forming an adhesive, strong, continuous, flexible paint *film*, possessed in a high degree by *linseed oil*, a medium which is fluid enough to be laid on with no thinner; therefore all the *medium* contributes to the film.

film glue [tim.] A thin, solid sheet of *phenol formaldehyde resin* laid between thin, costly, decorative, *face veneer* and the cheap, stronger, thicker, backing veneer. Film glue is easily applied. Since it does not wet the *veneer* and cause it to expand, it is the only *glue* possible with very thin veneers less than 1 mm thick. It is *thermosetting*.

fine solder [plu.] An alloy of $\frac{2}{3}$ tin, $\frac{1}{3}$ lead, more costly than *plumber's solder*, used in making the *blown joint*, for tinning, and for copper-bit soldering. It has the lowest melting point of all *solders*, about 185° C.

fine stuff [pla.] The material of the *finishing coat*.

fine-textured wood [tim.] Wood like birch or maple which has small pores which need no *filler* before varnishing, the opposite of *coarse-textured* wood.

finger-jointed [joi.] *See* **combed joint.**

finger plate [joi.] A plate fixed near the *latch* of a door to protect the surface from finger marks.

fingers [pla.] A *comb* or *drag* made by nailing together several laths with pointed ends.

finger slip [joi.] A small curved-edge *hone* for smoothing the inner surface of *gouges*.

finial A usually pointed ornament at the top of a *gable* or pinnacle or *newel*. *See* **hip knob.**

fining off [pla.] Applying the *finishing coat* of external rendering.

finish (1) [joi.] (USA) Fixed *joinery*.

(2) or **finishing coat** The final coat of *paint* or plaster.

(3) [pai.] The appearance of the final coat of paint, varnish or polish which may be *crystallized, gloss, flamboyant, hammer, polychromatic, textured,* or *wrinkle finish.*

finished floor or **finish floor** The visible, completed floor surface.

finisher or **trowel man** A *skilled man* who gives a smooth finish to precast concrete units with a wooden *float* or steel trowel, sometimes patching gaps with concrete of matching colour. A specialist architectural stone finisher smooths the surface with a stone and polishes it with an emery wheel or buffing wheel.

finish hardware [joi.] (USA) *Hardware* which is seen and therefore given a good finish, such as *door furniture*, clothes hooks, and so on, as opposed to *rough hardware*.

finishing carpentry [joi.] American term describing what is called *joinery* in Britain, that is doors, *skirtings, architraves,* etc.

finishing coat or **setting coat** or **skimming coat** or **fining coat** or **white coat** [pla.] A final layer of plaster about 3 mm thick, of different material (fine stuff) from the *coarse stuff* of the previous coats. With *lime putty* for the best interior finish, the mix would be 1 lime:$\frac{1}{4}$ to 1 *hemihydrate plaster*; for ordinary finishes up to one volume of sand may be added to this mix. With '*dry hydrate*' the following mixes are used: 1 of lime:1 to 1$\frac{1}{2}$ of sand, or for a rapid hardening surface, 2 lime:1 cement or hemihydrate plaster:3 sand. *Anhydrous* plasters may be used neat to give a hard, glossy finish or may be mixed with sand in the proportions 1 plaster:$\frac{3}{4}$ sand. *Hemihydrate plaster* without lime may be mixed with up to 3 times its volume of sand. These mixes are for inside plaster; for outside plastering *see* **stucco.**

finishing off [joi.] Preparing the finished surface of *joinery*.

finishings or (USA) **finish** The fixed *joinery* in a building; also the *plaster, paint,* or other final details to the walls, doors, etc.

finishing tools Plasterer's or *composition floor layer*'s trowels or floats for shaping curved or other difficult surfaces.

fir [tim.] A loose term which should properly be confined to the abies family, that is *whitewood, silver fir,* and others. It is also often used for *redwood* and *Douglas fir.*

fireback The wall behind a fireplace.

firebars Cast iron bars on which coal or wood or other solid fuel is burnt in a domestic or industrial furnace.

fire block [carp.] (USA) *Solid bridging* in floors or wooden walls which may be regarded as a *fire stop*.

fire breaks The fire-protecting doors, *closed stairs*, *concrete* floors, *division walls*, etc., which reduce the risk of fire in a building to an amount acceptable to insurance companies or the law. *See* **fire stop.**

firebrick *Brick* made from any clay difficult to fuse, generally one with a high content of quartz. It can usually be used up to temperatures between 1500° and 1600° C. *See* **refractory linings** (*C*).

fire cement Refractory cement such as the fireclay from which firebrick is made, or *high-alumina cement* (*C*).

fire check door A British Standard door (BS 459) which provides half an hour's protection against fire, with 3 mm ($\frac{1}{8}$ in.) flush plywood over 10 mm ($\frac{3}{8}$ in.) plasterboard on each face.

fireclay Clay which is rich in minerals containing much silica SiO_2, and alumina Al_2O_3, used for making *firebrick*.

fire cracks [pla.] Cracks in a plastered surface caused by exposure to the sun or other heat during drying. *See* **cracking, crazing.**

fire division wall (USA) A wall which divides up a building to resist the spread of fire. Unlike a *division wall* it does not necessarily rise through more than one *storey*.

fire door (1) A door to a furnace.

(2) A generally metal-plated door designed to slide shut when a *fusible link* melts so that it holds back a fire for an hour or more.

fire escape stair A *stair* required by law for escape in case of fire from buildings above a certain height. It may be inside or outside the building.

fire-extinguishing equipment *Drenchers, fire hydrants, sprinklers, emulsifiers,* hand-operated foam sprays and so on.

fire grading or **fire-resistance grading** In Britain structural members such as roofs, beams, columns, walls or doors can be tested by fire in accordance with BS 476 and graded for their fire resistance (or endurance). Fire resistance is usually measured by the time for which the member continues to satisfy three criteria: collapse, flame penetration, or excessive temperature rise on the 'cool' face. If it satisfies all three criteria for four hours it is described as having four hours' fire resistance. Thirty minutes is often adequate for small houses, while 6 hours could be needed for a warehouse with a *fire load density* of 380 000 Btu/ft² (4000 MJ/m²). The fire resistances needed by buildings and their elements are laid down in *Building Regulations*.

fire hazard Danger of fire. An internal hazard arises from the structure or the contents of the building. An external hazard arises outside the building. *See* BSCP 3, chapter 4, Fire precautions.

fire hydrant An outlet from a water main to which a fireman can connect his hose and control the flow as he wishes. It may be of 63, 38 or 19 mm ($2\frac{1}{2}$, $1\frac{1}{2}$ or $\frac{3}{4}$ in.) nominal bore and can be provided inside or outside a building, in colliery yards, timber yards, etc.

fire load Originally the total amount of combustible material in a room and its structure, expressed in heat units, kilojoules or Btu. There has also been a tendency to express it in kilograms or pounds of wood.

fire load density The *fire load* per unit area (BS 4422). This refers to floor area but others have recommended that fire load density should be calculated on wall area. A 'low' fire load density is less than 1135 MJ/m² (100 000 Btu/ft²) of floor area and this is usual for dwellings and offices. A 'moderate' one is from 1135 to 2270 MJ/m² (100 000 to 200 000 Btu/ft²). Warehouses often have a 'high' fire load density, 2270 to 4540 MJ/m² (200 000 to 400 000 Btu/ft²). *See* **fire grading**.

fire point (1) The lowest temperature at which a substance ignites and continues to burn when a flame is put to it. *See* **flash point.**

(2) A place where fire-extinguishing equipment is kept.

fireproof A term which should not be used since no practical construction can withstand fire indefinitely. In USA the term means a construction which will safely withstand the complete burning of the contents of the building. *See* **fire resisting.**

fire protection Every measure for the prevention, detection and extinction of fire, as well as the reduction or prevention of fire damage or loss of life and the design of buildings to resist fire (structural fire protection).

fire protection of structural steelwork Bare steel frames are liable to fail rapidly in fires, much more quickly even than thick wood joists. Fire protection is given by covering the frame with material which delays the arrival of heat to the steel. This covering may be brick, concrete, hollow clay, foamed slag or gypsum blocks, sprayed asbestos, plaster on *metal lathing*, etc. The thicker the cover, the more protection it provides. 6·3 cm (2·5 in.) of solid concrete or 2·5 cm (1 in.) of sprayed asbestos generally give the same protection. *See also* **fire grading, vermiculite-gypsum plaster.**

fire-resistance grading *See* **fire grading.**

fire-resisting door, floor, or **wall** Any door, floor or wall that can satisfy for a stated period of time the three criteria of *fire grading*. Any openings in it must be *protected* for the same period and the period chosen has to be suitable to the occupancy and *fire load* of the building.

fire-resisting finishes or **fire-retardant finishes** [pai.] Paints based on *silicones, polyvinyl chloride, chlorinated rubber, urea formaldehyde resins, casein,* borax, and other flame retardants which form a coating about 0·02 to 0·05 mm thick and thus considerably reduce the rate of *flame spread* of a *combustible* material.

fire-resisting glazing In London, wired glass or *electro-copper glazing* built into lights not exceeding 61 cm (2 ft) square is usually considered to have half an hour's fire resistance if it is at least 6 mm ($\frac{1}{4}$ in.) thick and the panes do not exceed 10 cm (4 in.) square. The frame of the light must have at least half an hour's fire resistance also.

fire stop or **draught stop** A barrier to fire, such as a brick wall built across an attic space at intervals or a horizontal barrier through the hollow part of a *stud* wall or *partition*, or *fire blocks*. These fire stops must be incombustible, or of timber 5 cm (2 in.) thick at least. *See* **beam filling**.

fire terms Apart from those beginning with 'fire', *see also* **compartmentation, flame spread, ignitability,** and **roof screen**, but many more are printed in BS 4422.

fire testing of materials Small specimens of building materials can be tested by the methods of BS 476, in particular for fire propagation, *ignitability*, non-combustibility, and *flame spread*.

fire tower (USA) In tall buildings, a stair designed as a fire escape with entries at each floor, protected by fire doors so that smoke cannot enter the stair.

fire venting Inducing hot air and smoke to leave a building as quickly as possible by vents, so that firemen can see to fight the fire.

fire wall (USA) A *division wall*.

fir fixed [carp.] Unplaned timber fixed by nailing only.

fir framed [carp.] Unplaned timbers fixed by preparing their joints as in roof *trusses*.

firing Exposure of *bricks* and other clayware to heat in a kiln.

firmer chisel [carp.] A carpenter's or joiner's ordinary *chisel*, stronger than a *paring chisel*, less strong than a *socket chisel*. It should not be struck with a hammer or mallet.

firmer gouge [carp.] A *chisel* with a blade curved like a *gouge*.

firring [carp.] *See* **furring**.

first fixer [carp.] A *carpenter* who is cutting or fixing *joists, rafters,* floorboards, *stairs*, or window frames.

first fixings (1) [joi.] *Grounds, plugs*, and so on which carry the *joinery*.
   (2) [carp.] Structural timber, *joists, rafters*, floors, etc.

first floor (1) (Britain) The floor which is next above the floor at ground level and is therefore about 2·75 m (9 ft) above ground. This definition is also accepted in USA for houses with neither *basement* nor *cellar*.
   (2) (USA) In buildings with a basement or cellar the first floor is the first above ground level (which would in Britain be called the *ground floor*).

first storey The space between the *first floor* and the floor above.

fished joint [carp.] *See* **Splice**.

fish glue or **isinglass** [tim.] A *glue* like *animal glue* but prepared from fish bladders and skins.

FITCH

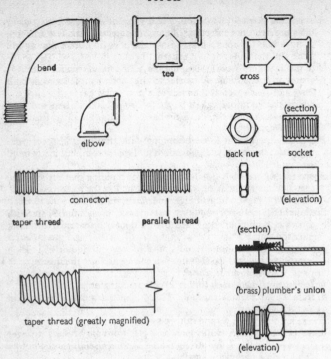

Fittings for screwed gas or water pipe, mainly of dead mild steel or malleable cast iron.

fitch [pai.] A long-handled small *brush* bound with tin, with which nearly inaccessible details are painted. *See* **lining tool**.

fitment or **fitting** Furniture fixed, often by the builder, as opposed to the loose furniture bought by the occupier.

fitter's hammer [mech.] An *engineer's hammer*.

fittings or **pipe fittings** [plu.] *Bends, couplings, crosses, elbows, unions,* etc. which for screwed pipe are screwed on to pipe ends. For screwed gas pipe the fittings are usually of *malleable cast iron* (*C*) and therefore cheap. *See* **copper fittings and above**.

five X or **5 X** Western red cedar *shingles* cut to 41 cm (16 in.) lengths. When placed together, five butts measure 5 cm (2 in.) thick.

fixed light or **fixed sash** A *dead light*.

fixed-price contracts [q.s.] These may be of three types, *lump sum, schedule of prices,* or *measure-and-value contracts.* Fixed-price con-

138

tracts are generally recommended by consultants and preferred by clients, since the contractor has an incentive to work fast and economically. Compare the slow, expensive *cost-reimbursement contract*.

fixer (1) or **fixer mason** or (Scotland and Cornwall) **builder mason** A *mason* who sets prepared stones in walls, whether the stone be only facing or to the full wall thickness.

(2) [pla.] A plasterer who fixes *fibrous plaster* as opposed to a 'shop hand'.

fixer's bedding *Lime putty* used by *fixer masons*.

fixing Glass panes are fixed when they are secured to ceilings or walls for such purposes as flush lighting fittings. Otherwise the word *glazing* is used.

fixing bracket In *patent glazing*, a fitting on a *glazing bar* for securing it to a structural member.

fixing brick or **fixing block** or **nailing block** or **nog** or **wood brick** A brick made from wood, or sawdust and clay, or from *diatomite*, or *lightweight concrete*, or other nailable material. It is used for fixing joinery. *Fixing fillets*, being thinner, shrink less and are a better fixing.

fixing fillet or **fixing slip** or **pad** or **pallet** or **pallet slip** [joi.] A piece of wood, the thickness of a mortar joint $23 \times 11 \cdot 5$ cm ($9 \times 4\frac{1}{2}$ in.), inserted into a *joint* as a fixing for *joinery*. *See* **fixing brick**.

fixings [joi.] *Common grounds*, *plugs*, *fixing fillets*, etc., for holding joinery, sometimes called *first fixings*.

fixing slip [joi.] A *fixing fillet*.

fixing strip A steel, non-ferrous metal or plastics device for fixing board or sheet *covering* to *partitions* or for fixing *cladding* to the wall frame or *sub-frame*.

fixture Anything fixed to a building. It becomes the landlord's property if its removal would damage part of the building. Examples are plumbing, wash basins, ceiling lamps.

flagstone or **flag** or **flagging** A slab of concrete, or cast or natural stone used for paving footways or gardens, originally sandstone which splits into flat sheets.

flaking (1) A ground on which *thatch* is laid, consisting of reeds woven over the rafters. No *battens* are used.

(2) The detachment of stone, brick, paint or plaster.

flamboyant finish [pai.] A glossy transparent *varnish* or *lacquer* over a bright metallic surface, often seen on new bicycles.

flame cleaning [pai.] Removing mill scale and water from weathered structural steelwork by a very hot flame. The surface is painted immediately after flame cleaning.

flame-retardant treatment A treatment, such as painting, which reduces the rate of *flame spread* on a surface, particularly on wall linings like *fibre board* or timber. *See* **fire-resisting finish**.

flame spread Building materials in Britain used for wall or ceiling lining are tested in accordance with BS 476 under intense radiant heat, together with a gas flame, to determine how quickly their surface ignites. They·are classified as follows: 1. very low, 2. low, 3. medium, 4. rapid.

flammable The term now used in Britain and USA by fire authorities instead of inflammable, and meaning that which burns with a flame. *See* **non-combustible.**

flanking window A window beside an outside door, with its sill at the doorsill level.

flanks The *intrados* of an arch, near its *springings.*

flank wall A wall at one side of a building.

flared header or **flare header** A brick which is dark at one end through being close to the fire during burning. It can be used for patterning the face of brickwork in *diaper* work. *See* **English cross bond.**

flash To make a weathertight joint, called a *flashing.*

flash drying [pai.] Rapid drying by exposing paint or varnish to radiant heat for a short time. *See* **stoving.**

flashing (1) A strip of impervious material usually *flexible metal* (such as zinc 0·8 mm thick or copper 0·56 mm thick or lead 1·8 mm thick) which excludes water from the junction between a roof covering and another surface (usually vertical). Flashings, at their upper end, are usually wedged tightly into mortar joints which have been raked out to receive them. They are sometimes made of *roofing felt. See also* **apron flashing, cover flashing, raking flashing, stepped flashing, burning in.**

(2) Burning bricks alternately with too much and too little air to give them varied colours.

(3) [pai.] The defect of patches in a *finish* which are glossier than the remainder, particularly at joins or laps.

flashing board or **lear board** or **layer board** A board on which *flashings* are fixed. *See* **layer board.**

flash point The lowest temperature at which a substance momentarily ignites when a flame is put to it. *See* **fire point.**

flat (1) A level platform, generally a roof, particularly a lead-covered roof (lead flat).

(2) (USA **apartment**) One floor of a multi-storey building or a dwelling unit on one floor. *Compare* **maisonette.**

(3) [pai.] Matt, *see* gloss.

(4) [pai.] *See* **flatting down.**

(5) [tim.] *See* **flat cutting.**

flat arch or **straight arch** or **French arch** An arch with a level *soffit* and *extrados* made usually of wedge-shaped, *gauged*, or moulded bricks which radiate from one centre. It is used over doors, windows, and fireplaces. In USA called a *jack arch* (*C*). *See* **soldier arch.**

flat coat [pai.] A coat of *filler.*

flat cost The cost of labour and material only.

flat cutting or **flatting** or **ripping** or **ripsawing** [tim.] The *re-sawing* of timber parallel to the edges (BS 565). *Compare* **deeping**, *see* **sliced veneer**.

flat-drawn sheet glass Ordinary *glass* for windows. *See* **sheet glass**.

flat grain [tim.] The grain of *flat-sawn* timber, most of which has annual rings at less than 45° with the face of the piece.

flat interlocking tile A *single-lap* standard British clay roofing tile normally measuring $39 \times 20 \times 1 \cdot 9$ cm ($15\frac{1}{2} \times 8 \times \frac{3}{4}$ in.) provided with two nail holes and no *nib*. It has a $7 \cdot 6$ cm (3 in.) side lap.

flat joint or **flush joint** A mortar joint whose surface is flush with the brickwork.

flat-joint jointed A *flat joint* in which a narrow groove has been cut with a *jointer*.

flat paint brush or **flat enamel brush** [pai.] A metal-bound *brush* used by the *housepainter*. It is of black bristle which is stiff enough to carry heavy *varnish*, *paint*, or *enamel* and is from 1 to 15 cm ($\frac{1}{2}$ to 6 in.) wide.

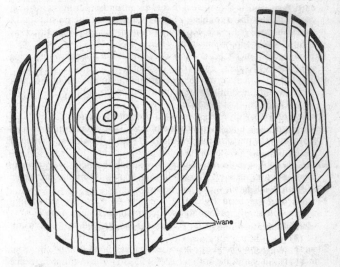

Flat-sawn log. The log has been seasoned after sawing, and the illustration shows the effects of shrinkage on it. The curved surfaces are caused by the radial shrinkage being half the tangential shrinkage. On the right is part of the same log showing the (concave outwards) warp which the shrinkage causes. The two central planks are fully quarter-sawn.

flat-pin plug [elec.] A common type of fused plug suitable for insertion into a *shuttered socket* outlet on a *ring main*. Its three contact pins are rectangular in cross-section.

flat pointing Pointing of brickwork to make *flat joints*.

flat roof (Scots, **platform roof**) A roof which slopes at less than 10° to the horizontal.

flat sawing [tim.] Sawing logs with parallel cuts, a method of *conversion* which wastes less timber than any other.

flat-sawn timber or **plain-sawn** or **slash-sawn** or (particularly USA) **bastard-sawn** [tim.] (1) Logs sawn with parallel cuts.

(2) Timber converted so that the annual rings meet the face at an angle less than 45°. (1) is a *conversion* method which produces some boards in which the annual rings meet the surface at less than 45°, the remainder being quarter-sawn (roughly the central one-third, including the centre planks).

flatting *See* **flat cutting** [tim.], **flatting down** [pai.].

flatting down or **rubbing** [pai.] *Sanding* with powdered pumice and felt, cuttle fish, glass paper, or other *abrasives*.

flatting varnish or **rubbing varnish** [pai.] A *varnish* which contains much hard *resin* and is therefore suitable as an undercoating varnish. It can be flatted down to make a smooth surface for the finish.

flat varnish [pai.] A *varnish*, *lacquer*, or *enamel* with its *gloss* reduced by adding wax, soap, *pigment*, or *filler*.

flat wall brush [pai.] A *brush* like a *distemper brush* but narrower, being only about 12·5 or 15 cm (5 or 6 in.) wide.

flaunching (1) A cement mortar *fillet* round the top of a *chimney stack* to throw off the rain, surrounding the *chimney pot*.

(2) The placing of this fillet.

fleaking *Flaking* with reed, to make a ground for *thatch*.

fleam [carp.] The angle between the file, while it is sharpening a *saw-tooth*, and the plane of the saw blade. It is therefore the side angle of the cutting face of the tooth.

Flemish bond A seventeenth-century *bond* which shows, in every course, *stretchers* and *headers* alternately. *See* **double Flemish bond, single Flemish bond, monk bond,** *and below*.

Flemish diagonal bond By laying a *course* of *stretchers* alternating with a course of headers and stretchers alternately, a face *bond* is obtained in which, as in *English cross bond*, a diagonal pattern can be seen.

Flemish garden-wall bond or **Sussex garden-wall bond** A *bond* showing, in each course and on both faces of a 23 cm (9 in.) wall, a sequence of three *stretchers* and one *header*. In thicker walls, one face is formed in English bond. Like *English cross bond* and *Flemish diagonal bond*, the face shows a diagonal pattern.

Fletton A cheap brick used in and near London, made from the shale of the Peterborough district.

flex or **flexible cable** or (USA) **lamp cord** [elec.] Flexible copper *conductors* enclosed in rubber and a textile or *polyvinyl chloride* binding. It is used for the final connection to an electrical fitting in a house, for example a lamp or heater. *See* **flexible cord.**

flexible-bag moulding [tim.] A process for making *moulded plywood*. Pressure, during gluing up, is applied to the hot *veneers* through a rubber bag, which forces the veneers on to a mould. In the vacuum process, the veneers are placed inside the bag, which is evacuated. In other methods, the bag is under strong internal pressure and is outside the plywood. *See* **skin.**

flexible cord [elec.] A flex in which each *conductor* has an area not larger than 4·5 mm^2 (0·007 in.^{2}).

flexible metal Sheet metal, originally zinc, lead, copper, or painted tinplate, but now also aluminium. Unlike corrugated sheet, flexible metal cannot span a gap, and so, like roofing felt, it must be laid on close boarding or plywood, covered with an underlay.

flexible-metal conduit [elec.] *Conduit* made from spirally wound steel strip.

flexible-metal roofing Roof coverings or *flashings* of flat *flexible-metal* sheet. Like *roofing felt*, these roof coverings need a smooth roof such as close-boarding underneath, covered with an *underlay*. Painted tinplate is used only in dry climates. *See* **seam,** *also* BSCP 143, Sheet Wall and Roof Coverings.

fleximer A cement mortar or other jointless flooring material containing some *elastomer* to improve its properties by reducing cracking and increasing adhesion. *See* **cement-rubber latex.**

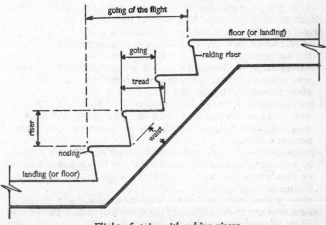

Flight of stairs with raking risers.

flex plug and socket [plu.] *See* **plug-in connector.**

flier or **flyer** (1) A rectangular *tread* in a *stair* (not a *winder*).

(2) [carp.] A *flying shore.*

flight A series of *steps* which joins one floor or *landing* to the next floor or landing (*see* illus. p. 143).

flint wall A wall built of flints showing their dark-grey broken faces, with brick or stone *quoins*. The modern way of building flint walls is as decorative and probably stronger than the old method of laying them by hand in mortar. The quoins and piers are first built to full height in brick or stone. Rough formwork is then fixed all over the face, the back of the wall being rough shuttered as the wall rises. Several courses of flints are laid dry against both shutters, and the space between filled with concrete which is sliced into the joints between the flints. When the concrete reaches the middle of the top-most course, several more courses are laid and concrete is again cast between them.

flitch [tim.] (1) A large timber, intended for conversion.

(2) Timber from which *veneers* are to be cut, usually with *wane* on one or more edges.

(3) The stack of veneers after cutting, piled in the order in which they were before cutting. *See* **swatch.**

flitched beam or **flitch beam** or **sandwich beam** [carp.] A beam built from two wooden beams sandwiching a vertical steel plate. The three are bolted together through the plate.

flitch plate [carp.] The steel plate which reinforces a *flitched beam.*

float (1) [pla.] A wooden plastering tool such as the *cross-grained float* or *Darby float*. *See* **blade, plasterer, trowel.**

(2) [plu.] A metal drain pipe, usually cast iron, hung just below floor level, taking waste from the floor above.

floated coat [pla.] A plaster coat smoothed by a *float.*

floater (USA) A tool used for finishing mortar *screeds* (3).

float glass Thick glass sheets which are competitive with *plate glass*, and are made by floating the molten glass on a surface of molten metal, thus producing a polished, smooth surface.

floating (1) [pla.] Levelling with a *floating rule* the second coat, the *floating coat* of *three-coat plaster*. *See* **coarse stuff.**

(2) [pai.] or **flooding** The re-arrangement or separation at the surface of a paint film of *pigment* grains. This sometimes happens when two pigments are mixed and may be caused by insufficient wetting, different densities or particle sizes, etc. Although it may in other paints be a defect because the final colour is not known before the coat is dry, floating is aimed at in aluminium and other metal paints. *See* **leafing.**

floating coat or **browning coat** or **topping coat** [pla.] The second of three coats in *three-coat plaster*. It is levelled with a *floating rule* between *screeds* and consists of *coarse stuff* like the first coat (rendering coat or pricking-up coat). Like the first coat it is not more than 1 cm (⅜ in.) thick.

floating floor *Discontinuous construction* for sound insulation, by separating the wearing surface of the floor from the loadbearing part, whether concrete or wooden. In both cases, a *glass-wool* quilt is laid on the rough floor, 5 × 5 cm (2 × 2 in.) battens laid on it without nailing, and finished floor boards nailed to the battens. If the finished floor is to be a concrete screed, this is laid direct on the glass wool. The ceiling below is fixed in the usual way.

floating rule or (USA) **rod** [pla.] A long wooden *rule* with which a floating coat is levelled to a plane surface between *screeds*. *See* **feather-edge rule**.

floatstone A porous opal or an iron block used for rubbing *gauged bricks*.

float valve [plu.] A *ball cock*.

flocculent gypsum *See* **gypsum insulation**.

flock spraying or (USA) **flocking** [pai.] Blowing soft, fluffy fibres from cotton, silk, or other textiles on to a sticky surface to form textile effects, such as suedes or felts of various colours.

flogging [carp.] Smoothing a timber floor with hand tools, now more often done with *sanding machines*.

floor-and-wall tiler [pla.] A *tradesman* who chooses, matches, cuts, and trims *wall tiles* and beds them in *mortar*, *glue*, or *plaster*. He also builds up hearths and fireplaces and sets very large tiles called *faience*, and may be a composition floor layer. *See* **tile slabber, floor-layer's labourer**.

floorboards The most usual boards for floors are planed square-edged wooden boards, nominally 15 × 2·5 cm (6 × 1 in.), that measure in fact about 14·5 × 2·2 cm (5¾ × ⅞ in.). Tongued and grooved boards provide a better floor because they are more fire-resistant, draught-proof and strong for the same thickness. Recently large flooring panels made of plywood or 19 mm (¾ in.) wood chipboard have become available which are quicker to lay and need fewer nails than boards. Wood chipboard is also cheaper than wooden boards. If tongued and grooved all round it will be superior to board floors in fire resistance and strength. If the plywood is also glued both to the joists and in the groove, the floor is further strengthened and made more airtight, and even fewer nails are needed, only 30 cm (12 in.) spacing in fact. Elastomer glues, developed in the USA since 1963, can be used in any weather except intensifying frost, and these it is that have made this site gluing practicable. *See* BSCP 201, 209.

floor chisel A *bolster* about 5 cm (2 in.) wide for pulling up floorboards.

floor clip or **bulldog clip** or (USA) **sleeper clip** Strips of 0·9 mm thick *sherardized* steel sheet pushed into and anchored in the surface of a *concrete* floor slab or *screed* just after it has been levelled. When the concrete has hardened, the two edges of the strip can be pulled up and bent into a U-shape and nailed to flooring *battens*, to which the finished wooden floor is fixed. A pad of asbestos or rubber sometimes forms part of the clip to reduce the transmission of sound from the floor to the room below (acoustic clip).

floor cramp or **flooring cramp** [carp.] A *cramp* for forcing floorboards together before nailing them down.

floor framing [carp.] The *common joists*, their *strutting* and support. *See* **double floor.**

floor guide A groove in a floor to guide a sliding door.

flooring brad *See* **brad.**

flooring tiles [pla.] Usually *concrete* or clay tiles set in cement mortar or in bituminous or other *adhesive*. For a more sound-absorbent, heat-insulating, decorative, or generally comfortable surface, tiles of *linoleum*, glass, *cork*, rubber, asphalt, or plastics are used. More durable floors are made of brick, or of steel *anchor plates* filled with concrete, or of cast-iron anchor plates. *See* **slab floor, thermoplastic tiles.** *See* BSCP 202, 203.

floor joist [carp.] A *common joist.*

floor-layer's labourer A helper to the *floor-and-wall tiler* who loads and unloads tiles or other materials for him, soaks tiles, mixes mortar or *jointless flooring* composition, sometimes helps in rubbing a *terrazzo* floor, and cleans up when the floor is laid.

floor line A pencil mark on a wall, stanchion, etc. to show finished floor level (F.F.L.).

floor lining [carp.] *Building paper* laid over a *rough floor* before the finished floor is placed on it.

floor plan A plan of a building showing the layout of all rooms, wall thicknesses, and other building information needed for completing the *carcase* of the building.

floor plug [elec.] An electrical connection in the floor.

floor sander A *sanding machine* used for smoothing wood floors.

floor spring or **spring hinge** [joi.] A sprung pivot housed in the floor to control the opening and closing of a *swing door*. *Compare* **helical hinge.**

floor stop [joi.] A *door stop* set in the floor.

floor strutting [carp.] *Herring-bone* or *solid strutting* between floor *joists* at midspan.

floor tile *See* **flooring tile.**

floor varnish [pai.] A *varnish* which is put on floorboards. It must therefore be quick drying, tough, abrasion-resistant, washable, and must take wax polish.

floor warming *See* **underfloor heating.**

flow or **level** [pai.] To spread into a smooth film, a property of liquid paints or varnishes which ensures that they will show neither brush marks nor *orange peeling*, though they may *run* badly.

flow pipe A pipe by which hot water leaves a boiler and enters the hot-water *cylinder* in a water heating system. *See* **primary flow and return.**

flue A passage for smoke in a chimney. For an open fire it measures at least 23×23 cm (9×9 in.) (*but see* **flue block**) and be faced with *flue lining*.

flue block A precast *hollow block* which, with others, forms a *flue*. For a gas fire a typical block has a *flueway* 38×5 cm (15×2 in.) and outside dimensions 48×22 cm ($19 \times 8\frac{3}{4}$ in.). Many other sizes exist.

flue gathering or **throating** That part of the chimney opening at the foot of the *flue* which contracts or changes direction by corbelling.

flue lining (1) *Terra cotta* or fireclay pipes of 23×23 cm (9×9 in.) or 24 cm ($9\frac{1}{2}$ in.) dia. inside for an open fire, and 30 or 60 cm (12 or 24 in.) long. They have superseded *pargetting*. *See* **flue block, chimney block.**

(2) When a modern gas- or oil-fired furnace is installed at an old flue, a pliable flue lining tube should be inserted along its full length. Because these boilers are so efficient, they have cool flue gases liable to leave condensation water in the flue. The condensate is likely to pass through the brickwork and to stain the plaster in the rooms severely. These flue linings, made of aluminium, asbestos, and paper, are sealed to the chimney at top and bottom so that they are surrounded by a jacket of insulating still air. Alternatively the space between brick and lining may be filled with lightweight concrete. They are produced in sizes from 3 in. diameter upwards.

flue pipe A metal or *asbestos-cement* pipe which leads smoke from a slow-combustion stove to the flue. To prevent flame striking asbestos cement (which would break), the first $1 \cdot 8$ m (6 ft) next the stove should be of metal.

flueway The clear space for flue gases within a *flue* or *flue lining*.

fluorescent lighting [elec.] *Cold-cathode lamps*.

fluorescent pigments [pai.] Metallic tungstates, borates, and silicates, and some organic dyes. They form brilliant paints when effectively used, particularly in dull surroundings, because they convert invisible radiation to visible, and therefore are brighter than neighbouring surfaces. Some of them deteriorate quickly on exposure. *See* **phosphorescent paint.**

fluorocarbon resin *See* **polytetrafluorethylene.**

flush (1) In one plane, for example, a flush door.

(2) To flake off, said of the face of walling stone. *See* **hollow bed.**

(3) [plu.] To send a quantity of water down a pipe or channel to clean it.

flush door [joi.] A smooth-surfaced door, built of *plywood, hardboard,* or *coreboard,* either a *hollow core* or *solid door*.

flushed joint A joint above or below which the surface of stone has flaked off.

flushing trough [plu.] A long water tank extending above and across a range of WCs installed together, and supplying them with flushing water. It has the advantage that any WC can be flushed at short intervals without the waiting period needed for the filling of a *cistern*.

flushing valve or (USA) **flushometer** [plu.] A valve which supplies a precise quantity of water to flush a WC and therefore replaces the usual *cistern*. It has the advantage that there need be no wait while the cistern refills, but not all water authorities allow it. The London Metropolitan Water Board forbids it because it wastes water.

flush joint A *flat joint*.

flush panel or **solid panel** [joi.] A panel which is flush with its framing.

flush soffit A smooth under-surface, particularly of *spandrel steps*.

flush valve [plu.] A *flushing valve*.

fluxes In *soldering, brazing,* and *welding* (*C*), fusible substances like borax which cover the joint, prevent oxidation and so help the molten metal to stick. *See also C.*

flyer A *flier* (a rectangular stair tread or a flying shore).

flying bond (1) *Monk bond.*

(2) *American bond.*

flying scaffold A scaffold hung usually by ropes from an *outrigger*. *See* cradle.

flying shore or **flier** A horizontal strut fixed between two walls above ground level, often placed between two houses in a street when the intermediate house has been demolished. *See* **straining piece, needle, raking shore** (illus.).

fly rafter [carp.] A decorated *barge board*.

fly wire Fine wire mesh used as *scrim* between the joints in *wallboards*.

foam (1) *Lightweight concrete* made from foamed cement.

(2) *See* **foamed slag.**

(3) Foamed *synthetic resins* used as a lightweight infilling in *sandwich construction*.

foamed slag Blast-furnace slag which has been treated during cooling to give it a foamy texture. It is a useful *insulator* with a thermal *conductivity* of 0·10 to 0·11 W/m deg. C. (0·7 to 0·8 Btu in./ft²h deg. F.) for the loose (3 to 13 mm) aggregate weighing 528 kg/m³ (33 lb/ft³). Concrete made of it weighs 800 to 1900 kg/m³ (50 to 120 lb/ft³) and has a K-value between 0·17 and 0·32 W/m deg. C. (1·2 to 2·2 Btu in./ft²h deg. F.) according to its density. *Pumice concrete, no-fines concrete* and other *lightweight concretes* have rather similar properties. *See* BSCP 877.

foil A description of copper, zinc, aluminium, or lead which is thinner than *sheet* or *strip*, that is, less than 0·16 mm.

folded flooring [carp.] Floorboards forced into place by springing instead of by using a *floor cramp*.

folding casements or **doors** [joi.] (1) A pair of casements or doors with rebated *meeting stiles*, hung in a single opening.

(2) Two or more casements or doors hinged together so that they can open and fold in a confined space (BS 565).

folding rule A *fourfold rule* or a *zigzag rule*.

folding shutters [joi.] *Boxing shutters*.

folding stair A *loft ladder*.

folding wedges or **easing wedges** or **striking wedges** or **lowering wedges** [carp.] Wedges, often of hardwood, used in pairs to tighten up or slack off *dead shores*, *flying shores*, *raking shores*, and *falsework* (*C*), and *centers* of all sorts. *See* **easing the wedges, sand box.**

follower [plu.] *See* **boxwood bobbin.**

foot block An *architrave block*.

foot bolt [joi.] A strong *tower bolt* set vertically.

foot cut or **plate cut** or **seat cut** [carp.] (USA) The horizontal saw-cut in the *birdsmouth* at the foot of a *common rafter*. *See* **plumb cut.**

footing A foundation to a wall. *See also C.*

foot plate [carp.] (1) A horizontal timber laid over and crossing the *wall plate*. It joins the foot of the *rafter* to the foot of an *ashler* piece.

(2) A *sole plate*.

footprints or **pipe tongs** or (USA) **combination pliers** [plu.] A pipe fitter's adjustable wrench with serrated jaws, made in three sizes to grip pipes from 7·6 cm (3 in.) dia. down to a few mm. It also turns nuts but burrs the flats so that they cannot afterwards be turned with a spanner. *See* **gas pliers, plumber** (illus.).

footstone A *gable springer*.

force cup [plu.] A cheap tool for unblocking wastes or drains. It is a rubber cup fixed on the end of a short stick, and pushed up and down over the waste plug so as to move the water above the blockage, thus driving down, and sucking back alternately.

forced circulation [plu.] A pumped circulation, as opposed to a *gravity circulation*.

forced drying [pai.] *Drying* at a temperature not above 65° C. (150° F.). *Compare* **stoving.**

foreman An experienced *tradesman* who is placed in charge of other men working at his trade. He may work with the tools or merely supervise others. *See* **general foreman.**

foreman bricklayer An experienced *bricklayer* who gives out instructions concerning dimensions, levels, and *bonds* for brickwork.

foreman carpenter and joiner An experienced *carpenter* and joiner who, in Britain, may be further qualified by a certificate of the City and Guilds of London Institute. He is generally specialized either as an outside foreman, or as an inside foreman (shop foreman) who supervises bench workers and wood machinists. The outside foreman works on a building site, supervising carpenters and joiners, working under the *agent* or *general foreman* and sets out work from *architect*s' or engineers' drawings. The shop foreman also sets out work but on a smaller scale, and his work is usually finer but not so heavy.

foreman glazier An experienced *glazier* who measures work, estimates cost, and may choose glass and plan the cutting of it. He can prepare *templates* for curved glass and may be able to work with stained glass.

foreman plasterer A capable *plasterer*, usually with experience in both *fibrous* and solid plaster, able to work to the instructions of an architect. In Britain he may have the additional qualification of the technical certificate of the City and Guilds of London Institute. The best foremen can give a price for work before doing it.

fore plane [joi.] A plane between the *jointer* and the *jack plane* in length.

forked tenon [joi.] A joint in which a *tenon* cut in the middle of a long *rail* is inserted into an *open mortise*.

formaldehyde *See* synthetic resin.

form of tender [q.s.] In *tendering*, documents including a *bill of quantities*, sent out by an architect to contractors, on which they state their price for doing the work.

Forstner bit [carp.] A bit with a sharp ring at its outer edge, for sinking *blind holes*.

fossil resin or **fossil gum** [pai.] *Resins* such as *copal* which have become hard through ageing in the ground.

foul-air flue A ventilating duct which draws air out of a room.

foul water Sewage, also called *soil*.

foundation stone A large stone set in a building near ground level, generally well above the foundation level. Carved outside it are the date of setting, usually also the *building owner*'s, the builder's, and the *architect*'s names. A hollow within the stone may also be filled with contemporary objects such as coins, and, in England, a copy of *The Times* newspaper of the date when the foundation stone was laid.

fourfold rule or **folding rule** A two-feet (or in Scotland three-feet) long, wooden pocket rule divided into inches and eighths, hinged at three joints, to fold into one quarter of its length. *See* push-pull rule, zig-zag rule.

four-piece butt matching or **diamond matching** [tim.] A method of joining four adjacent sheets of figured, *sliced veneer* so that the richest *figure* is at the centre of the panel. Each sheet is cut square, near the edge of the best figure. These four right-angles join at the centre of the panel, forming a richly figured veneer nearly four times as big as on each separate sheet. The method gives a vivid effect with strongly striped veneers.

fourteen-inch wall The name sometimes given in Britain to a wall which is 1½ bricks, that is about 34 cm (13½ in.) thick.

foxtail wedging or **fox wedging** [carp.] *See* secret wedging.

foxy timber Timber which is beginning to decay and becoming dull red in colour.

frame [carp.] The timber members of *joinery* or a building, which are connected by *halving*, by *mortise-and-tenon*, or by similar joints. However the tendency in American *frame construction* is to have very few carpentry joints and nearly all the joints are nailed. *See also C.*

frame construction [carp.] (USA) Wood house building. Typical methods are *balloon framing, braced frames, platform frames*.

framed and braced door or **framed ledged and braced door** or **framed braced and boarded door** [joi.] A door which shows, on the face side, vertical boards as well as the two *stiles* and the top *rail* which is rebated to take the ends of the boards. On the back can be seen, in addition, the horizontal *ledges* at the bottom and middle of the door, as well as the two diagonal *braces* (*C*).

framed and ledged door [joi.] A door like a *framed and braced door* without the diagonal *braces* (*C*).

framed door [joi.] A door with a rigid frame (generally with tenon and mortise joints) consisting at least of top, bottom, and *lock rails*, *hanging stile*, and *shutting stile*.

framed floor or **double-framed** or **triple floor** [carp.] A floor of *common joists* carried by *binders* which are carried by beams. This type of floor is more laborious and costly than concrete floors and is therefore no longer built in Britain.

framed ground [joi.] *Grounds* framed like a door frame round openings with the head tenoned to the posts, used as a fixing for *joinery*.

framed partition A *partition* built up on its own frame of timber or less commonly of other material such as reinforced concrete or metal. *See* **common partition, coverings, double-tier partition, head, nogging, sill, stud, trussed partition.**

frame house [carp.] The American sawn timber house of *frame construction*, sheathed usually with *weather-boards* or *shingles* outside.

frame-saw [tim.] (1) A power saw for wood or stone with one or several vertical or horizontal blades held tight in a frame.

(2) [carp.] A heavy *bow saw.*

framing square [carp.] A *steel square.*

freemason A term which in the Middle Ages meant a skilled *mason*, capable of cutting *freestone*, that is of carving. The term is not now used in building but refers to members of certain associations.

free-standing A description of a part of a building which does not touch other parts and thus supports itself, for example, a *chimney stack* or a *column.*

freestone Building stone which is fine-grained and uniform enough to be worked in any direction and can thus be carved. Freestones are generally limestones or fine-grained sandstones.

free stuff [tim.] *Clear timber.*

freight elevator or **trunk lift** American terms for the British goods lift, used for hoisting furniture and other heavy loads in a building, but not for carrying passengers. *See* **dumb waiter.**

French arch *See* **Dutch arch**.

French casement or **French door** or **French window** [joi.] A *casement door*.

French fliers The *fliers* round an *open-well stair* which has *quarter-space landings*.

Frenchman A kitchen knife with the end bent over, used with the *jointing rule* for trimming mortar *joints*.

French polish [pai.] *Shellac* dissolved in *methylated spirits*. *See* **spiriting off**.

French roof A *mansard roof*.

French stuc Plasterwork imitating stone.

French tiles Clay *interlocking tiles*, occasionally imported to Britain.

French varnish [pai.] Varnish made of *shellac* and *methylated spirit*.

French window [joi.] A *casement door* is the most exact description of this door, but French window is the commonest term for it.

Freon Trade name for a non-toxic non-inflammable American refrigerating medium.

fresco [pai.] Painting on wet lime plaster with pigments mixed in water. The *pigments* are limited to those few which are sufficiently *alkali resistant*. Considerable skill is needed, and plasterer and painter must work together. For these reasons fresco painting is now rarely practised.

fresh-air inlet A pipe connected to the open air, admitting fresh air to the lower end of a house drain, near its connection to the *sewer* (*C*). It is fitted with a hinged mica flap at its upper end to prevent it working as a foul-air outlet, because in country or suburban districts it is fixed near ground level. Since these flaps are not reliable, many authorities do not now insist on a fresh-air inlet. The drains must in this case be fully ventilated by the *ventilation pipe* at the upper end of the house drain. *See* **intercepting trap**.

fret saw [joi.] A saw for cutting round sharp curves. It consists of a thin, narrow, replaceable blade held in a frame. The blade can be released from the frame to insert it into a drilled hole in the wood from which sawing begins. Many sizes are obtainable, from very small handsaws to power-operated *jig saws*.

fretted lead [plu.] H-shaped lead *cames* for *leaded lights*.

fretwork (1) [joi.] Work done with a *fret saw*.
(2) *See* **leaded lights**.

friction latch or **friction catch** [joi.] A small spring catch mortised into the edge of a door like a *ball catch*.

frieze The part of the wall of a room above the picture rail.

frieze panel [joi.] The highest *panel* in a door of more than four panels.

frieze rail [joi.] The rail below the *frieze panel* in a door.

frig bob saw A long handsaw used for cutting Bath stone at the quarry.

frit Treated and finely ground sand, glass, and flint used for glazing bricks and other *ceramics*.

frog An indentation on a bed face or faces of a brick to reduce its weight. At least for V-shaped frogs, walls built frog down (frog empty) are stronger than walls built frog up.

frontage The length of a site in contact with a road.

frontage line A *building line*.

front hearth That part of the floor of the fireplace which projects into the room.

front lintel The *lintel* supporting the visible wall over an opening.

front putty *Face putty*.

frost cracks [tim.] Cracks in wood caused by frost splitting the growing tree. They are closed by *calluses* but remain a *defect*.

frost heart [tim.] A defect of *heartwood* shown by a deepened colour caused by frost during the growth of the tree.

frosting [pai.] A description of a translucent, finely wrinkled finish formed during *drying*, characteristic of *tung* and other oils which have not been properly heat-treated. *See* gas-checking.

froststat [plu.] An automatic control, a type of *thermostat* that turns on the heat when the *circulating water* temperature falls nearly to 0° C.

frowy [tim.] Brittle, or soft.

fugitive pigment [pai.] A *pigment* which fades quickly.

full [mech.] A mechanical or *joinery* term describing a part that is slightly oversize, the opposite of *bare*.

full coat [pai.] As thick a *coat* of *paint* or *varnish* as can be properly applied.

full gloss [pai.] The highest grade of *gloss*.

full size (1) The *tight size* of an opening for glass.

(2) Full size slates in Scotland are 36×20 cm (14×8 in.) or larger.

full-way valve or **gate valve** [plu.] A cock which does not impede the flow of water, therefore used where the water pressure is barely enough for the flow required.

fungicidal paint [pai.] A *paint* which discourages fungus even in tropical conditions. Many disinfectants can be blended with paint for this purpose.

fungus [tim.] The cause of the *decay* of wood, *dry rot* being the commonest in buildings in Britain. *See* moisture content.

furniture [joi.] *See* door furniture.

furred (1) [plu.] Said of pipes and boilers which have become encrusted with hard lime or other salts deposited from the water heated in them.

(2) *See* furring.

furring or **firring** (1) Lathing fixed to *common grounds* and plastered, leaving an air space between brick and plaster.

(2) (USA) A cavity within an outer wall to keep out damp and for insulation. It may be formed with lath and plaster or with *hollow blocks* or bricks.

(3) [carp.] Timber strips laid, for example, on uneven joists to

pack them out, and make a plane surface for floor boards, or for the close boarding of a roof or wall.

(4) [plu.] *See* **furred** (1).

furring strip [carp.] A *common ground*. *See* **furring**.

furrowed surface Horizontal or vertical flutings at about 1 cm ($\frac{3}{8}$ in.) centres on *ashlar*, often on a face projecting between *drafted margins*.

fuse or **fuse element** [elec.] A small piece of wire in an electric circuit which melts when the current exceeds a certain value. It is a fire protection for the house and the wiring. If the fuse is properly designed no *short circuit* can last more than a fraction of a second. *See* **cartridge fuse**.

fuse box or **fuse board** [elec.] A *cutout box*.

fuse element [elec.] The fusible wire in a *fuse*.

fuse link [elec.] A container, generally of porcelain, which holds the *fuse element*. It can be pulled out of the *cutout box* to insert a new element.

fuse switch or **switch fuse** [elec.] A switch containing a *fuse*. *Compare* **switch and fuse**.

fusible link A metal part which, until it melts, holds open a *fire door*. It then releases the fire door, which closes.

fusible plug [mech.] A metal plug of low melting-point, screwed into the part of a boiler just above the furnace, under the water. If the water level drops below the fusible plug, this will melt and steam will blow down into the fire and put the fire out. A similar plug is used in *sprinklers* and *drenchers*.

fusible tape [tim.] *See* **joint tape**.

G

gabbart scaffold (Scotland) A sturdy *scaffold* of squared timber. The *standards* are of three *deals* bolted together, the middle deal being cut off at platform levels to let the *ledger* pass between the two extreme deals of the standard. The ledger thus rests on the middle deal.

gable or **gable end** The triangular part of the end wall of a building with a sloping *roof*, between the *barge boards* or *rafters*. A gable may be of any material – weatherboards, brick, stone, hung tiles, etc.

gable board A *barge board*.

gable coping A coping to a gable wall which projects above the roof and is coped mainly with *kneelers*.

gable end A *gable*.

gable post [carp.] A short post at the apex of a *gable* into which *barge boards* are *housed*.

gable roof A roof with a *gable* at one or both ends. *See* **roofs** (illus.).

gable shoulder The projection formed by the *gable springer* at the foot of a gable coping.

gable springer The stone overhanging at the foot of a gable coping, also called a foot stone, skew block, etc. It is below the lowest *kneeler*.

gable wall A wall crowned by a *gable*.

gable window A window built into a gable or looking like one.

gaboon or **gaboon mahogany** or **okoume** [tim.] An *African mahogany* (Aucoumea klaineana).

gain [carp.] A *mortise* or notch to receive another timber or a timber *connector*.

gallery apartment house (USA) An *apartment house* in which access to the dwellings is obtained from an open corridor.

gallet or **garnet** A *spall*, a chip of rock.

gallows bracket [carp.] A triangularly-framed bracket which projects from a wall.

galvanized pipe [plu.] Galvanized *screwed pipe* for water. The only difference from gas pipe is that it is galvanized, and gas pipe is black.

gambrel roof (1) or **half-hipped roof** A roof having a *gablet* near the ridge and the lower part hipped. *See* **roofs** (illus.).

   (2) US term for a *mansard roof*.

gamma protein [pai.] Protein from soya bean meal. It is used as an *extender* for *casein* in water paint or *distemper*.

gang-boarding A *cat ladder*.

gang saw [tim.] A reciprocating mechanical saw such as a *frame saw*.

gangway A path of *scaffold boards* laid for men to walk on or to wheel barrows along.

gap-filling glue [tim.] A *glue* for joining surfaces which cannot be fitted closely enough or pressed tightly enough for a *close-contact glue*. *See* **casein glue**.

Garchey sink A *waste-disposal unit* installed in some modern flats. All ordinary household rubbish can be emptied through it and no dustbins are needed except for tins, bottles, and newspaper.

garden-wall bonds Brickwork usually one *brick* thick, built to show a *fair face* each side. Since garden walls are not heavily loaded, the number of stretchers can be high. *See* **American bond, Flemish bond**.

garderobe (USA) A small bedroom or a small room in which clothes or other articles are kept.

garnet hinge [joi.] A *cross garnet hinge*.

garnet paper *Abrasive* paper covered with finely powdered garnet.

gas-checking or **gas-crazing** [pai.] *Crystallizing*, webbing, crows-footing or *frosting* of *paints* or *varnishes* containing vegetable *drying oils*, so called because it occurs when these oils dry in heating gas.

gas circulator [plu.] An appliance for heating water by gas, which is connected to a container for hot water, normally a *cylinder*. In warm weather, coal- or oil-fired boilers are not needed for *central heating*. An *immersion heater* or gas circulator can take the place of the boiler for supplying domestic hot water. Both are more expensive to run than the boiler, but give less work to the householder.

gas concrete A *lightweight concrete* foamed by aluminium powder in the mixing water reacting with it to form tiny hydrogen bubbles.

gasket [plu.] (1) Hemp fibres wound round the threads of a screwed joint in a water pipe before it is screwed up. When wetted, the hemp expands and makes a watertight joint.

(2) or **gaskin** In jointing stoneware pipes with cement mortar, a ring of old rope pushed into the foot of the socket before any mortar is inserted. *See* **jointing material**, *also* **C**, and **Neoprene**.

gas pliers Strong pliers with concave jaws for gripping gas pipe, serrated like the jaws of *footprints*.

gas proof [pai.] Description of a film which does not *gas check* in heating gas.

gate hook or **gudgeon** A metal bar driven by its point into a wooden post or built into a *masonry* or *brick* gate pier. It has an exposed, upstanding pin on which the *hinge* of the gate is dropped to hang the gate. *See* **band-and-hook hinge**.

gate pier A gate post of *concrete*, *brick*, or stone.

gate post A post, generally of wood, on which a gate hinge is fixed or on which the gate shuts.

gather (1) To bring *flues* together in a stack.

(2) *See* **flue gathering**.

gauge (1) The proportions of different materials in a *mortar or plaster* mix.

(2) [pla.] To accelerate the hardening rate of a plaster by adding *cement* or *gypsum plaster* as appropriate.

(3) or **margin** The exposed depth of a *slate* or *tile*, the distance between the bottom edge of one row of slates or plain tiles, and the bottom edge of the next row above or below. This is the same as the distance between the centres of the *battens* or the fixing nails. *See* **lap**. The following rules apply to *sized*, not to *random* slates, nor to *single-lap tiles*. For *centre-nailed slates*, gauge $= \dfrac{\text{length} - \text{lap}}{2}$. For plain tiles or *head-nailed* slates, gauge (cm) $= \dfrac{\text{length} - \text{lap} - 2 \text{ cm}}{2}$, nailing 2 cm from the head.

(4) A wooden or metal boundary strip used in asphalting to show the correct thickness of asphalt, like a *screed* in plastering.

(5) [joi.] *See* **mortise gauge**.

(6) *See below*.

gauge board (1) or **spot board** or **mortar board** [pla.] A board about 90 cm (3 ft) square for carrying *plaster* and tools. It is usually placed on a stand about 68 cm (27 in.) high. *See also* **gauging board**.

(2) [carp.] A *pitch board*.

gauge box A *batch box*.

gauged arch An arch built from *gauged bricks*, that is, soft bricks sawn to shape or rubbed smooth on a stone or another brick. They are laid with very fine joints, often of *lime mortar* without cement or sand (pure *lime putty*), which may be 3 mm ($\frac{1}{8}$ in.) thick or even less.

gauged bricks or **rubbed brickwork** Brickwork like that used for a *gauged arch*. It was in fashion in England from 1650 to 1750 and in Colonial Virginia, then sporadically in fine Victorian buildings and even more rarely in the present century.

gauged mortar (1) Cement-lime *mortar*.

(2) *Gauged stuff*.

gauged stuff or **gauged plaster** or **putty and plaster** [pla.] *Lime putty*, usually for the finishing coat of interior *cornices*, *ceilings*, and *mouldings*, to which *gypsum plaster* or cement has been added, to counteract shrinkage and hasten the set.

gauge pot A watering can or other container for pouring liquid *grout* (*C*).

gauge rod A *storey rod*.

gauging (1) [pla.] Adding *Portland cement* or *gypsum plaster* to a mix to hasten its set.

(2) [carp.] Marking timber with a *mortise gauge* or *marking gauge*.

(3) Sawing and rubbing bricks to size and shape for *gauged brickwork*.

gauging board A *banker* for mixing mortar, plaster, etc.

gauging box A *batch box*.

gauging plaster *Gypsum plaster*, or *plaster of Paris*.

gauging trowel *See* **plasterer's tools** (illus.).

gaul [pla.] A hollow in a *finishing coat* caused by bad trowelling.

G-cramp [joi.] A steel, G-shaped screw *cramp* used by *joiners* when gluing wood.

gelatin moulding [pla.] A method of making jelly moulds for complicated undercut *fibrous plaster* castings.

gelling [pai.] (1) Conversion of liquid to jelly.
  (2) *See* **feeding**.

general contractor A *main contractor*.

general foreman The *main contractor*'s representative on a site, in charge of all labour under the *agent*. He coordinates the work of trades *foremen*, whether employed by the main contractor or by a *sub-contractor*. He has usually graduated from a trade and been a trade foreman. *Compare* **ganger** (*C*).

geometric stair A stair with a *string* which is continuous round a semicircular or elliptical well, and thus has no *newel posts* and often no *landings* between floors.

Georgian glass Thick *glass* with a square mesh of steel wire embedded in it.

German siding or **novelty siding** [carp.] *Weather-boards* concavely rounded on the top edge and rebated on the inner face of the bottom edge.

gesso [pai.] A brilliant white composition of *whiting* and *glue*, or *plaster of Paris* and *size*, or of other materials, used as a background for painted designs on wood or plaster, sometimes in relief.

gib [carp.] An iron or steel packing piece which clasps together the parts of a *cottered joint*, passing through a hole in the *king post*. It is wedged tight with steel wedges, the *cotters*.

gig stick [pla.] A *radius rod*.

gimlet [carp.] A small tool with a wooden handle at right angles to the point, used for boring holes smaller than 6 mm ($\frac{1}{4}$ in.) dia. in wood. It was used by the ancient Greeks. *Compare* **awl**.

gimlet point [joi.] A description of the point of wood *screws* or *coach screws* which are intended to form at least part of their own hole and to grip the wood at their point.

gin A tripod and *gin block* or other simple *lifting tackle* (*C*).

gin block or **gin wheel** or **jinnie wheel** or **rubbish pulley** A single pulley for fibre rope carried in a steel frame with a hook at the top from which it can be hung.

girder casing (1) Material which covers and protects from fire the part of a steel *girder* (*C*) that is below ceiling level. It may be *concrete*, *brick*, *terra cotta*, *sprayed asbestos*, etc.
  (2) *Formwork* (*C*).

girt (1) [carp.] A *rail* or intermediate beam in wooden-framed buildings, often carrying floor *joists*, a small girder.
  (2) [tim.] *Girth*.

girth or **girt** [tim.] The circumference of a round timber, measured by tape. Timber is paid for at a price per m³ ft calculated from the *quarter-girth*.

girt strip [carp.] A *ribbon board*.

gland (1) [plu.] or **olive** A compressible copper or brass ring which in *non-manipulative compression joints* is slipped over the copper tube and under the screwed fitting. When the fitting is tightened, the soft metal of the gland is compressed and deformed, thus closing the gap between the tube and the fitting.

(2) [elec.] A seal used at the end of a cable to prevent water entering. A similar gland is used at water taps and stop valves to prevent leakage outwards.

gland joint A joint on a copper hot-water or soil pipe which allows temperature movement.

glass *See* **heat-absorbing glass, obscured glass, plate glass, sheet glass,** and *below*.

glass blocks or **glass bricks** Hollow translucent blocks of glass obscured by patterns moulded on one or both faces. When used as *partitions* they give a pleasant diffused light, but they have low heat insulation value and low *fire resistance*, though they are non-combustible. Glass blocks should be built as non-loadbearing walls (with *expansion strips* along the edges and top) in areas not exceeding 11 m² (120 ft²) or 6 m (20 ft) in height. In pavement lights solid glass is used. *See* BSCP 122, Hollow Glass Blocks.

glass-concrete construction (1) Reinforced concrete *pavement lights* or floor or roof slabs with glass lenses cast in.

(2) Walls with loadbearing concrete *mullions* between which hollow *glass blocks* are built in. These glass walls are generally without *opening lights*.

glass cutter A tool for cutting glass, either by a diamond or by a hard, sharp, metal wheel.

glassed surface *See* **polished work.**

glassfibre-reinforced resin *Synthetic resin* reinforced by glass fibre in the way that concrete is reinforced by steel, used for building caravan roofs, boat hulls, large pipes, sports car bodies, *corrugated sheet* for roofing, and many other things which cannot easily be built any other way. In building, one important use for the do-it-yourself householder is the jointing and repair of cast iron rainwater gutters, the jointing of metal to wood, and other water-proof, non-rusting welds. Translucent, as a shell roof, it has been used for spanning 15 m (49 ft).

glasspaper or **sandpaper** or **garnet paper** or **emery cloth** *Abrasive* paper made from glass, flint, garnet, corundum, or similar powders glued to cloth or paper. About fifteen different finenesses exist and unfortunately there are different names for each grade of fineness by each different maker. Emery cloth can be used wet and thus produces no unpleasant or poisonous dust.

glass silk or **glass wool** Flexible fibres formed from molten glass and used as an *insulator* for heat and sound. It is highly fire-resisting and can be obtained made up into *blanket* between waterproof papers, or resin-bonded or bitumen-bonded or loose. It weighs from 48 to 96 kg/m³ (3 to 6 lb/ft³) and its thermal *conductivity* (K) = about 0·033 W/m degree C. (0·23 Btu in./ft²h deg. F.) in the above states. *Compare* **mineral wool**.

glass size The *glazing size*.

glass slate or **glass tile** A piece of glass made to the same size as a *slate* or *tile* and laid among the slates or tiles to give light to an attic. Several glass slates must be used together if any considerable amount of light is to be let in. Apart from glass some translucent plastics are allowed: unplasticized PVC, *polyester resin* with glassfibre reinforcement (both stabilized against ultra-violet light), or acrylic material. For roof coverings in towns, fire hazard must be carefully considered, and some plastics have poor fire resistance.

glass stop (1) A device at the lower end of a *patent glazing* bar to prevent panes sliding down.

(2) A *glazing bead*.

glass substitutes Several synthetic resins can be used in place of glass and most of them are much less easily damaged. Some of these are: methyl methacrylate (*Perspex*), cellulose acetate, polystyrene, and vinyl polymers. Perspex and polystyrene can be used for optical purposes.

glass tile *See* **glass slate**.

glass wool *Glass silk*.

glaze (1) To instal glass in any sort of *light*. *Compare* **fixing**.

(2) A glass-like waterproof protection fired on to the surface of pottery, bricks, walls, and occasionally, roofing tiles. It may be transparent, coloured, or white.

(3) [pai.] A *glaze coat*.

glaze coat [pai.] A nearly-transparent, thin, coloured *coat* put on to enhance the colour below. The process of putting on the coat is called glazing.

glazed brick or **enamelled brick** A brick with a shiny surface from the *glaze* fired on to it.

glazed tile Earthenware *wall tiles* mainly for interior use, obtainable with a cream or white earthenware glaze or an *enamel* glaze of many possible colours. All glazed wall tiles *craze*, particularly where the temperature is variable. The only tiles which do not craze are unglazed tiles such as *quarry tiles*.

glazed ware or **glazed stoneware** Pipes and drain fittings made of vitrified clay (sometimes called earthenware) glazed by the vapour of common salt thrown into the kiln during firing.

glazier A *tradesman* who cuts glass and fixes it in a window or door frame with putty and *glazing sprigs* or spring-clips or *glazing beads* of wood, metal, or plastics. He also removes old putty.

glazier's chisel A glazier's *putty knife* shaped like a chisel.

glazier's points *Glazing sprigs.*

glazier's putty or **painter's putty** A plastic material used for bedding glass in *lights* and for making a weatherproof fillet of *face putty* outside the glass, holding it to the frame. It is made of *whiting* mixed with *linseed* oil with, occasionally, some white lead added. *Compare* **lime putty.**

glazing (1) To fit glass into lights. *See above and below; also* **fixing, patent glazing.** *See* BSCP 145, 152, 153.

    (2) [pai.] *See* **glaze coat.**

glazing bar or **sash bar** or **astragal** A rebated wood or T-shaped metal bar which holds the *panes* of glass in a window. The term 'glazing bar' is often kept for *roof lights* or for *patent glazing.* Metal windows usually have no glazing bars and thus let in more light.

glazing bead or **glass stop** A small hardwood strip, mitred and nailed or screwed round a rebate, to hold the glass instead of *face putty.* The glass is often bedded in wash leather or asbestos rope, etc. *See* **patent glazing.**

glazing size or **glass size** The size of a piece of glass cut for *glazing.* The clearance between glass and window should be 1·5 mm ($\frac{1}{16}$ in.) all round. The glass size both ways should thus be 3 mm ($\frac{1}{8}$ in.) less than the distance between the extreme edges of rebates in window bars, or *tight size.*

glazing sprig or **glazier's point** or **brad** A small headless *nail* buried in the *face putty* round a pane of glass. It holds the glass while the face putty is being placed.

gloss [pai.] The reflection of light by a painted surface. The stages of gloss in *finish* recognized by the British Standards Institution are:

    (1) flat (or matt), practically without *sheen,* even from oblique angles.

    (2) eggshell flat.

    (3) eggshell gloss.

    (4) semi-gloss.

    (5) full gloss, that is, a smooth, almost mirror-like (specular) gloss from any angle.

glossing up [pai.] The defect of a *gloss* which develops on a matt surface when it is handled.

glue [tim.] Liquid used for sticking materials to each other, particularly wood to wood. *Animal glue* has been known for hundreds of years, and is still in use for interior work, but is being superseded for outside work by the *synthetic resins* which were not used before 1930. Other glues are *casein glue,* blood albumen or soluble dried blood, and glues of vegetable origin such as *cassava glue,* or protein glues such as *soya glue.*

glue block [carp.] An *angle block.*

glue kettle [joi.] A cast-iron pot with a water jacket outside it held in another cast iron pot, preferably tinned or galvanized. It is used for heating *animal glue*.

glue line [tim.] The thin surface of *glue* between two parts.

glue spread [tim.] *See* spread.

glu-lam [tim.] *Laminated wood* used structurally, e.g. to build a roof *truss*.

glycerin, glycerine or **glycerol** [pai.] An alcohol which, like other alcohols, mixes in all proportions with water and is used in preparing *synthetic* and natural *resins* for *paints* and *varnishes*. It is also added to *distemper* to make it more flexible.

going The horizontal distance between two successive *nosings* is the going of a *tread*. The sum of the goings of the treads is the going of the *flight*. *See* going rod.

going rod A rod for setting out the *going* of a *flight* of stairs. *Compare* storey rod.

gold bronze [pai.] A copper or copper alloy powder for *bronzing*.

gold size [pai.] (1) An *oleo-resinous varnish* which dries quickly to the tacky state but hardens slowly and is used for fixing gold leaf to a surface.

(2) An oleo-resinous varnish with a high proportion of *driers* which hardens quickly and is used for making *filler*.

gold stoving varnish [pai.] A transparent *varnish* which forms a yellow film on tinplate or other silvery surfaces, either by means of a dye or by discoloration of the film during *stoving*.

goods lift The British equivalent of *freight elevator*.

gore A *lune* for covering a dome.

gorge A *throat*.

gouge (1) [joi.] A chisel with a curved cutting edge for hollowing out wood, more used by carvers than *joiners*. Carving gouges are very numerous. Joiner's or turner's gouges are less varied and are generally shaped like part of a cylinder, except for the *parting tool*.

(2) A *mason*'s tool for carving stone. It may be wood-handled and struck with the *dummy*, or all-steel and struck with the mallet.

gouge bit [carp.] A drilling bit with a rounded end.

gouge slip or **slipstone** [carp.] An *oilstone slip*.

grade [tim.] A quality class, both in Britain and USA, for timber. *See* stress graded timber.

gradiograph A *levelling rule* for checking the slopes of drains.

graduated courses *Diminishing courses*.

graffito or **sgraffito** or **scratch work** [pla.] A *plaster* surface decorated by scoring a pattern on it while it is soft, and exposing a lower coat of a different colour. The upper layer is often white, the lower layer black or dark red. More than two colours can be used if required.

grain [tim.] The general direction, size, and arrangement of the fibres and other elements in wood. *See* figure.

grainer [pai.] A painter who can paint wood or stone to imitate wood *grain* and knot marks, marble veins, and so on.

graining [pai.] Painting a surface to look like the *grain* of wood or marble, etc., by manipulating a wet coat of semi-transparent 'graining colour' with graining combs, brush, rags, and other implements.

granitic finish A *face mix*, resembling granite, on precast concrete.

granulated cork *See* **cork, corkboard.**

grappler The wedge-shaped, eyed spike for the top end of a *bracket scaffold*. It is driven hard into a brick joint.

grass table An *earth table*.

gravel board or **gravel plank** A horizontal board fixed to the underside of a *close-boarded fence* to prevent the vertical boards from reaching the ground. It is more easily replaced than a vertical board and is less easily rotted than is the end grain of the vertical boards.

gravel roof (USA) A roof made waterproof with *roofing felt*, sealed or bonded, and covered with a layer of gravel to improve its insulation value and protect it from the sun.

gravity circulation of hot water [plu.] A circuit which works only by virtue of the difference in densities between hot and cold water, not by a pump. Nearly all systems which supply hot water only (and no central heating) are gravity systems.

greasiness [pai.] A greasy surface on a paint film, caused by lack of *compatibility*.

green (1) *See* **green concrete.**

(2) [tim.] A description of unseasoned timber, which, unless otherwise specified, can be taken to be of 50% *moisture content*.

green brick A clay brick before burning, during drying.

green concrete or **green mortar** Concrete or cement mortar after its initial set and before it has begun to harden properly, when it has a dark green colour. It remains green up to about seven days or more in cold weather, less in hot weather.

greenheart [tim.] (Ocotea rodiaei) A hardwood from Guyana which is remarkable for its very high *modulus of elasticity* (*C*), 23 500 N/mm² (3·4 × 10⁶ psi) at 12% *moisture content* with a density of 1020 kg/m³ (64 lb/ft³). Owing to this high density, which is, like all timber densities, only an average and very variable, it often sinks in water even when dry. It varies in colour from black to brown, resists insect attack, is easily split, is harder to work than oak, and cannot be screwed without boring owing to its tendency to split. It is used for docks, piles, and similar structures where high strength is more important than low cost of timber and labour. *Compare* **Douglas fir.**

greystone lime Lime burnt from chalk containing enough silica and alumina to make it hydraulic. *See* **semi-hydraulic lime.**

grid plan A plan in which setting-out lines called grid lines coincide with the most important walls and other *building components*.

Prefabricated buildings are usually designed to fit a grid plan. *Compare* **planning grid**.

grillage A metal frame which carries *slates*, *shingles*, or *tiles*, and thus replaces slating battens. The horizontal bars can be set at the gauge (3) of the tiles, slates or shingles, *See also* (*C*).

grille (1) An open, often decorative screen of metal or wood.

(2) A grating or screen through which air passes usually into (rather than from) a ventilating duct. A grille has no damper and connects to a return air duct, unlike a *register*.

grinder [mech.] An *abrasive* wheel turned by hand or by electric motor, used for sharpening tools and removing metal from other surfaces. *See* **grindstone**.

grinding slip [joi.] An *oilstone slip*.

grindstone [carp.] An *abrasive* wheel of natural sandstone which is turned at a low speed, often by hand or treadle, so as to enable a *carpenter* or *joiner* to sharpen his edge tools. Electrically driven, fast-turning grinders have the disadvantage that they heat the tools and may thus spoil the *temper* (*C*) of the steel.

grinning through The showing through plaster of lathing beneath, or the showing of a lower coat of paint through a top coat. *See* **pattern staining**.

grog Broken pottery or *brick* used in making bricks, *refractory mortar*, etc.

groin The curved line at which the *soffits* of two *vaults* are seen to intersect. For **groyne** *see C*.

grommet or **grummet** [plu.] A hemp washer soaked in jointing compound and fitted between the back nut and the socket of a *connector* to make a tight joint. Many other grommets exist.

gross features [tim.] In considering the stress to which a timber part may be subjected, gross features are *knots*, *wane*, *slope of grain*, *shakes*, *checks*, *splits*, each of which is defined by certain allowable limits in BSCP 112. Rate of growth is also limited. *See* **stress-graded timber**.

ground (1) [carp.] A *common ground* or framed ground.

(2) [elec.] (USA) *Earth* or earth connection.

(3) [pai.] Any surface which is or will be painted.

(4) The first of several coats of paint. *See* **ground coat**.

ground brush [pai.] A wire- or string-bound paint *brush*, round or oval in section, used for painting large areas (grounds). *See* **house painter** (illus.).

ground casing [joi.] (USA) A *blind casing*.

ground coat [pai.] An opaque coat put on under a *glaze coat* or *scumble*.

grounded work [joi.] *Joinery* fixed to grounds.

ground floor The floor which is nearest the ground level. It is generally about one foot above ground level. *See* **first floor**.

ground plan A drawing of the ground floor of a building. It may also show the foundations.

ground plate [carp.] The lowest horizontal timber of a building frame, often called the *sole plate*.

grounds [carp.] *Common grounds* or *framed grounds*.

ground sill [carp.] A *sole plate*.

ground storey The space in a building between *ground floor* and *first floor*.

ground table An *earth table*.

groundwork Battens over close boarding or over *roofing felt* on *rafters*, fixed as a base for slating or tiling.

growth ring [tim.] Usually the same as an *annual ring*, though it is possible in a year of exceptional weather for two growth rings to occur.

GRP or **glassfibre reinforced plastics** *See* **glassfibre-reinforced resin.**

grub saw A saw for cutting stone by hand.

grub screw or **set screw** [joi.] A short *screw* fitted into threads on the metal part of a door handle to grip the spindle tightly. It has a slotted head which is small enough to pass completely into the hole and thus should not scratch the finger if properly fitted.

guard bead [joi.] (1) or **guide bead** or **inner bead** or **window bead** or (Scotland) **baton** or (USA) **inside stop** or **stop bead** A *bead* mitred round the inner edge of a *sash window* to prevent the inner *sash* from swinging into the room. *See* **sash stop.**

(2) A bead mitred round a shop window to protect it from rain.

guard board A *scaffold board* placed on edge at the outside of a gantry to prevent objects dropping and injuring people below.

gudgeon (1) A metal dowel for locking neighbouring stones together. (2) A *gate hook*.

guide bead [joi.] A *guard bead*.

guide coat [pai.] A very thin *coat* of loosely bound paint applied over a *surfacer* before it is rubbed down. It is removed during rubbing, but it guides the person doing the rubbing, since it shows him the high places.

guillotine [joi.] A *trimming machine*.

gullet [joi.] The gap beside a file tooth or *saw-tooth*; also the length of a saw-tooth from point to root.

gum arabic [pai.] A fine, white powder obtained from certain acacia trees, used in making transparent paints.

gummy [pai.] A description of the heavy *drag* on a *brush* from a sticky paint (gummy paint). It may be due to cold weather, excessive evaporation of the *solvent*, etc. *Compare* **slip.**

gum vein [tim.] Local accumulation of *resin* as a streak which occurs in some hardwoods.

gunstock stile [joi.] A *diminishing stile*.

gusset piece In *flexible-metal roofing*, a piece of metal soldered over an external corner between a roof sheet and two intersecting upright surfaces. *See* **dog ear.**

gutter A channel along the edge of a road or an *eave*, to remove rain-water. *Eaves gutters* may be of cast iron, pressed steel, concrete, asbestos cement, lead, copper, aluminium, zinc, timber, plastics, etc. A channel at the intersection of two roof slopes is a *valley gutter*. *See* **box gutter, secret gutter, open valley.**

gutter bearers [carp.] (1) Short 5×5 cm (2×2 in.) timbers under a lead gutter which carry the boards (*layer boards*) on which the lead is laid.

(2) Timber bearers which carry *snow boards* and rest on each side of a *box gutter*.

gutter bed A *flexible-metal sheet* laid over the *tilting fillet* behind an *eaves gutter* to prevent overflow from entering the wall.

gutter board A *gutter bearer*.

gutter plate [carp.] (1) A wall plate below a lead gutter.

(2) A side of a valley gutter. It is lined with flexible metal and carries the feet of the rafters. *Compare* **pole plate.**

gymnosperm [tim.] Trees with naked seeds, that is conifers (fir, pine, yew), all of which are called *softwoods*, though some may be very hard.

gypsum $CaSO_4 \cdot 2H_2O$, which occurs as the minerals alabaster, satin spar, and selenite. It is the raw material for *gypsum plasters* and is also the final stage of these plasters when they have set. It is an *extender* in *water paints* or *distempers*, rarely in *oil paints* because it is slightly soluble in oil and not very opaque.

gypsum baseboard [pla.] Square-edged plasterboard which is made for plastering with gypsum or anhydrite plaster. The edges must be *scrimmed* before plastering.

gypsum blocks *Building blocks*, usually hollow, made of *gypsum plaster*.

gypsum insulation or **flocculent gypsum** A loose, *gypsum* material weighing 288 to 320 kg/m³ (18 to 20 lb/ft³) and having a thermal *conductivity* (K) of 0·065 W/m deg. C. (0·45 Btu in./ft²h deg. F.).

gypsum lath [pla.] Sheets of gypsum plasterboard in relatively small sheets, nailed to a wall or ceiling for plastering. It is usually plastered two coats, like *fibre board*, gypsum wallboard, or base board. The first coat is a mix of 1 volume of *gypsum plaster* with $1\frac{1}{2}$ of sand, the finishing coat being any neat gypsum or retarded *hemihydrate plaster* gauged with lime.

gypsum plank or (USA) **board lath** [pla.] *Gypsum plasterboard* 19 mm ($\frac{3}{4}$ in.) thick, 60 cm (2 ft) wide, either for direct decoration or with a surface to take plaster.

gypsum plaster [pla.] Plaster made from *gypsum* by heating it to drive off water, for example, anhydrous gypsum plaster ($CaSO_4$), *Keene's* ($CaSO_4$ with borax or alum as an *accelerator*), *plaster of Paris* (the hemihydrate), and retarded *hemihydrate plaster*. Unlike cement, it expands on setting and therefore does not crack unless there are faults in the backing. It must never be mixed with *Portland* cement.

gypsum plasterboard [pla.] *Building board* made of a core of *gypsum* or *anhydrite plaster*, usually enclosed between two sheets of heavy paper. When plastered it requires two-coat work about 9 mm ($\frac{3}{8}$ in.) thick, since a skim coat alone, 3 mm ($\frac{1}{8}$ in.) thick, nearly always cracks. There are many types, including wallboard and *insulating plasterboards*. Its K-value is about 0·06 W/m deg. C. (0·42 Btu in./ft²h deg. F.).

gypsum-vermiculite plaster *See* **vermiculite-gypsum plaster.**

gypsum wallboard Gypsum plasterboard with its *face* self-finished or designed to be decorated directly. The *back* may be plastered.

H

hacking (1) A course of *rubble walling* which is composed alternately of single stones and of stones arranged two to the height of the course.

(2) (USA) Laying bricks in such a way that the bottom edge of each course is set in from the line of the course below.

hacking knife A knife with which old *putty* is removed from a light before reglazing it. *See* **house painter** (illus.).

hacksaw A handsaw or mechanical saw for cutting metal, consisting of a steel blade stretched tight in a frame. The blade is replaceable and wears out quickly by breakage of the teeth, which must be hard and are therefore brittle since hard steel is brittle.

haft The handle of a light tool such as a knife or an awl. *Compare* **helve.**

hair [pla.] Bullocks' or goats' hair was used in lime undercoats (*coarse stuff*) as reinforcement in the proportion of 1/kg of hair to 120-180 litres of *coarse stuff* to reduce cracking. Jute fibre, manila, or asbestos fibre are also used but are less effective.

hair beater [pla.] Two *laths* fixed together with wire or string to beat out *hair*, and thus clean and separate it.

hair cracking [pai.] Fine, erratic, random cracks which do not penetrate the top coat.

haired mortar [pla.] *Coarse stuff* containing *hair.*

hair hook [pla.] A tool like a hoe, but having, instead of the hoe blade, two bent tines for mixing hair into coarse stuff. *See* **larry.**

half bat or **snap header** A half-brick, cut in two across the length.

half-bed [q.s.] The *labour* of laying a stone on its bed is called a half-bed since the upper *bed-joint* surface is thus laid for the stone above. *See* **half-plain work, half-sawn stone.**

half-brick wall A wall the width of a brick, entirely built of *stretchers*, therefore in Britain, 10·2 cm (4 in.) thick in standard bricks.

half-hatchet (USA) A *carpenter*'s hatchet with a notch for drawing nails, like a plasterer's *lath hammer.*

half-landing (1) or **landing** A platform intermediate between two floors of a building, and joined to them by stairs.

(2) A *half-space landing*.

half-lap joint [carp.] A joint formed by *halving.*

half-plain work [q.s.] The *labour* of laying *ashlar.*

half-principal [carp.] A *rafter* which does not reach the ridge.

half rip-saw [carp.] A *rip-saw* with closer teeth than the usual rip-saw.

half-round Semicircular, such as a semicircular drainage channel, or a *moulding*, or a file flat on one face, curved on the other.

half-round veneer [tim.] *Veneer* cut on a lathe from a *flitch* which is roughly semicircular. The figure is intermediate in character between *sliced* and *rotary veneer*, since the cut radius can be made very large and the curvature very flat. The curve of the face of the flitch from which the veneers are cut may therefore be much less than a half-circle. *See* stay log.

half-sawn stone [q.s.] Stone which has been sawn only, the term half being used because half of the sawing cost is charged to the stones each side of the cut. *See* half-bed.

half-space landing A *landing* whose length is the width of both flights plus the well. *See* quarter-space landing.

half-span roof A *lean-to roof*.

half-timber A piece of timber measuring not less than 13×25 cm (5×10 in.) in cross section, made by halving a baulk along its length.

halved joint [carp.] A joint made by *halving* two timbers.

halving [carp.] A general way of forming an *angle joint*, or occasionally a *lengthening joint*, between two timbers of the same thickness. One half of each is cut away, the cut surfaces placed together, and the result is a joint in which the outer faces are flush. *See* **dovetail halving, lap joint, end-lap joint, splice.**

hammer [mech.] A steel head held by a wooden handle at right angles to it, fitted into a central eye. One end of the head is provided with a *face* for driving nails, the other end of the head is called a *peen* and may be hemispherical (*ball peen*) or wedge-shaped (*cross peen* or *straight peen*). The commonest hammers used in building are the *joiner*'s, *engineer*'s, *club*, *lath*, *claw*.

hammer axe [mech.] A hammer like the *lath hammer*.

hammer-dressed stone or **hammer-faced stone** Stone which has been only roughly faced, that is with the hammer at the quarry.

hammer finish (1) [pai.] A *finish* like hammered metal, produced by coloured *enamels* containing metal powder applied with a spray gun.

    (2) *See* **hammering.**

hammer-headed chisel Any *mason*'s chisel with a flat conical steel head, which is struck by a hammer and not by a mallet. *Compare* **mallet-headed chisel.**

hammer-headed key (1) A *double-dovetail key.*

    (2) A stone *cramp* like a *slate cramp* for locking stones together.

hammering Bending sheet metal such as copper into decorative shapes with a hammer. *Compare* **bossing.**

hammer pitching or **hammer pinching** Working the surface of *pitch-faced stone.*

hand (of doors) [joi.] The hand (right or left) of doors and windows is important when hinges and some door springs, latches or rim locks are ordered, but mortise locks, butt hinges, H-hinges, and some others are not handed. In Britain a door is right-hand hung if the hinges are on the right of a person opening the door towards himself, but its rising butts or lift-off hinges would be left hand. American terminology is different, but both in Britain and USA a right-hand lock fits a right-hand hung door, a right-hand screw tightens clockwise, and a right-hand stair has the handrail on the right going up.

hand brace [carp.] A carpenter's *brace*. *Compare* **hand drill**.

hand drill [mech.] A small boring tool usually of 6 mm ($\frac{1}{4}$ in.) or smaller capacity, like the *breast drill* but lighter and lacking its upward extension. It is more suitable for drilling in confined spaces or for drilling metal than is the carpenter's *brace*.

handed A description of building parts, including *joinery*, which match each other as an object matches its image in a mirror. When two objects are handed, they form a matching *pair*, one left-handed, the other right-handed. *See* **hand** (of doors).

hand electrical tools [mech.] Electric drills, screwdrivers, *circular saws*, nut-runners, *sanders*, *grinders*, *hand rotary planes*, etc.

hand float [pla.] A wood tool for laying on the *finishing coat* of plaster. It measures about $30 \times 10 \times 1$ cm ($12 \times 4 \times \frac{3}{8}$ in.), the *grain* being parallel to the length (also **straight-grained float** or **skimming float**).

handrail or **guard rail** [joi.] A rail forming the top of a *balustrade* on a balcony, a bridge, stair, etc.

handrail bolt or **joint bolt** [joi.] A bolt threaded at both ends. A square nut at one end is gripped in a *mortise* in an end of one handrail. In the other handrail, a similar mortise is provided, but the nut is circular and notched. This nut can be turned by striking the notches with a *handrail punch* inserted into the mortise from beneath the handrail. The tightening of the nut brings the two ends of the handrail closely together.

handrail punch [joi.] A small tool inserted into a *mortise* under a wooden handrail to tighten a nut on a *handrail bolt*.

handrail screw [joi.] (1) A *dowel screw*.

(2) A *handrail bolt*.

handrail scroll [joi.] A spiral ending to a handrail.

hand rotary electric planer [joi.] An electrically driven planing machine, held and handled like a bench *plane*, except that no force is required, and the cutting is done by the adzing action of a *cutter block*. Planing is therefore quick and relatively effortless. The motor needs to be of 1 hp for a 6·3 cm ($2\frac{1}{2}$ in.) wide cutter.

handsaw [carp.] Any joiner's or carpenter's saw held in the hand, such as a *rip saw*, *cross-cut*, or *tenon saw*, generally a cross-cut saw, never a power-operated saw. *See* **setting** (illus.).

handscrew or **screw clamp** [joi.] A *cramp*, generally with wood jaws and screws.

hang [joi.] To fit a door or a window to a building by its hinges.

hangar A building which shelters aircraft.

hanger (1) Generally a steel member from which other parts are hung, for instance a vertical steel bar which carries the weight of *walings* (*C*) in a deep excavation.

(2) A *stirrup strap*.

hanging gutter (USA) A metal *eaves gutter* fastened to *rafter* ends or to a *fascia*.

hanging post or **hingeing post** The post from which a door or gate hangs.

hanging sash [joi.] A *sash window*.

hanging shingling or **weather shingling** or **vertical shingling** *Shingles* fixed to vertical or nearly vertical slopes.

hanging stile [joi.] The *stile* of a door or casement to which *hinges* are fixed.

hardboard Usually *fibreboard* formed under pressure to a density of 480 to 800 kg/m³ (30 to 50 lb/ft³) (medium hardboard). Standard hardboard is denser than 800 kg/m³ and tempered hardboard is even denser, more than 960 kg/m³ (60 lb/ft³). It is also more resistant to water. Most hardboards have one smooth and one textured face and sometimes the smooth face is covered with plastics or metal or wood veneer or it may be embossed with a pattern to represent leather, tile, etc. Medium hardboard is made in thicknesses from 6·4 to 19 mm and in sheet sizes up to 1·8 × 3·6 m (6 × 12 ft). Standard and tempered hardboards are made from 2 to 12·7 mm thick and in sizes up to 1·6 × 3·6 m.

hard-burnt (1) A description of a burnt clay *brick*, *tile*, etc., which has been burnt at high temperature. This gives the material high compressive strength and durability and low *absorption*.

(2) [pla.] Plasters like *Keene's cement* are called hard-burnt.

hard dry [pai.] A stage in the drying of a paint film when it is nearly free from tackiness and is dried throughout its depth. It can therefore be flatted or another coat can be safely applied.

hardener (1) An *accelerator* for a *synthetic resin*.

(2) [pla.] A material used to harden plaster casts or gelatine moulds, such as alum, dextrine or polyvinyl acetate.

(3) A solution in water of sodium silicate, zinc silico-fluoride, etc., applied to a concrete floor to strengthen it and so reduce *dusting*.

hard finish [pla.] A smooth *finishing coat* containing *gypsum plaster*. *See* **hard plaster**.

hard gloss paint [pai.] A popular *oil paint* like *enamel*. It obtains its hard glossy finish from a proportion of *resin* in the oil *medium* (like varnish). It is hard by comparison with *oil gloss paint*.

hard-metal sheathed cable [elec.] A cable covered with a tube of hard metal, such as copper but not lead, over the cable insulation.

hard plaster [pla.] Plasters resist knocks in the following order, the hardest first: (1) cement and sand mixes, (2) plasters like *Keene's*, (3) retarded *hemihydrate plasters*, (4) *gauged* lime plasters. Generally the greater the lime content the softer is the plaster. Soft plaster has better sound *absorption*, does not show *condensation* so badly, and has a texture which many architects prefer to that of hard plaster. Hard plaster generally means a *hard finish*.

hard putty [pai.] *Hard stopping*.

hard solder [plu.] *Solder* containing copper, which therefore melts at a higher temperature than *soft solder*. *See* **brazing**.

171

hard stopping or **stopper** [pai.] A stiff paste made with water and powder, applied by *stopping knife* to fill deep holes in a ground. Unlike *glazier's putty* it sets hard quickly, since it contains *plaster of Paris*. Compare **filler**.

hardware Originally builder's iron and steel supplies, now also plastics and non-ferrous metal parts. *See* **finish hardware**.

hard water Water containing calcium or magnesium salts in solution which react with soap and thus prevent a lather forming easily in the water. These waters also give up their salts on heating and the salts are easily deposited in pipes, boilers, or channels, which become *furred up*. *See* **water softener**.

hardwoods [tim.] Wood from broad-leaved, usually *deciduous trees*. They belong to the botanical group of *angiosperms*. Not all hardwoods are hard though most British ones are. *See* **softwood, timbers**.

hardwood strip floor *See* **parquet floor layer**.

harl (Scotland) *Rough cast*.

hasp [joi.] A hinged, slotted arm or plate. *See* **hasp and staple**.

hasp and staple [joi.] A fixing for doors, gates, or box lids, in which the *hasp* is locked over a *staple* by a padlock or peg which passes through the staple.

hatchet [carp.] A small axe for dressing timber, held in one hand. Some, like the plasterer's *lath hammer*, are called hammers.

hatchet iron [plu.] A *soldering iron* shaped like a hatchet.

haunch or **hauncheon** [joi.] In a *tenon* made the full width of the wood from which it projects, but narrowed near the point, the wide part near the root is called the haunch. *See* **haunched tenon**, *also C*.

haunched tenon [joi.] A *tenon* which is narrower at the tip than at the root. The wider part is called the haunch.

hauncheon [joi.] A *haunch* at the root of a *tenon*.

haunching (1) Concrete round the sides of a buried *stoneware* pipe to support it above the bedding concrete.

(2) [joi.] A *mortise* for a hauncheon.

hawk or **mortar board** [pla.] A small pinewood square about $30 \times 30 \times 2$ cm ($12 \times 12 \times \frac{3}{4}$ in.), with a handle below it for carrying *mortar* or *plaster*. *Light alloy* hawks are also made. *See* **mason** (illus.).

hay band A straw rope left inside the cavity of a *cavity wall* to collect mortar droppings. It is laid on the wall ties and drawn up when the next course of ties is about to be fixed.

haydite (USA) *Expanded clay* used as an *aggregate* for lightweight concrete. *See* **perlite**.

head The upper horizontal member of a door frame, window frame, *sub-frame*, *partition* frame, or *column*. It is also the larger end of a bolt or hammer or the upper end of a slate, rainwater pipe, etc.

head block [carp.] A block bolted to the end of a timber tie to take the thrust of a *rafter*. It may also be keyed into the tie.

head board A board carried on the head of the *bricklayer's labourer* in the north of England, and used by him instead of a *hod* for carrying mortar. *See also C.*

head casing [joi.] (USA) The part of the *architrave* outside and over a door. It may be topped by a *weathering* such as a *drip cap* or by a hood.

header (1) A *brick* laid across a wall to *bond* together the different parts of a wall, and, by extension, the exposed end of a brick.

(2) or **header joist** [carp.] American term for a *trimmer joist* used in floors or a similar member in the walls of *frame construction.* For the confusion between American and British senses *see* **trimmer.**

header joist [carp.] (USA) A *header* (2).

head flashing A *flashing* like a small *gutter* round the edge of a projection through a roof.

head guard A lead gutter which crosses the cavity in a *cavity wall* and protects the head of a wooden *lintel* or frame from water.

heading bond Brickwork *bond* of headers only, used for footings and for all curved walls.

heading course A course of *headers.*

heading joint (1) A *cross joint.*

(2) [carp.] The line on which two boards butt. *See* **splayed heading joint.**

head jamb [joi.] (USA) A *door head.*

head joint (USA) A *cross joint.*

head moulding A *moulding* over an opening.

head nailing The nailing of *slates* at about 2 cm from the head. Each nail is thus covered by two slates and a better covering is formed than with *centre nailing,* but the method is unsuitable for exposed windy sites. Since the *gauge* is slightly smaller than with centre nailing, the cost and weight per unit area of roof are slightly more.

headroom or **headway** The clear height from floor to ceiling. In *stairs* it is measured from the *nosing* vertically upwards, and is usually not less than 2 m ($6\frac{1}{2}$ ft.).

head weather moulding [joi.] A small member framed into the *head* of a window frame, to throw rainwater clear of the window.

heart [tim.] The centre of a log. *See* **boxed heart, heart centre.**

heart bond A *bond* for walls which are too thick for *through stones.* Two *headers* meet within the wall and their joint is covered by another header.

heart centre or **pith** [tim.] The core of a tree, also called *parenchyma. See* **balsa wood, burr.**

hearting The infilling of a wall faced on one or both sides with better-looking material.

heart plank [tim.] A plank containing the heart of a *hardwood* log and thus most of its defects. *See* **centre plank.**

heartshake [tim.] A radial *shake* originating at the heart (BS 565).

heartwood or **duramen** [tim.] Sometimes the best timber in a log, to be distinguished from the *sapwood* which, before the tree was felled, was living tissue, and is softer, and paler in colour. *Compare* **heart centre.**

heat-absorbing glass or **anti-actinic glass** A rolled, faintly green *glass*, which allows only one sixth of the infra-red rays (heat rays) to pass through it and about half the visible spectrum. It must be so placed that sunlight falls uniformly on it without irregular shadows since the sun heats it quickly and may thus crack it.

heat exchanger A device containing one fluid, usually air or water, that is heated (or cooled) by a hotter (or cooler) fluid passing through special tubes inside it. A *calorifier* is one type.

heating [plu.] *See* **central heating, coil heating, district heating, unit heater.**

heating and ventilation engineer A mechanical engineer concerned with the design and maintenance of heating and ventilation systems, hot and cold water supplies, refrigeration plant, *sprinkler* and *drencher* systems, and centralized *vacuum-cleaning plant*. His most specialized knowledge deals with such heat exchangers as boilers and refrigerators. Many are members of the Institution of Mechanical Engineers or at least of the Institution of Heating and Ventilation Engineers. *Compare* **hot-water fitter.**

heating battery The heating surface of a *calorifier*.

heating element [elec.] That part of an electric heater which consists of a wire which is heated by an electric current.

heat insulation The ability of a material to impede heat flow. It is inversely proportional to the *air-to-air heat transmission* coefficient of the material. *See* **insulating materials.**

heat-recovery wheel A wheel that is mounted across the incoming and outgoing air ducts of an air-conditioning system and, by slowly rotating, transfers heat in winter from the outgoing to the incoming air. In summer it can function in reverse to cool the incoming air. High efficiencies are claimed but stationary *heat exchangers* working on the same principle also exist.

heat-resistant paint or **enamel** *Paint* or *enamel* which can be *stoved* at high temperatures or used on *radiators* or similar equipment. They often contain *silicone resins*.

heat-resisting glass A glass with a low *coefficient of expansion* (*C*), like borosilicate glass, that has superseded mica for the windows of *roomheaters*.

heavy-bodied paint (1) A viscous *paint*.

(2) A paint which makes a strong film.

heel [joi.] (1) The rear end of a *plane*.

(2) The lower end of the *hanging stile* of a door.

(3) The part of a beam or *rafter* resting on a support.

heel strap [carp.] A U-shaped steel *strap* (*C*) bolted to the *tie-beam* of a wooden *truss* near the *wall plate*. The strap passes over the back of the *principal rafter*, joining it to the tie-beam, and transmitting the rafter thrust to the tie-beam.

height board [carp.] A *storey rod*.

height money Additional pay for working more than 12 m (40 ft) above the ground or above a building.

helical hinge [joi.] A hinge for a *swing door* which is hung from its frame. *Compare* **floor spring**.

helical stair The correct but not the usual name for a *spiral stair*.

helve The handle of an axe, pick, sledge hammer, or similar heavy tool usually made of *ash* in Britain, or *hickory* in USA. *Compare* **haft**.

hemihydrate plaster [pla.] *Plaster* obtained by heating *gypsum* ($CaSO_4 \cdot 2H_2O$) which loses part of its water and becomes half-hydrated, the hemihydrate ($CaSO_4 \cdot \frac{1}{2}H_2O$). It is the very quick-setting plaster called *plaster of Paris*. To make it suitable for plasterer's work a retarder of set, usually *keratin*, is added to form retarded hemihydrate plaster. Hemihydrate plaster in contact with iron or steel should have 5% of its weight of hydrated lime mixed with it to prevent corrosion. The plaster is sometimes sold with the lime mixed into it. *See* **gypsum plaster**.

herring-bone bond *Diagonal bond*.

herring-bone matching [tim.] *Book matching*.

herring-bone strutting or (USA) **cross bridging** [carp.] A method of stiffening floor *joists* at their midspan by fixing a light strut from the bottom of each to the top of its neighbours and the top of each to the bottom of its neighbours. *See also* **bridging, solid bridging**.

hessian or (USA) **burlap** Strong coarse material woven from jute or hemp for making sacks, brattice cloth wrapping for electric power cables, or reinforcement for *fibrous plaster*.

hew [carp.] To shape timbers with hatchet or axe. *See below*.

hewn stone Stone with a good finish (dressed stone).

H-hinge or **parliament hinge** or **shutter hinge** [joi.] A *hinge* screwed through holes on two lengthened parts (legs of the H) away from the joint. The knuckle (the crossbar of the H) projects beyond the face of the closed door or shutter but this projection allows the door to clear *architraves* and lie flat against the wall when opened. It is used for outside shutters.

hickey A portable steel tool used in USA for bending steel tube, conduit, or reinforcement.

hickory [tim.] A strong north American timber used like English *ash* for tool handles, ladder rungs, etc., where high shock resistance and bending strength are needed. In East Anglia, hazel is sometimes called hickory, though it is much less strong.

hiding power or **opacity** [pai.] The ability of a paint to obscure any colour beneath it. (The term *covering power*, meaning both hiding power and *spreading rate*, is not now used.)

high-calcium lime or **fat** or **rich** or **white** or **white chalk** or **non-hydraulic lime** [pla.] A relatively pure lime (mainly CaO) giving a very plastic putty. It can safely be mixed with *Portland cement*.

high-density plywood or **compregnated wood** or **lignified wood** [tim.] *Plywood* or *laminated wood* formed, often with *film glue*, at pressures of 500 psi or more. The density is more than twice that of ordinary plywood but the mechanical strength is about sevenfold. It generally cannot be nailed or bent, and must be worked with metal-working tools. It is used for airscrews and for the ribs in plywood floors of aircraft. Also called compreg or impreg.

highlighting [pai.] Emphasizing the relief of a surface by making certain parts of it paler than the general colour.

high-pressure system A method of *central heating* with small pipes and rapid circulation of water at about 205° C. (400° F.) and 16 atmospheres. It was preferred to the low-pressure system when this required awkwardly large pipes before **small bore** systems were used.

hinge [joi.] A metal, pinned connection between a door or gate and the jamb or post on which it swings. The commonest hinges in building are *butt, back-flap, band and hook, counter-flap, cross garnet, helical, H-hinges, lift-off butts, rising butts*. The following hinges are symmetrical and therefore not handed: backflaps, butts, counter-flaps, H-hinges. All other types of hinge are *handed* and the *hand* must be stated.

hinge-bound door [joi.] A door which is hard to close because the hinges have been screwed in with the door set too close to the frame. With *butt hinges* this can easily be remedied. The door is unhung, slips of cardboard are inserted behind the butts, and the door rehung. The process is repeated until the door closes easily.

hip or (Scotland) **piend** The outstanding edge formed by the meeting of two roof surfaces, provided on the ends of roofs which do not finish with a *gable*. Rainwater flows away from a hip and towards a *valley. See* **hipped roof**.

hip capping (1) The uppermost strip of *roofing felt* or other protection over a hip.

(2) *Weaving.*

hip hook or **hip iron** or (Scotland) **piend strap** A metal bar fixed to the *hip rafter*. It projects and is seen at the foot of the *hip*. Its function is to hold the lowest *hip tile* in place.

hip knob A *finial* to a *ridge* where it meets a *hip* or *gable*.

hipped end The sloping triangular end of a *hipped roof*.

hipped gable roof A *jerkin-head roof. See* **roofs** (illus.).

hipped roof A roof which has four slopes instead of the two slopes of the ordinary gabled roof. The shorter sides are roofed with small sloping triangles, called the hipped ends, which are bounded by two hips above (meeting at the ridge) and an *eave* below. Normally the eaves are at the same level all round. *See* **roof** (illus.).

hip rafter or **angle ridge** or **angle rafter** A rafter forming a *hip*. The *jack-rafters* meet on it.

hip roll (1) or **ridge roll** [carp.] A round timber with a V-cut beneath to cover a hip.

(2) The *flexible-metal* covering which fits over a wooden hip roll.

hip tile (1) Clay or concrete *tiles* which cover those roofing tiles which meet at a hip. *Angular*, round, and *bonnet* hip tiles are the types standardized in Britain. Hip tiles do not generally overlap each other. *See* **ridge tile**.

(2) An asbestos-cement hip covering.

hitch A loose knot used on a hemp rope when hoisting material.

hoarding High fencing erected by a *contractor* round his building site, mainly to prevent theft.

hod A tray measuring $40 \times 23 \times 23$ cm ($16 \times 9 \times 9$ in.) shaped like a $40 \times 23 \times 23$ cm box cut diagonally in two. It is fixed to a long handle by which it can be held on the shoulder with one hand and thus carried up a ladder. It contains twelve bricks or 9 litres (2 gallons) of mortar and is therefore very heavy when loaded. It is made of light alloy or wood. *See* **bricklayer's labourer**.

hod carrier A *bricklayer's labourer* who carries a *hod*.

hogsback tile A *ridge tile* which is not quite half-round.

holderbat A fixing for holding a pipe to a wall or *soffit*. It consists of a steel bar built or wedged into brickwork or concrete and forged into a semicircular shape at the outer end. Another semicircular piece screws down on to it and clamps the pipe. *See* **fasteners** (illus.).

holdfast A steel or cast-iron spike driven into a joint of brickwork. At its outer end is a flattened eyed piece, through which a screw can be driven to fix *joinery*, etc. *See* **fasteners** (illus.).

holding-down clip A folded clip of *flexible-metal roofing* sheet shaped like a *capping*, but fixed to the roof boarding so as to anchor and join adjacent lengths of capping

hole saw or **tubular saw** or **hole cutter** or **annular bit** [mech.] A drilling tool which cuts a ring-shaped sinking and can if necessary cut out a complete cylinder of wood or metal. In carpentry it is used for fitting the *shear plate* and the split ring timber *connectors* and consists of a pipe with one serrated end. It greatly increases the maximum diameter of hole which can be made with any given drill.

holidays or **skips** [pai.] Areas which have been left unpainted.

holing Punching holes in slates by hand or machine before fixing them.

hollow (1) A concave surface.

(2) or **hollow plane** [joi.] In Britain a plane for forming convex surfaces. *Compare* **round**.

hollow-backed flooring [carp.] Floorboards hollowed out on the underside so as to improve the ventilation and bedding of the boards on the *joists*.

hollow bed A *bed joint* which is not filled in the middle, a method of laying *sills* so as to prevent them breaking if the *masonry* settles unequally. If other stones are laid like this, they may flake off.

hollow blocks or **hollow tiles** Concrete or burnt clay hollow *building blocks* are used for making partitions or external walls, or for forming reinforced concrete *hollow-tile floors* (*C*). Lightweight, thermally-insulating, hollow blocks are also made of *foamed slag* concrete, diatomite, gypsum, etc. *See* **solid masonry unit, flue block.**

hollow chamfer A concave *chamfer.*

hollow clay tile *Hollow blocks.*

hollow-core door [joi.] A *flush door* in which the plywood or *hardboard* of both faces is glued to a skeleton framework (or core). It is less heavy and cheaper than a *solid door.*

hollow glass blocks *See* **glass blocks.**

hollow partition (1) A *partition* built of hollow blocks.

(2) A partition built in two leaves with a gap between for sound insulation, thermal insulation, or to accommodate a sliding door. *See* **discontinuous construction.**

hollow plane [joi.] *See* **hollow.**

hollow roll or **seam roll** A method of jointing two pieces of *flexible-metal* sheet in the direction of the fall of a curved or sloping roof. The edges of adjacent sheets are laid together, lifted up, and bent round to form a cylindrical roll without a *wood roll*, but sometimes with a stiffening sheet of copper, etc., called a tack.

hollow wall (1) A *cavity wall.*

(2) (USA) A wall built of two leaves which are bonded together, not with wall ties, but with bricks, such as a wall built in *rat-trap bond*.

hollow-wood construction [carp.] Wood construction with *plywood* (decorative or structural or both) glued on both faces as in hollow-core doors.

homogeneous fibre wallboard A wallboard made of the same material as *insulating fibreboard* but generally thinner. It is used as a cheap wall and ceiling lining where high insulation of sound and heat are not needed. It can be used as a base for plaster and as a permanent shuttering for concrete or as an underlay below linoleum or carpets. It is made in thicknesses of 10 mm ($\frac{1}{2}$ in.) or less, up to 1·2 m (4 ft) wide and 3·6 m (12 ft) long.

Honduras mahogany [tim.] (Swietenia macrophylla) A Central American mahogany of the same family as *Cuban mahogany* but slightly paler, softer, less dense, and easier to work.

hone or **oilstone** [carp.] A very smooth quartz stone used to give a final, uniform, long-lasting polished edge to a cutting tool which has been rubbed on a coarser stone such as a *grindstone*. The surface is oiled before the blade is rubbed on it. Hones are also used for rubbing *terrazzo* or interior stonework.

honeycombing [tim.] Separation of the fibres in the interior of timber due to drying stresses. *See also C.*

honeycomb slating *Diagonal slating* in which the *slates* have three corners cut off, the *tail* as well as the two edges.

honeycomb wall (1) A *half-brick wall* built of *stretchers* with gaps between, so that they are only held by bed joints at their ends, above and below.

(2) or **sleeper wall** A half-brick wall under the wooden ground floor of a house with no basement, supporting the floor *joists*, with a few bricks omitted for ventilation.

honing gauge [joi.] A small clamp with one wheel attached. The clamp holds a *chisel* while it is being rubbed on a stone and thus keeps the same angle for the chisel edge throughout the rubbing process.

hood A canopy to throw rainwater off a window or door or other opening.

hook [joi.] The extension of the *cutting iron* past the *sole* of a plane.

hook-and-band hinge *See* **band-and-hook hinge**.

hook and eye [joi.] A fastening for a door or a casement window consisting of a *cabin hook* on the frame which engages with a *screw eye* on the door.

hook bolt A galvanized bolt bent into a U-shape at the end which is not threaded. A nut and washer fit over the other end, which passes through metal or *asbestos-cement* wall or roof sheet and holds it in place. The hook engages with the steel angle used as a *purlin* or *rafter*. The thread may be cut on the bar or rolled on it, in the latter case it stands up from the bar.

hook joint or **hooked joint** [joi.] A joint used between the meeting edges of doors, *casements*, and showcases which must be airtight. The *rebate* on one *stile* is cut to an S-shape, which fits into a similar groove on the opposite stile. A similar arrangement can be made for the *meeting stiles* of double windows.

hook rebate [joi.] The S-shaped rebate in a hook joint.

hook strip [joi.] (USA) A cleat fixed to a wall as a base to which clothes hooks can be screwed.

hoop iron Thin strips of iron or steel about 50×1.5 mm ($2 \times \frac{1}{16}$ in.) occasionally used as reinforcement in the bed joints of brickwork from about the seventeenth century in England. In good practice it is usually tarred and sanded.

hoop-iron bond Brickwork reinforced with *hoop iron* in the bed joints. *See* reinforced brickwork (*C*).

hopper The draught-preventing, triangular *deadlights* at the side of a *hopper light. Compare* **rainwater hopper**.

hopper head A *rainwater head.*

hopper light [joi.] (1) A *light* hinged at the bottom with draught preventing *hoppers* each side.

(2) A light hinged at the bottom between deep *reveals* where there is no space for hoppers at the side.

hopper window or **hospital window** [joi.] A window formed of one or more *hopper lights* above each other.

horizontal shore A *flying shore*.

horn [joi.] Originally an extension beyond the frame of a mortised member (usually a *stile*) to strengthen the *mortise* during wedging up. Since it protects the frame during transport, it is also used with dowelled joints. It helps to build a frame into brickwork.

horse (1) [carp.] Framing used as a temporary support, such as a *trestle*.

(2) [carp.] A *string* carrying the *treads* and *risers* of a *stair*.

(3) [pla.] A short board housed to receive the *stock* (or wood backing) to the shaped metal *templet* which forms a plaster *moulding* to the required profile. *See* **running mould,** *also below.* ·

(4) [plu.] A wood *finial* which is to be covered with lead.

horsed mould [pla.] A wooden stock (carrying zinc plates cut to the profile of the desired plaster *moulding*) housed firmly into a *horse* and fixed to it by wooden stays. The horsed mould is pushed along the angle between ceiling and wall by one plasterer. It is held in position by a *running rule* nailed to the wall and a *nib guide* nailed to the ceiling. A second plasterer feeds the plaster on to the moulding. *See* **muffle.**

horsing up [pla.] Building up a *horsed mould* for running a plaster *cornice*, etc.

hose cock or **hose bib** or **sill cock** (USA) A tap at sill height outside a building, with a *fitting* which will take a hose.

hospital door [joi.] A *flush door*.

hospital window [joi.] A *hopper window*.

hot-air heater An air heater which warms a room by driving warm air into it through holes in the floor or walls.

hot-air seasoning [tim.] Drying timber in a *kiln*.

hot pressing [tim.] Gluing in a *press* between heated *platens*, usual for *thermo-setting* glues and common for other *glues* which set more quickly when heated. Used in making *plywood*. *See* **caul.**

hot spraying [pai.] Spraying of *paints* and *lacquers* which have been heated to reduce their viscosity instead of adding volatile *thinners*. In this way a thicker coat can be formed.

hot surface [pai.] An abnormally absorbent surface (BS 2015).

hot-water cylinder A *cylinder* for storing hot water.

hot-water fitter or **heating and domestic engineer** A *tradesman* who instals and maintains small boilers, furnaces, etc., sometimes also *air-conditioning* systems, usually working under a *heating and ventilation engineer*. He may hold the technical certificate of the City and Guilds of London Institute but with further qualifications may eventually become a heating and ventilation engineer himself.

housebreaking The demolition of buildings.

housed joint [joi.] or (USA) **dado joint** A shallow sinking in the face of one board to enclose (or house) the end of another timber in the way that steps are housed into a *close string*.

housed string [joi.] A *close string*.

house painter A craftsman who paints buildings and prepares them for painting with wire brush, scraper, pumice stone, blow lamp, or chemicals, filling in cracks with *hard stopping*, *glazier's putty*, *plastic wood*, or plaster.

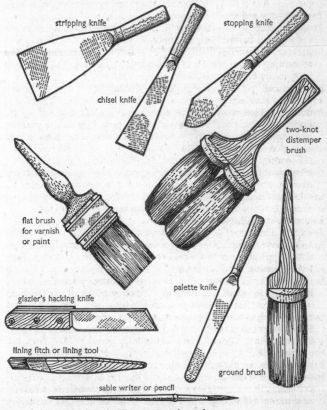

stripping knife

stopping knife

chisel knife

two-knot distemper brush

flat brush for varnish or paint

glazier's hacking knife

palette knife

lining fitch or lining tool

ground brush

sable writer or pencil

House painter's tools.

181

house painter and decorator A *housepainter* who can also hang paper and do gilding and marbling. He may hold a technical certificate of the City and Guilds of London Institute. *See* **painter's labourer.**

house service cutout [elec.] A cast-iron or plastics *cutout box* in which the *service cable* ends. It is sealed by the electricity undertaking.

housing (1) [joi.] A *housed joint*.

(2) A quantity of houses or flats.

housing society *See* **cooperative housing society.**

hovelling Extending the walls of a *chimney stack* upwards with openings all round to improve the draught in high winds.

hub [plu.] (USA) The enlarged end (or bell or *socket*) of a cast-iron pipe.

hue [pai.] *See* **colour.**

humidifier Plant such as washers or sprays for regulating air moisture content and temperature in *air-conditioning*.

hungry or **starved** (1) [pai.] Description of a surface which is too absorptive for the amount or kind of paint put on it. The paint film is therefore thin and patchy.

(2) [pla.] A plastering mix of poor workability.

hydralime [pla.] *Hydrated lime.*

hydrant A connection to a water main, usually a *fire hydrant*.

hydrated lime [pla.] *Slaked lime* ($Ca(OH)_2$) formed by adding water to quicklime (CaO). It is known as *lime putty* when it has been kept in water for some weeks.

(2) or **dry hydrate** Lime bought as $Ca(OH)_2$ from the manufacturer as powder in bags. Though more expensive than quicklime, its use may be unavoidable where there is no space for a *maturing pit* or where time is not available for maturing the putty. Dry hydrate should be soaked for some hours before use, to make it plastic.

hydraulic A description of *limes* or mortars which, like *Portland cement*, set and harden under water because of their content of burnt clay (aluminium silicate). They also harden more quickly than lime mortar but cannot all be mixed with Portland cement in plastering. Plasterers consider some of them unworkable, partly because the most hydraulic limes cannot be allowed to mature, since they would harden in so doing. Non-hydraulic limes (pure CaO) are the most workable or 'fat'. *Eminently hydraulic* limes are 'lean', the least workable, and set under water in three days whereas *feebly hydraulic* limes require twenty-one days. *See* **semi-hydraulic lime.**

hydraulic cement *Cement* which hardens under water, like Portland, *see* **hydraulic.**

hydraulic glue An old term for glue which remains hard under water. *See* **synthetic resins.**

hydraulicity The property of a mortar (not possessed by *high-calcium lime* mortar) of setting in the absence of air or in excess water. *See* **hydraulic.**

hydraulic lift A passenger lift raised by a hydraulic ram underneath it. Electric lifts were superseding this type but they are returning to fashion.

hydraulic test [plu.] The *water test* for drains.

hyperbolic paraboloid roof A shell roof which looks like a butterfly in elevation. It began to be fairly widely used in the 1950s, built in timber or concrete. *See* **hypar** (*C*).

I

ignitability The ability of a material to be ignited by a small flame. For a building material this can be tested by a method described in BS 476.

imbrex In *Italian tiling*, the over-tile which is semicircular and fits over the *tegula* or under-tile.

imbricated A description of any surface which looks like a tiled roof.

immersion heater [elec.] An electric resistance heater submerged in a water tank. *See* **automatic immersion heater.** *Compare* **electrode boiler.**

imperfect manufacture [tim.] Defects in conversion or planing such as variation in sawing, *mismatching*, torn or *chipped grain* or other tool marks, *skips* in planing, machine burn, insufficient *tongue* or groove, and so on. It is less weakening than a *defect*.

impregnated flax felt *Roofing felt* made of felted jute or flax or hair, waterproofed with fluxed coal-tar pitch, brown wood tars, or wood pitches.

impregnation [tim.] This usually means the impregnation of timbers with preservatives under pressure or with alternating vacuum and pressure. It is much more effective than application by brush, but needs special plant and is therefore best done by a timber wholesaler. *See* **preservatives** (*C*).

improved wood [tim.] A *high-density plywood*.

improvement line A *building line*, the line of an improved road.

inband (Scotland) A *header* stone visible in a *reveal*.

inbark [tim.] Bark embedded in wood, ingrown bark.

incandescent lamp (1) [elec.] An ordinary filament lamp.

(2) A gas or oil lamp with a mantle impregnated by thorium and cerium oxides and made incandescent by the heat of the flame.

incentive system of wages A wage system in which men receive more pay if they do more work, for example, *bonus* or piece rates. However if, as may occur in Britain, the extra pay increases disproportionately the amount of tax paid, the incentive may not be an incentive to work steadily. The effect may be to encourage a man to work hard for a short time and then to stay away from work so that some tax is paid back to him.

incise To cut or carve stone, glass, wood etc.

inclinator [joi.] An armchair which carries the occupant upstairs, installed in his home by the celebrated comedian Groucho Marx, who also invented the word.

inclined shore A *raking shore*.

incombustible building material In USA any material which does not burn in a 2½-hour standard test in a furnace is considered incombustible. In Britain the term had a similar meaning but fire authorities now prefer *non-combustible*. Compare **combustible, non-inflammable**.

incompatibility [pai.] The opposite of *compatibility*.

increaser [plu.] A taper pipe increasing in diameter in the direction of flow.

incrustation (1) or **fur** [plu.] Hard lime or other materials deposited in water pipes or conduits.

(2) A layer of corrosive material which collects on a wall surface in an industrial district and should be removed periodically to prevent stone wearing away.

indent A gap left in a course of brickwork or stone between *toothers* to *bond* with future work.

indented joint [carp.] A joint in which the wooden fishplates and the main timbers are cut with mating notches (which may be wedged). The *fishplates* (C) are bolted to the main timbers.

indenting *Toothing*.

indenture A legal agreement used, for example, between an *apprentice* and his master.

indicating bolt [joi.] A door bolt often installed on a WC door to show whether it is vacant or engaged.

indirect cylinder [plu.] *Direct cylinders*, in the past, *furred* up quickly in districts where the water was hard. To prevent furring, the domestic water supply now often has two circuits for hot water. The *primary flow-and-return* pipes are at least 19 mm ($\frac{3}{4}$ in.) diameter, and are connected to the boiler and the cylinder only. The water in the primary circuit is never changed, and the pipes therefore do not fur up. The water which is drawn from the hot taps passes from the cylinder out of pipes known as the draw-off pipes. Long draw-off pipes cool quickly and hot taps connected to them must be run for a long time before the water comes out hot. But the water in these *dead legs* may be kept hot by connecting them to the cylinder through a secondary circulation. This is a pipe connected to the bottom of the cylinder that allows the water in the draw-offs to pass back into the cylinder and heat up again.

indirect heating Heating of rooms by a distant source of heat, the heat being brought to the room by steam, water, or hot air. Generally means *central heating*. Compare **direct heating**.

indirect lighting Lighting a room by any means which hides the lamp. It generally involves hiding the lamps behind a *cornice* so that the light is thrown down into the room by the white ceiling. It is generally more restful to the eyes than direct lighting. *See* **cove lighting**.

industrialized building methods Not industrial building, but a high degree of prefabrication applied to domestic or other construction so as to reduce site work to the minimum; this involves careful planning, and the maximum of standardization. The quantity of factory work on the *building components* is deliberately increased so as to reduce the cost, and improve the quality and speed of construction. *See* large-panel.

inert pigment [pai.] (1) A *pigment* which, unlike a *drier*, does not react chemically in a paint.

(2) A paint *extender*.

infilling Rigid material placed within a *frame* (particularly a building frame or *partition* frame) for fire resistance, insulation, weather protection, or stiffness. Brickwork is the usual infilling in Britain, but *insulating materials* are also used. *See* coverings.

inflammable *See* flammable.

inflated structure *See* air-house.

infra-red drying [pai.] *See* stoving.

ingle-nook A corner by a fireside, usually with a seat built in.

ingo or ingoing (Scotland) A reveal to a window, fireplace, etc.

ingo plate [joi.] (Scotland) A *reveal lining*.

ingrown bark [tim.] *Inbark*.

inhibiting pigment [pai.] Zinc or other chromates, red lead, zinc, aluminium, or graphite powders. They are added to paints, particularly *priming coats* to prevent corrosion of a metal surface.

inhibitor [pai.] A material such as arsenic or antimony compounds which delays a chemical action in *pickling* (*C*) acids, or small proportions of anti-oxidants which reduce *skinning* or stabilize the paint in a tank in which objects are painted by dipping. *Compare* corrosion inhibitor, inhibiting pigment.

inlaid parquet [joi.] *Parquet* flooring glued in blocks about 60 cm (2 ft) square on to a wood backing and then fixed to floor boards.

inlay A decorative design which has been cut into a surface of *linoleum*, wood, or metal and filled with material of a different colour, often by gluing.

inner bead [joi.] A *guard bead*.

inodorous felt A brown, bitumen-impregnated flax *roofing felt* used as an *underlay* to roofs covered with *flexible metal sheet*.

input system A method of ventilating by sucking air into an electric fan from the roof to the rooms through *ducts* and louvres. A simple air cleaner is usually included, together with an automatically controlled air heater. *See* combined extract and input system.

insert or patch or shim [tim.] An inlay of *veneer* which fills a knot hole or other hole in *plywood*. *See* joint tape.

inside-angle tool [pla.] (USA) A float for shaping internal angles.

inside facing [joi.] (Scotland) *Inside lining*.

inside glazing *External glazing* placed from within the frame. *See* internal glazing, outside glazing.

inside lining [joi.] In a *cased frame*, a jamb or head member that faces into the building. In USA it may be called the inside casing or box casing. *See* window bead.

inside stop [joi.] (USA) A *guard bead*.

inside trim or **casing trim** [joi.] (USA) The *architrave* within a door or window.

inspection certificate A note from a Local Authority certifying that the drains of a building are satisfactory, and that it is ready for occupation.

inspection fitting or **inspection eye** An *access eye*.

inspection junction A special length of drain pipe with a branch, consisting of a short pipe leading up to ground level, through which the flow can be inspected.

installation or **internal installation** The gas pipes and appliances on the consumer's side of the control cock at the board's gas meter. *See* BSCP 331.

instantaneous (sink) water heater A small, non-storage gas water heater usually with its own taps and swivel spout for fitting over a sink. *See also* **multi-point water heater.**

insulate To protect a room or building from sound or heat (or heat loss) usually by the use of *discontinuous construction* to break up the sound paths, combined with *insulating materials*. *See* **conductivity, insulator.**

insulated metal roofing Insulated panels up to 0·6 by 3·6 m (2 × 12 ft), faced with *flexible metal* about 0·6 mm thick. Various grades of insulation exist, but the weight is about 10 kg/m² (2 lb/ft²); a *U-value* of 2·0 W/m² deg. C. (0·35 Btu/ft²h deg. F.) can be further improved to 0·97 W/m² deg. C. by adding 13 mm of *insulating fibre board* ceiling. It is a type of *roof decking*.

insulating board Any board used for insulating, of which there are many different types. They may be used partly as a wall or ceiling facing or purely for insulating, placed next to roof boarding, or built into *insulated metal roofing*.

insulating fibre board Lightweight *fibre board* about 11 mm or more thick, with a *K-value* between 0·05 and 0·065 W/m deg. C. (0·34 to 0·35 Btu in./ft²h deg. F.) and weighing up to 400 kg/m³ (25 lb/ft³). Sizes are up to 1·2 m by 4 m (4 × 13 ft).

insulating fibre board lath [pla.] *Insulating fibre board* prepared with special edges as a base for plaster.

insulating materials Materials for insulating rooms or buildings from heat or cold or sound. They are numerous but can be divided into several classes, that is (a) *insulating fibre boards*, *plasterboard*, asbestos or *asbestos-diatomite sheet*, *corkboards*, (b) *loose fills*, (c) *blankets*, (d) *woodwool* or *compressed straw slab*, (e) partition blocks of asbestos diatomite, diatomite brick, foamed slag, etc., (f) *double glazing*, not properly a material but a special use of a material. *See* **insulators, expanded polystyrene,** BSCP 3, chapter 2.

insulating plasterboard *Plasterboard* backed with brightly polished aluminium *foil* (total weight 7 kg/m² (1·5 lb/ft²)) having a *U-value* through a one-inch air space of 2·3 W/m² deg. C. (0·4 Btu/ft²h deg. F.).

(When the foil is applied to both faces the U-value is correspondingly reduced.) It is made as wallboard in sizes up to 1·2 by 3·6 m (4 by 12 ft), and as a base for plaster in smaller sizes.

insulating strip An *expansion strip*.

insulators (1) *Insulating materials*, that is materials having a low heat *conductivity*. (For sound insulation, different types of material are needed, since those that insulate from heat do not always insulate so well from sound.) The effectiveness of a sound insulation increases with its weight, but heat insulators generally improve with a reduction in weight and a corresponding increase in the number of air cells enclosed. *See* **discontinuous construction, loose fill.**

(2) [elec.] Materials which have low electrical conductivity, such as dry paper, timber, or cotton, certain resins or varnishes, rubber, many plastics, mica, porcelain, glass, and so on.

intagliated Engraved or stamped in, for example, *intaglio tiles*.

intaglio tiles Tiles decorated with patterns pressed into them.

intake belt course A projecting *string course* at a place where the wall thickness is reduced.

intarsia or **tarsia** [joi.] An *inlay* like *marquetry*.

integral waterproofing The *waterproofing* of concrete by including an *admixture* with the mixing water or cement. It is not certain that any admixture can make a bad concrete waterproof. Concrete can, however, be made waterproof without admixtures, merely by accurate proportioning of aggregates, cement and water, and care in placing and curing. British civil service departments have used no tanking whatever round all concrete basements, however wet, built since the early 1950s, and many of these basements are completely watertight.

intercepting trap or **interceptor** or **disconnecting trap** A water-sealed *trap*, fitted between a house drain and a sewer, to disconnect the air in the two. In many towns, these traps are no longer being installed, as it is believed that they are unnecessary for well-laid modern house drains and that they hinder the ventilation of the *sewer* (*C*). Where a fresh air inlet is used, it is connected to the upstream side. *Compare* **petrol intercepting chamber.**

intergrown knot [tim.] A *live knot*.

interim certificate A payment for building work, but not the final *certificate*.

interior-type plywood [tim.] *Plywood* made with *glue* which does not resist moisture well. Many of the *synthetic resins* and some *casein glues* are moisture-proof and can thus be used without difficulty out of doors (unlike *animal glue*). This moisture-resistance of course does not prevent the wood from rotting.

interlaced fencing or **woven board** or **interwoven fencing** Fencing made by weaving together straight, very thin boards so that no space remains to be seen through.

interlocked grain or **interlocking grain** or **twisted fibres** [tim.] A *grain* in which the fibres slope one way in one series of rings, then slowly reverse their slope in the next growth rings, and so on. *Spiral grain* is one example. *Ribbon grain* is shown when the wood is quarter sawn. Wood with this grain is difficult to work or to split.

interlocking joint A joint in *ashlar* in which a projection on one stone beds in a groove on the next one. *See* **joggle**.

interlocking tile *Single-lap tiles* developed from the *pantile*, measuring about 38×20 cm (15×8 in.), nailed and often also held by *nibs*. The side overlaps of the tiles are specially grooved, the left-hand edge of each tile (viewed from the ground) fitting under a groove in the right-hand edge of the next tile in the same course. *See* **concrete interlocking tile, flat interlocking tile, Poole's tiles**.

intermediate rafter A *common rafter*.

internal dormer A vertical door or window in a sloping roof, within the general line of the roof so that the flat surface in front of the door or window must be waterproofed (in the past often with lead). *See* **dormer**.

internal glazing Glazing on internal walls. *See* **external glazing**.

internal hazard *See* **fire hazard**.

International SfB (Samarbetskommittén för Byggnadsfrågor) An international classification of building information subjects. The system originated in Sweden but is now sponsored and administered by CIB (International Council for Building Research Studies and Documentation). The CI/SfB English language version is the responsibility of the *RIBA*.

intertie [carp.] An intermediate horizontal member in a *framed partition*, to stiffen it at a door head or elsewhere between floor levels.

intrados or **soffit** The visible under-surface of an arch or *vault*.

intumescent paint A coating that swells to foam when heated and can thus be used to seal and so increase the fire resistance of building elements such as doors.

iron [joi.] A *cutting iron. See also C*.

iron cement [mech.] A mixture of iron turnings (80 g), sal ammoniac (1 g), and flowers of sulphur (1 g) used for joining cast-iron pipes and for mending cracked cast-iron parts. It withstands the action of water and is strong. *See* **rust joint**.

iron core [joi.] A steel bar covered by a wooden handrail, connecting the tops of *balusters*.

ironmongery Cast or wrought iron, or *hardware*.

iron oxide [pai.] Iron oxides are used in *pigments* both manufactured and natural, for instance Venetian red (haematite). Magnetite (Fe_3O_4) or a mixture of oxides make black or purple pigments. The *umbers, siennas*, and *ochres* are oxides or hydrated oxides of a yellow to chestnut or dark brown colour and were among the earliest pigments used by man.

iron sand Fine chilled shot, fed with water into a cut during the sawing of hard stone.

ironwork Wrought or cast iron, usually decorative.

irregular coursed rubble *Rubble walling* built to *courses* of various depths.

isolating membrane An *underlay*.

isolation strip An *expansion strip*.

Italianized roofing Roofing of Italianized zinc sheet, that is zinc sheet with three or more equally spaced, parallel, half-round corrugations running down the slope.

Italian tiling or (USA) **pan and roll tiles** *Single-lap tiles* which form a roof covering with two different sorts of tiles, the curved over-tile or imbrex and the flat, tray shaped under-tile or tegula. *Compare* **Spanish tile.**

item [q.s.] A description of a volume of material and work supplied, or of one *labour* in a *bill of quantities* followed by its unit (metres, ft, m², ft², kg, lb). When *tendering*, the *contractor* states his price opposite each item.

J

jack arch (USA) A *Welsh arch* or a *flat arch*. *See also* C.

jack plane [carp.] A *bench plane* used for cleaning wood from the saw, or from any preliminary rough work. A wooden jack plane is about 43 cm (17 in.) long; steel jack planes are shorter.

jack rafter A short *rafter* between *hip rafter* and *eave* or between *valley* and *ridge*.

jack rib or **cripple** or **jack timber** [carp.] A curved *jack rafter* used in a small dome roof.

jamb The vertical flank of an opening, to the full thickness of the wall, often also the joinery covering the flank. *See* **reveal**.

jamb lining [joi.] A timber facing covering a jamb.

jamb post or **jamb stone** A post or stone, forming a door jamb.

Japan [pai.] *Black Japan*.

Japanese lacquer [pai.] A glossy coating obtained by tapping the sap from the Japanese varnish tree (Rhus vernicifera) or sumach. So durable was the old Japanese *lacquer* of 200 years ago that a cargo of it wrecked in the sea was recovered undamaged after lying submerged for two hundred years.

jedding axe A *cavil*.

jemmy A *pinch bar* about 38 cm (15 in.) long.

jenny A *gin block*.

jerkin-head roof or **hipped gable** or **shread head** or (USA) **clipped gable** A roof which is *hipped* from the *ridge* halfway to the *eaves* and gabled from there down, the contrary of a *gambrel roof*. *See* **roof** (illus.).

jerry builder A builder who builds with poor materials and workmanship.

Jetfreezing There are several methods of temporarily freezing a central heating or water supply pipe, that enable the pipe to be opened without draining it. One of these functions by the use of liquid nitrogen at $-196°$ C., another, Jetfreezing, uses liquid carbon dioxide that changes into the solid 'dry ice' as soon as it leaves the cylinder. A closely fitting jacket around the pipe, injected with the cold gas, enables pipes up to 10 cm (4 in.) diameter to be frozen, the smaller diameters usually within 15 minutes. The method has also long been used for maintenance work on oil-filled electrical power cables.

jib door or **gib door** A door whose face is flush with the wall and decorated so as to be as little seen as possible.

jig A clamp or other device for holding work or guiding a tool so that repetitive jobs can be accurately worked without repeating the marking out.

jig saw or **scroll saw** [tim.] A reciprocating, power-operated saw like the *fret saw*, used for cutting sharper curves than those which can be cut by the *band saw*.

jinnie wheel A *gin block*.

jobber or **builder's handyman** A semi-skilled man who can do any sort of repair to small houses such as bricklaying, plastering, painting, plumbing, *joinery*, roofing. *Compare* **builder's labourer.**

job made (USA) Made on a building site.

jog (USA) An offset or change in the direction of a line or a surface. *See* **joggle.**

joggle [carp.] (1) A small projection at the end of a mortised piece to strengthen it. *See* **horn.**

(2) A *stub tenon*.

(3) or **groutnick** In *blockwork* or *masonry* walls, a recess on one block which fits a projection on another block, or which forms, with a similar recess on the other block, a cavity into which mortar is poured, forming a cement *joggle* (*C*). A secret joggle in an *ashlar* arch stone is an *interlocking joint* which cannot be seen on the face.

(4) A metal cramp for locking together stones in the same course. *See below.*

joggled Generally this means: shaped with an indentation or projection, but *see* **joggle.**

joggle piece [carp.] A post shouldered like the foot of a *king post* to form an abutment for a strut.

joggle post [carp.] American term for a *king post*.

joiner (1) A man who makes *joinery* and works mainly at the bench on wood which has been cut and shaped by the machinists. His work is finer than the *carpenter*'s, much of it being highly finished and done in the good conditions of the joinery shop where it is not exposed to the weather.

(2) (Scotland) A *carpenter-and-joiner*.

joiner's gauge A *marking gauge*.

joiner's hammer or **Warrington hammer** A *hammer* with a *cross peen* head weighing from 110 to 570 g (4 to 20 oz.).

joiner's labourer or **carpenter's mate** A helper to a *carpenter* or *joiner*, who may be known as a gluer-up if he prepares *glue* and joints. He can generally stack timber.

joinery (1) The making and fixing of wood finishes to a building such as doors, *skirting boards*, *architraves*, linings, *windows*, picture rails.

(2) (USA) The joints in woodwork whether for heavy construction (English *carpentry*) or for *cabinet maker*'s or other fine work. *See* **finishing carpentry.**

joint (1) The *mortar* between adjacent bricks or stones, *bed joints*, *cross joints*, and *wall joints*. *See* **flat, concave, keyed, weather-struck.**

(2) [carp.] A connection between two members to form a corner (*angle joint*) or for lengthening or widening a timber surface.

(3) Expansion joint *see* (*C*).

joint bolt [joi.] A *handrail bolt*.

jointer (1) Bricklayer's tools, used after bricklaying, for putting various surface finishes on to mortar *joints* in *pointing* or *jointing*. *See* **mason's tools** (illus.).

(2) [carp.] or **jointing plane** A *plane* longer than the *try plane*, and used for smoothing long edges to be joined. Steel jointers are about 60 cm (2 ft) long, wooden ones may be 76 cm (2·5 ft) long. *See* **rubbed joint.**

jointer saw A machine for sawing stone.

joint fastener [carp.] A *corrugated fastener*.

joint filler A sort of putty inserted between abutting ends of *gypsum plasterboard*.

jointing Working the surface of mortar *joints* to give a finished face while green (rather than raking them out and refilling them as in pointing). *See* **tooling, weather-struck joint.**

jointing compound [plu.] Paste, old paint, *iron cement*, *ptfe*, etc., put into a joint between steel or iron pipes to prevent leakage.

jointing material [mech.] A material obtainable in sheets from which *gaskets* or washers can be cut to shape for insertion in the joints of flanged pipes, pumps, etc., to make them watertight. It is often of rubber for cold water or asbestos for steam.

jointing plane [carp.] A *jointer*.

jointing rule A long *straight edge* used by bricklayers with the *jointer* in pointing.

jointless flooring or **composition flooring** Many types of flooring exist which can be laid without close joints, e.g. asphalt, anhydrite, cement-bitumen, *cement-rubber latex*, *cement-wood*, *granolithic screed* (*C*), *magnesite* flooring, *pitch mastic*, *terrazzo*. In spite of the name, many jointless floors are best laid with joints at about 2·5 m apart, particularly those containing cement, to allow for *shrinkage* (*C*). Jointless floors are laid by *plasterers* or by composition floor layers. Those mentioned so far are not expensive in materials but they are all of mortar screed thickness, 2·5 to 4 cm (1 to $1\frac{5}{8}$ in.) minimum. Many new types made of relatively costly plastics, including fleximers – rubbery materials sometimes mixed with Portland cement – can be applied thinly to new or old concrete sub-floors or screeds by brush or spray or trowel in thicknesses down to that of a paint film. Most of them are as wear- and corrosion-resistant and rapid-hardening as good concrete (some of them much better) but pva (polyvinyl acetate) will not resist permanent moisture. Some of the less expensive materials are acrylic emulsions, used over industrial floors of concrete, steel plate or wood. They often fill in gaps before the final, more expensive polyester or epoxide resin floor is laid. Another fine, glossy, flexible surface, very effective on smooth sanded wood is made from polyurethane (polyol/isocyanate) resins. *See* BSCP 204, In-situ floor finishes.

joint mould or **section mould** [pla.] A zinc, plywood, or cardboard *template*, shaped for a plaster member.

joint rule [pla.] A steel *rule* from 5 to 46 cm (2 to 18 in.) long, used by *plasterers* in forming the *mitres* at the junctions of cornice *mouldings*. One end is cut at an angle of 45°. *See* **plasterer** (illus.).

joint runner or **pouring rope** [plu.] Asbestos rope or similar material, used for packing the outside of a pipe joint which is to be filled with molten lead. *See* **pipe-jointing clip**.

joint tape (1) Paper or paper-faced cotton tape (sometimes embossed) which is fixed over the joints between wallboards. *Compare* **scrim**.

(2) [tim.] Gummed brown paper with which the *face* of *veneer* is reinforced before it is glued to its backing, mainly to hold *inserts* in place. Fusible tape is converted to glue in *hot pressing*. *See* **film glue, tapeless splicer**.

joist (1) In Britain and USA a wood or steel beam directly supporting a floor. It is usually a *common joist*. Steel joists are often distinguished by calling them RSJs or *rolled-steel joists* (*C*).

(2) In USA, rectangular *lumber* from 5 cm (2 in.) up to (but not including) 12·7 cm (5 in.) thick and 10 cm (4 in.) or more wide, graded for its bending strength loaded on edge. When graded for its bending strength loaded on face it is a *plank*.

joist anchor A *wall anchor*.

joist hanger A steel plate or *strap* (*C*) or cast-iron shoe which carries the end of a wooden *joist*.

journeyman or **craftsman** A skilled worker, competent in his trade, who has served an *apprenticeship* of the recognized number of years. Usually called a *tradesman*.

jumbo brick (USA) A *brick* of larger size than usual, whether intentionally or by mistake.

jumper (1) In *snecked* or *squared rubble* a *stretcher* which covers more than one *cross joint*.

(2) [elec.] (USA) A temporary electrical connection made round part of a circuit.

(3) [plu.] A brass, mushroom-shaped part in the domestic water tap, whether for sink or basin or bath, which carries the washer on the lower face of the mushroom disc. The stalk of the mushroom points upwards into a hollow guide, the screw-down part of the tap. When the tap opens, the force of the water lifts the jumper into this guide, raising the washer with it. Washers need replacing every year or two when the tap starts to drip.

junction [plu.] A *special* drain pipe made with a *socket* to take a *branch*.

junction box [elec.] A box which covers the joints between the ends of conductors in house wiring. It also joins up the ends of their metal sheaths.

K

kauri [pai.] A fossil *copal resin* from the largest conifer of the North Island of New Zealand which is believed to live 3,600 years. It was at one time much used in hard-drying *varnishes*.

Keene's cement or **Parian plaster** or **hard-burnt plaster** [pla.] An anhydrous gypsum ($CaSO_4$) plaster with an *accelerator* of set, a *hard plaster* with a smooth, vermin-proof finish. It is used for *finishing coats*, often over a Portland cement rendering and a coarse variety of Keene's in the *floating coat*. It should not be mixed with lime. It has a gradual set which enables it to be trowelled smooth without danger of spoiling it.

keeper [joi.] (1) A metal loop over the fall bar of a *thumb latch* to limit its travel.

(2) A *striking plate* or other guide for a bolt in a door.

keeping the gauge Maintaining the spacing of the brick *courses*, usually four per 30 cm (12 in.) height for a metric English *brick* of 6·5 cm (2⅝ in.) depth.

keeping the perpends Laying *bricks* or stones or *slates* or *tiles* accurately so that the *cross joints* (perpends) or the visible edges of slates and tiles in alternate *courses* shall be in the same vertical line.

keratin A *retarder* for *plaster of Paris*, obtained from the horns, hoofs, nails, or scales of animals.

kerf A saw-cut in wood or stone etc. *See* **setting** and illustration.

kerfed beam [carp.] A beam cut with several saw-cuts to allow it to be bent.

kerfing [carp.] Making saw-cuts on one side of a piece of wood so as to bend it towards that side, a convenient way of making the risers of curtail or bullnose steps.

kettle Various sorts of open vessel for melting glue, containing paint, etc.

kevel (1) *See* **cavil**.

(2) A strong piece of wood used for securing a rope's end.

key (1) The roughness of a surface, the texture which enables *plaster*, *mortar*, *glue*, etc., to grip it with a *mechanical bond* (*C*).

(2) [carp.] A special *hardwood* piece let into a joint to strengthen it, such as a *double-dovetail key* or a *feather*, or a small hardwood slip let into a *mitre* joint.

(3) [carp.] A *counter batten* dovetailed across boards to prevent them warping.

(4) [mech.] A bar driven between a drive shaft and the hub of a wheel to ensure that the shaft drives the wheel.

(5) A *cotter*.

(6) A bricklayer's pointing tool for making a *keyed joint*.

key block A *keystone*.

key course A course of stones instead of a keystone at the crown of a *vault* or wide arch.

key drop [joi.] An *escutcheon* cover.

keyed [joi.] (1) Held or locked by a *key*.

(2) Said of a dowel which is grooved, to allow air and excess *glue* to be driven out.

keyed beam [carp.] A beam with a *lap joint* into which *joggles* have been cut in each member. *Hardwood* or metal rectangular wedges (keys) are driven into these holes, and increase the bending strength of the joint. *See also* **compound beam.**

keyed joint Concave *pointing* of a mortar *joint*.

keyed mortise and tenon [carp.] A *tusk tenon*.

keyhole saw [joi.] A *compass saw* which tapers to a point.

keying-in *Bonding* a *brick* wall to another already built.

key plan A small plan to show the position of the units in a scheme.

key plate [joi.] An *escutcheon*.

keystone The central wedge-shaped *arch-stone* at the crown of an arch, put in last. It is no more important than the other arch-stones, but its insertion means the completion of the arch.

keyway [mech.] A slot cut along a shaft, and another along its matching hub, to receive a *key*.

khaya [tim.] *See* **African mahogany.**

kick (1) The difference in slope between *patent glazing* and the surrounding roof.

(2) A *frog* in a *brick*.

kicking plate or **kickplate** [joi.] A metal plate fixed to the bottom *rail* of a door to protect it.

kieselguhr *Diatomite*.

kiln (1) A furnace in which cement, brick, or lime is burnt.

(2) A chamber over a furnace, used for drying timber.

kiln dried or **kiln seasoned** [tim.] A description of timber which has been dried in a kiln. *Compare* **natural seasoning.**

king bolt or **king rod** A vertical steel rod hanging from the ridge to the *tie-beam* of a wooden roof truss, taking the place of a *king post*.

king closer A three-quarter brick used as a *closer*. A diagonal piece is cut off one corner by a vertical plane passing from the centre of one end to the centre of one side. (It is actually $\frac{7}{8}$ of a brick but is usually called a $\frac{3}{4}$ brick.)

king post or (USA) **joggle post** [carp.] A vertical timber which hangs from the ridge of a king-post roof truss and carries the tie-beam of the *truss* at its foot. Its foot is joggled to carry two struts (*see* **joggle piece**) and the tie-beam is often held to the king post by a *cottered joint*.

king-post truss [carp.] A wooden roof *truss* consisting of a pair of principal rafters held by a horizontal *tie-beam*, a vertical *king post* between tie-beam and *ridge* and usually also two struts to the rafters from a thickening (joggle) at the foot of the king post. It is suitable for spans up to about 11 m (36 ft) but like the *queen-post truss* is now not used in Britain.

kiss marks Marks where *bricks* have been touching in the kiln.

kite winder Of three *winders* turning a right angle, this is the central one, so called because in plan it looks like a kite.

Knapen system Patented method of drying out damp walls by drilling holes into them from outside at about the level of the internal floor. The holes do not pass right through the wall and are left open so as to drain outwards and to be kept ventilated.

knapped flints Flints broken across the middle, therefore dark coloured, often grouped to form a square on the wall face.

knapping hammer A *hammer* for shaping stones.

knee (1) or **elbow** [plu.] A sharp right-angle bend in a pipe.

    (2) [joi.] A curve in a handrail which is convex on the upper side like a human knee. *Compare* **ramp.**

    (3) or **crook** [tim.] A naturally-curved short timber.

kneeler or **skew table** or **skew** The sloping-topped, level-bedded stones in a *gable coping*, above the *gable springer*.

knob-and-tube wiring [elec.] An early American method of concealed wiring of buildings, still used in districts liable to be flooded. The conductors are carried in rubber-insulated cable on porcelain knobs fixed to floor *joists*. The cable passes, where necessary, through timber in a porcelain tube.

knobbing or **knobbling** or **skiffling** Dressing stones roughly in the quarry by removing protruding humps.

knocked down (mainly USA) Description of building components, delivered to a job completely cut and shaped, but not assembled.

knocker [joi.] A hinged metal bar, striking on a metal plate on an entrance door.

knocking up Mixing and making workable a batch of *plaster*, *concrete*, *mortar*, or *paint*. (The adding of fresh water to stiffened mortar or concrete is usually forbidden, since such material should be thrown away.)

knockings Stone chips smaller than *spalls*.

knot [tim.] A place in a tree trunk from which a branch has grown out. Knots which reduce the strength of wood are *dead*, *loose*, or *unsound* knots. Harmless knots are *tight*, *sound*, *live*, and *pin* knots. Some of these terms are synonymous.

knot brush [pai.] A thick *brush* with its bristles or fibres bunched in round or oval shapes and often used for distempering (one-, two-, or three-knot brushes).

knot cluster [tim.] A group of two or more knots such that the wood fibres are deflected round the entire group (BS 565).

knotting [pai.] *Shellac* dissolved in *methylated spirit*, used as a local *sealer* over *knots* in new wood so that it can be painted without danger of the knots exuding sap through the paint. Other quick-drying compositions are also used.

knuckle [joi.] That part of a *hinge* consisting of holes through which the pin passes.

knuckle joint [carp.] A *curb joint* in a *mansard roof*.

knuckle soldered joint [plu.] A right-angled joint made between two lead pipes.

kraft paper Strong brown paper used in building as the containing sheet of insulation *blankets*, in *building paper*, and in other ways.

K-value The thermal *conductivity* of a material. Compare **U-value.**

L

labour [q.s.] Work which is done with material already itemized and paid for elsewhere. A labour is often a separate *item* in a bill, although one item may contain several labours.

labour constant [q.s.] The amount of labour required to do a unit of work, e.g. 1 m² (ft²) of brickwork 22·5 cm (9 in.) thick.

labourer A man who has not been apprenticed to any trade and is therefore paid less than a *tradesman*. In practice he may have some skill in many trades, like a *jobber* or a *builder's labourer*, and thus be very useful.

laced valley A *valley* formed by *tiles* or *slates* without a *valley gutter*. A wide board is fixed in the *valley*, on which slates or tiles of 1½ times the normal width are laid. The two slopes intersect sharply and do not blend into each other as in a *swept valley*.

lacing course (1) A *course* of bricks, or stones, or several courses of *tiles* laid together, to strengthen a *rubble* or *flint wall*.

(2) A course of upright bricks or deep stones which cuts through two or more arch rings and bonds them together.

lacquer [pai.] A glossy finish which dries rapidly by *evaporation* of the *vehicle*. Lacquers are used for decoration (*Japanese lacquer*) and for coating tin cans and bright metal surfaces such as brass (transparent lacquer). These modern lacquers are distinguished from *varnish* and *enamel* by being based on cellulose compounds, including *nitrocellulose*. *See* **acetone, resins**.

ladder Ladders consist of two long wooden *stiles* fixed at a certain distance apart by *rungs* wedged tightly into holes mortised in the stiles at 20 to 30 cm (8 to 12 in.) centres. *Light-alloy* ladders are much lighter than wood ladders. Different types are the *builder's ladder*, *standing ladder*, *extending ladder*, *step ladder*, and *steps*.

ladder scaffold A *scaffold* quickly erected on ladders braced together, used for painting or other light work.

ladkin or **latterkin** A *hardwood* tool for opening the *cames* of *leaded lights*.

lag (1) To wrap hot pipes or tanks and thus reduce heat losses.

(2) *See* **lagging**.

lag bolt (USA) A *coach screw*.

lagging (1) Horizontal boards nailed across the *centers* and supporting an arch during construction.

(2) *See* **lag, moulded insulation**.

lag screw or **lag bolt** (USA) A *coach screw*.

laid on [joi.] *Planted*.

lake [pai.] A *pigment* consisting of a *dye* precipitated on to an inorganic base (or carrier or filler) such as aluminium hydroxide.

laminate (1) To impregnate many layers of paper, textile, or *veneer* with a *synthetic resin* and to compress them at high temperature to make a sheet of durable, strong material. *See* **laminated plastic.**

(2) Laminates include *laminated plastics*, glass-fibre reinforced plastics, plywood, *blockboard*, *battenboard*, *laminboard* and *glu-lam*.

laminated arch [tim.] A wood arch built of *laminated wood*.

laminated fibre wallboard *Fibre board* made in thin layers cemented together up to a thickness of 6 mm ($\frac{1}{4}$ in.). It is used for panelling walls, ceilings, etc. and is made in sizes up to $1 \cdot 2 \times 3 \cdot 6$ m (4×12 ft), with a surface which is smooth or pebbled, painted or prepared for painting. Preformed boards made to curves from 15 to 60 cm (6 to 24 in.) dia. can be used for clothing *columns* or as column *formwork* (C). These boards have been made in England since 1906.

laminated joint [joi.] A *combed joint*.

laminated lead sheet A weather-resisting covering to walls, consisting of thin layers of lead glued to other material.

laminated plastics or **synthetic-resin-bonded paper sheet** Sheets of paper or textile, soaked with a *synthetic resin*, and pressed together to make a stiff board or glossy-surfaced covering for a wall or board. (Warerite, Formica, and others.) In spite of the name they are obtainable not only as sheet but also as bars, cylinders and other sections. *See* **laminate.**

laminated wood Layers of *veneer* or wood glued or mechanically fastened together without *cross bands*. In practice many constructions intermediate between this (pure) laminated wood and *plywood* exist and it is rare to find one devoid of crossings. Laminated wood is built of *plies* which are thicker than the usual plywood veneers.

laminboard [tim.] Built-up board like *blockboard* but with core strips up to 7 mm wide and 7 mm thick.

lamp black or **vegetable black** [pai.] A *pigment* like *bone black* but made by burning coal tar products with very little air. *See* **carbon black.**

lamp cord [elec.] (USA) A *flex*.

lampholder or **lamp socket** [elec.] A wall fixture or fitting at the end of a *flex*, which carries an electric light bulb.

lampholder plug [elec.] *See adaptor*.

landing A platform between two flights or merely at an end of a flight, a *half-landing*, *half-space* or *quarter-space landing*.

Langman cylinder [plu.] An *indirect cylinder*, introduced in 1973 for use with *sealed systems* at temperatures up to 110° C., designed so that hot water above boiling point cannot reach the taps to issue dangerously as steam. The indirect heating tubes are located in the lower half of the cylinder (in cylinders of earlier type they are in the upper half) and are separated from the upper half by a baffle plate that delays the flow of water to the main draw-off pipe and so allows it to cool. The cold feed water flowing into the lower half further cools the water.

lantern or **lantern light** A frame raised above the general level of a flat roof, glazed all round to admit light and sometimes air. It occasionally has a skylight. *See* **saucer dome**.

lap (1) In *centre nailing*, the effective length of a *slate* or *tile* which is covered by two others. On plain tiles or *head-nailed* slates, the lap is measured from the centre of the nail hole, and therefore the lap is equal to the length covered by two others minus the distance from centre of nail hole to head of slate (about one inch). Lap should be greater on a flat than on a steep roof. For a steep roof of 60° slope, a lap of 6·3 cm (2½ in.) is enough, but for a 30° slope 10 cm (4 in.) lap is needed, and slates should also be at least 30 cm (12 in.) wide. *See* **weather, gauge**.

(2) [pai.] To place one coat of *paint* or *varnish* beside another and over its edge so as to make an invisible join.

(3) [pai.] The join so formed.

(4) [plu.] *See* **passings**.

lap cement *See* **sealing compound** (2).

lapis lazuli The sapphire of the Bible and the ancients, used now like marble for decorating walls, or ground to make the pigment *ultramarine*.

lap joint or **lapped joint** [carp.] A joint between timbers made by laying one over the other and bolting through them or clamping them with U-bolts. *Halving* is a development of the lap joint. *See also* **C**.

lapped tenons [carp.] Two *tenons* which enter a *mortise* from opposite ends and lap each other within it.

lap siding [tim.] *Clapboard*.

larch [tim.] (Larix) A hard, strong, resinous *softwood*, not easy to convert owing to knots and *resin* and therefore used in the round in mines, in piling, and in *carpentry*, although it is also used in *joinery* and flooring. Larch is one of the few *deciduous* conifers. The American larch is called tamarack, also a Larix.

large knot [tim.] No size is specified in BS 565, except for *pin knot* (max. 6 mm or ¼ in.) but *see* **standard knot**.

large-panel construction An *industrialized building method* which was first used in the USSR and consists of reinforced concrete panels one or two storeys high, sometimes made of lightweight insulating concrete. For large-panel timber walls which have been prefabricated for many years both in Europe and America, the term is not used.

larmier French word for an overhanging stone with a throat gouged in it to throw off water. Also **drip, weather moulding**.

larry (1) [pla.] A tool 2·1 m (7 ft) long, shaped like a hoe, used for mixing *plaster*. *See* **hair hook**.

(2) Fluid *mortar*.

larrying or **larrying up** Using fluid *mortar* in which the *bricks* are slid into position and not laid, after which mortar is poured in to fill the vertical joints.

lashing or **whip** or **bond** A short length of fibre rope or steel rope for tying *scaffold* timbers.

latch [joi.] A door fastening which may be of several types, the two most important of which are: (1) The *thumb latch*, a bar, pivoted on the door, which catches in a hook on the frame.

(2) The bevelled metal tongue operated by a door handle and controlled by a spring in the common *mortise lock*, *rim lock*, or *cylinder lock*. Unlike a lock, it engages by spring when the door is closed, without any key being turned.

latch bolt [joi.] (USA) A *latch* (2).

latchet [plu.] A *lead tack*.

latex emulsions [pai.] *Dispersions* in water, originally of rubber from a tree, now also of many *synthetic resins* or rubbers, the best known in Britain being polyvinyl acetate (PVA). *Polystyrenes* and poly-methacrylates are also made in Britain. *See* **emulsion paint**.

lath [pla.] A sawn or split strip of wood, 5 to 10 mm by 25 to 32 mm (about $\frac{5}{16}$ by $1\frac{1}{8}$ in.) and about 1 m (3 ft) long, formerly used in new work as a base for plaster. More generally it is any material such as *fibre board*, *plasterboard* or *metal lathing*, used as a base for plaster.

lathe [tim.] A machine on which wood or metal is turned to a circular shape, such as the powerful machine on which *rotary* and *half-round veneers* are cut.

lath hammer or **claw hatchet** or **shingling hatchet** [pla.] A plasterer's *hammer* for nailing laths. It has an axe edge as well as a hammer face, the axe side being nicked near the handle to form a claw for drawing nails. *See* **plasterer** (illus.).

lathing Any base for *plaster*. *See* **lath**.

lathing hammer [pla.] A *lath hammer*.

lath, plaster, and set [pla.] *Two-coat work*, a *floating coat*, and a *finishing coat* often used on modern backings such as plasterboard or *insulating board*.

lath, plaster, float, and set [pla.] *Three-coat work*.

latterkin A *ladkin*.

lattice window A *light* in which small *panes* of glass are bedded in metal *cames*. The usual type is a *leaded light*.

laundry chute or **clothes chute** A duct from a bathroom to a lower floor, into which dirty clothes are dropped.

laundry tray or **laundry tub** (USA) A deep, wide sink fixed to a wall for washing clothes.

lavatory A wash basin; the term has now come to mean the room containing the wash basin, and by further extension the room containing a WC.

lay bar A horizontal *glazing bar*.

layer board or **lear board** [carp.] A board on which the lowest sheet of a *box gutter* is laid.

laying trowel (1) A *brick trowel*.

(2) [pla.] or **laying-on trowel** A rectangular steel trowel about $10\frac{1}{2} \times 4\frac{1}{2}$ in. with which *plaster* is laid on to a surface. *See* **plasterer**.

lay light A *light* fixed horizontally in a ceiling.

lay panel [joi.] A door or other *panel* with its length horizontal.

leaching cesspool (USA) A *cesspool* which leaks. Cesspools also leak in Britain although this is usually forbidden by Local Authorities. *See* **tight cesspool**.

In all the terms below down to 'lead wool' (except where otherwise indicated) the syllable 'lead' is pronounced 'led' and refers to the heavy metal.

lead (pronounced 'led') [plu.] (1) A building material used for *plumbing* by the Romans and for roofing since early Saxon times. Lead rainwater pipes often of very beautiful design were used until the nineteenth century, when cast iron became common, and lead is now too expensive for any but the most extravagant *architect*. The sheet lead used for building is from 1 to 3 mm thick; *see* **lead flat**.

(2) A lead *came* of a *lattice window*.

lead (pronounced 'leed') (1) [elec.] A *conductor*.

(2) or **corner lead** (USA) A section of brickwork plumbed exactly and built up ahead of the remainder by steps in the courses called *racking back*.

lead burner [plu.] A specialized *plumber*.

lead burning [plu.] Welding lead without solder; it therefore requires a more directional flame than soldering.

lead-capped nail An American nail with a lead washer forming the underside of the head. When driven on to a roof sheet, it makes a watertight joint.

lead cesspool A lead *rainwater hopper* at the lower end of a parapet gutter to collect rainwater before it enters a downpipe.

lead chromes [pai.] Yellow to orange *pigments*, very opaque and of high *staining power*, consisting of basic lead chromate often with *chrome yellow*. Some primrose shades may contain a small percentage of precipitated aluminium compounds with up to 4% of Al_2O_3. *Compare* **zinc chromes**.

lead chrome green or **Brunswick green** [pai.] Composite *pigments*, prepared from *lead chrome* and *Prussian blue*. *See* **chrome green**.

lead-clothed glazing bar or **lead-covered glazing bar** *See* **lead sheath**.

lead damp course A *damp course* of sheet *lead*, an excellent material but costly and heavy.

lead dot [plu.] *See* **dots**.

lead driers [pai.] Lead compounds which quicken the hardening of *drying oils*, e.g. *litharge*, lead linoleate, lead resinate, and other organic salts of lead. *See* **drier**.

leaded light or **lead glazing** A *light* in which small diamond-shaped panes of glass are held in lead *cames*.

leaded zinc oxides [pai.] *White pigments* which are mixtures of ZnO (zinc oxide) and PbO·2PbSO₄ (*basic lead sulphate*). They are described in Britain as 15%, 25%, or 35% leaded. This means that the pigment contains 15% or 25% or 35% of PbO·2PbSO₄ ($\pm$ about 2%), the remainder being ZnO. These pigments are made from sulphide ores of lead and zinc, burnt without separating them.

leader head (pronounced 'leeder') (USA) A *rainwater head*.

lead flat A *flat roof* covered by lead sheet with *rolls* and *drips* at suitable intervals. Lead flats are best laid in areas smaller than 2·2 m² (24 ft²) and shorter than 3 m (10 ft), falling at 1·2 per cent (12 mm per m). The lead should be placed on waterproof *building paper* that does not stick to the wood or the lead, covering either planed boards or 2·5 mm plywood. Lead should never touch oak because its tannic acid attacks the metal. Lead is not now used for large areas because it is costly and heavier than other *flexible metal*. Lead is unsuitable for steep slopes because it creeps unless it is fixed to the boarding with lead *dots* 76 cm (30 in.) apart both ways. Lead was formerly described by its weight in lb/ft², the heaviest for flats being 7 lb/ft², which is 3·15 mm thick, used on the very best roofs and now known as code 7 lead. Code 3, the lightest for flat roofs, weighs 3 lb/ft² and is 1·25 mm thick. It is used on roofs with no traffic where the lead needs no *bossing*.

lead-free paint Paint without lead compounds, used in food packing, etc. *See* **lead restricted, lead paint**.

lead glazing *See* **leaded light**.

lead joint [plu.] A *spigot-and-socket joint* in large cast iron water pipes, made by pouring molten lead into the gap or caulking it with *lead wool*.

lead-light glazier or **lead-light maker** A *tradesman* who sets out and solders lead or copper *cames*, cuts glass, *glazes* the light, and fixes it. If he is specialized in stained glass he may be known as a fret glazier or decorative glass worker. If he is also *fixing*, he is called a stained-glass fitter and fixer. *See* **leaded light**.

lead nail A small copper alloy *nail* for fixing lead sheet to a roof.

lead paint Paint containing lead *pigment*, particularly *white lead*. All lead pigments are poisonous because they are soluble in the juices of the stomach and can thus be absorbed by the body. The greatest danger from white lead is the inhaling of its dust. Involuntary swallowing and absorption through the skin are less likely.

lead plug (1) A small cylinder of lead driven into a *joint* or hole in a wall. It is a tight fixing for a *screw*.

(2) A lead *cramp* cast between neighbouring stones in a course into a groove cut in each to hold them together.

lead-restricted paint Paint which contains less than 5% of PbO calculated according to the British *lead paint* regulations of 1927. Such paints are regarded as *lead-free*.

lead roof A *lead flat*.

lead safe or **drip sink** A shallow lead tray with a waste pipe, near a sink, bath, draw-off tap, or WC to catch any overflow. It is drained through a pipe trapped like a waste-water fitment. *See* **receptor**.

lead sheath An enclosure to the steel core of a *glazing bar* which embodies *lead wings* and *condensation grooves* (lead-clothed *glazing bar*). *See also* C.

lead slate or **copper slate** or **lead sleeve**, etc. A specially made lead, copper, or other *flexible-metal* flashing to fit round a pipe where it passes through a roof. It is shaped like a top hat with the top cut off. *Compare* **soaker**.

lead soaker *See* **soaker**.

lead spitter A short outlet from a lead *rainwater hopper* or gutter to a down pipe.

lead (or copper, etc.) tack or **ear** or **tingle** or **latchet** or **bale tack** A lead (or copper, etc.) strip about 6 cm (2½ in.) wide which fixes any free edge of lead *flashing*, coverings, *rolls*, *seams*, or pipes. One end is fixed to the roof, the other end to the roof covering. *See* **hollow roll**.

lead wedge or **bat** A tapered piece of lead made by beating folded scrap lead or by casting, used for fixing a *flashing* to a *raglet*. *See* **tag**.

lead wing A projecting lead fin forming part of the *lead sheath* of a lead-clothed *glazing bar* in *patent glazing*. It is dressed down on to the glass to hold it and to prevent water entering at the edge of the glass.

lead wool [plu.] Lead cut up into thin strands and used for caulking iron pipes (*spigot-and-socket joints*) when for some reason it is impossible to pour molten lead into the socket.

leaf (1) One of a pair of doors or windows or one of the slates at a slate ridge.

(2) One solid half of a *cavity wall*, in Britain generally a *half-brick wall* tied by *wall ties* to the other half. *See* **withe**.

leafing [pai.] The *floating* of metallic *paints* containing aluminium or bronze powder. The metal grains are flat (leaf shaped) and should also lie flat to give good *colour* and protect the paint film. Mica may sometimes be added to increase this property.

lean lime A lime of low workability and low *volume yield*. *Compare* **high-calcium lime**.

lean mix A *mortar* or *concrete* mix with little cement, or a plaster which is unworkable.

lean mortar [pla.] *Mortar* which is harsh, difficult to spread; the contrary of *fat*.

lean-to roof or **half-span roof** A roof whose summit is carried on a wall which is higher than the top of the roof. *Compare* **penthouse roof**.

lear board A *layer board*.

LECA Light Expanded Clay Aggregate, a *lightweight aggregate*.

ledge [joi.] One of the two or three horizontal timbers on the back of a *batten door* or on a *framed and braced door*.

ledged and braced door [joi.] A *batten door* which is diagonally braced between the *ledges*. It is an unframed door of medieval type, still used for many purposes. *See* **clenching**

ledged door [joi.] A *batten door*.

ledgement A *string course*.

ledger A horizontal pole, parallel to the wall in wooden *scaffolding*, lashed to the *standards* and carrying the *putlogs*. In *tubular scaffolding* similar terms are used, but different fixings.

ledger board [carp.] A *ribbon board*.

leggatt A *thatcher's* wooden tool for striking the butts of *reeds* to bring them into line.

Leighton Buzzard sand A clean, white closely graded sand, quarried for building purposes, and used as a standard sand for testing cements in England.

lengthening joints [carp.] Joints which increase the length of timbers, as opposed to *angle joints*. Some of these are *scarfed*, *lapped*, *halved*, fished, and *indented joints*. They are little used now compared with a century ago, when long steel or reinforced concrete members were not obtainable.

Lesbian rule [pla.] A lead strip pressed instead of a *squeeze* round a *moulding* to make a temporary record of it.

let in [joi.] *Housed*.

letter plate [joi.] A front-door fitting to receive letters, consisting of a slotted plate fixed to the outside of the door. The best types have a hinged cover plate with a spring to hold it shut, preventing draughts.

level or **mechanic's level** or **spirit level** A *level tube* (*C*) set in a *straight edge* from 23 to 90 cm (9 in. to 3 ft) long. *See* **mason's tools** (illus.).

levelling [pai.] *See* **flow, pulling over**.

levelling rule [pla.] A *straight edge* about 3 m (10 ft) long used with a spirit *level* for bringing *dots* and *screeds* to a uniform surface. *See* **rule**.

lever boards Adjustable *louvres*.

lever cap [joi.] The metal piece above the *back iron* of a metal *plane* which holds the back and *cutting irons* in place by a cam action.

lever lock [joi.] A good-quality *lock* in which the key must move several levers to shoot the bolt.

lewis or **lifting pins** An arrangement of sever wedges or of curved steel bars which grip into a dovetailed *mortise* the top of a heavy block of stone or concrete. Lewises are of severa different types. Some can be released at a distance by the banksman w.. ,... a string attached to a releasing wedge.

lewis hole A hole cut in stone or cast in concrete to enable it to be lifted with a *lewis*.

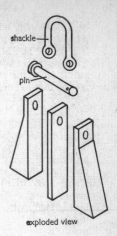

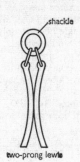

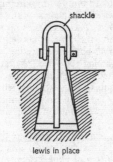

Lifting lewises.

lewising tool A *mason*'s chisel for cutting *lewis holes*.

lift or (USA) **elevator** [mech.] An enclosed platform for carrying goods or passengers from one level to another in a tall building. *See* **hydraulic lift**, *also* BSCP 407.

lifting or raising [pai.] Failure caused by the swelling of a dry film of *paint* or *varnish* when another coat is applied over it, usually manifested by wrinkling. *See also* **pulling up**.

lifting pins A *lewis*.

lifting shutters Shutters which drop behind the *window back* when not in use. They are balanced by cords and weights like a *sash*.

lift latch (USA) A *thumb latch*.

lift-off butts or **loose butt hinges** or (USA) **loose-joint butt** [joi.] *Hinges* which can be taken apart by lifting one leaf from the other. A door can thus be easily unhung without unscrewing its hinges.

lift shaft or **lift well** The vertical opening in a building, through which the lift and its counterweight travel.

ligger (1) A hazel or willow stick, 1·5 m (5 ft) long, held down by *spars* at the ridge of *thatch*.

(2) [pla.] Scots for a *spot board*.

light One glazed or unglazed window (usually of several panes), whether fixed or opening, for instance the part between two *mullions* or *transoms*. *See* **dead light**.

light alloys Alloys of aluminium (which have been used in building since the 1930s) and alloys of magnesium (which have only recently

207

been developed, but are even lighter). Their specific gravities are: Al alloys 2·7, Mg Alloys 1·8 (compared to steel 7·8). *See* **duralumin** (*C*), BSCP 118, 143.

light-gauge copper tube [plu.] Domestic *copper pipes* with a wall thickness down to 0·7 mm for 13 mm (½ in.) bore pipe. Since these pipes are too thin to have a screw thread cut on them, they are joined by *capillary* or *copper fittings*. The tolerances on outside dia. are only ±0·04 mm, which ensures the close fit needed for capillary joints.

lighting panel [elec.] A *cutout* box for protecting lighting circuits.

lightning conductor or **lightning rod** [elec.] A thick copper *lead* (pronounced 'leed') connected to *earth*, projecting above a building, and provided with sharp points which very much reduce the chance of the building being struck by lightning. *See* **air termination**, BSCP 326.

lightning shake [tim.] Compression failure of wood, seen as a cross break.

lightweight aggregate *Vermiculite* and *perlite* are lightweight aggregates used instead of sand in plaster, but also in concrete for roofs of large span. *Pumice* or *foamed slag* or *clinker* aggregates can also be used to make *lightweight concrete*. As early as 1919 *expanded clay* aggregates were in use in USA for shipbuilding, and they are now used very widely for the structures of multi-storey buildings.

lightweight concretes are of two main types: (a) *aerated concretes*, weighing around 960 kg/m³ (60 lb/ft³) or less, which are highly insulating and not very strong; (b) concretes made from *lightweight aggregate*. These can be used for structural purposes (columns, beams, and slabs), are less good insulators though better than dense *concrete* (*C*), and usually weigh less than 1760 kg/m³ (110 lb/ft³). (BS 2787:1956 says 1440 kg/m³ (90 lb/ft³).) *See also C*.

light well An *air shaft*.

lignified wood [tim.] *High-density plywood* impregnated with lignin at high pressure.

lignin [tim.] The most important constituent of wood after cellulose, consisting of resins which cement the wood fibres together.

lime Chalk and other forms of calcium carbonate burnt in a kiln are called quicklime (CaO). When soaked in water this becomes *hydrated lime*. Both forms are lime. *See* **high-calcium lime, hydraulic.**

lime ashes *Small lime*.

lime concrete A mixture of gravel, sand, and *lime* which sets hard and was used in Roman times and later before *Portland cement* was made.

lime mortar *Lime* and sand mixed with water as a bricklaying *mortar*, sometimes with *cement*. If the quantity of cement is equal to the quantity of lime (or nearly so) the mortar can be considered to be a *cement mortar*, because cement is more powerful and more quick acting than lime. But the terms are not precise.

lime paste *Slaked lime* with water (which may be *lime putty*).

lime plaster [pla.] A mixture of *lime*, sand, and *hair*. *See* **coarse stuff.**

lime powder The powder obtained when lime is *air slaked*. It cannot be used in building.

lime putty [pla.] When *quicklime* is soaked in water, passed through a 3 mm ($\frac{1}{8}$ in.) mesh sieve, and allowed to settle for a month in a *maturing bin*, the material dug out is lime putty. *Compare* **glazier's putty**; *see* **coarse stuff, fattening up.**

lime tallow wash [pai.] *Limewash* mixed with tallow, used for roof and wall surfaces; one of the few coatings which do not damage bituminous roofing.

limewash or **whitewash** or **whitening** [pai.] A milk formed with *quicklime* by soaking in excess water. It is made more durable by adding tallow, *size*, *casein*, alum, or other binder.

limpet asbestos *See* **sprayed asbestos.**

limpet washer A conical *washer* (*C*) fixed under the nut of a *hook bolt* to hold down a corrugated sheet. It is shaped to fit the top of the corrugation. A *diamond washer* has the same function.

line A cord used for setting out building work, particularly by *bricklayers*. It is also used in tunnelling by mine surveyors, who chalk the line, then, when it is properly set, flick it against the roof. The resulting line marked on the rock shows the direction of the tunnel. *See* **fibre rope** (*C*).

line level A small spirit *level* which, suspended at the middle of a taut bricklayer's line, can be used to level it to within about 1 mm per m ($\frac{1}{8}$ in. in 10 ft).

linen tape A tape used for rough measuring and setting out. It is light and easily handled but less accurate than a steel *tape* (*C*), particularly if no steel wire is interwoven in it.

line pins Steel pins about 8 cm (3 in.) long inserted in the mortar *joint* at the ends of a wall and used for holding a bricklayer's *line*.

liner (1) [pai.] A *lining tool*.

(2) [plu.] A *sleeve piece*.

lining (1) [joi.] An *architrave*, a cover to structural work, in particular the wooden surround to a door opening. A lining (or door case) is not strong enough to carry a door without proper anchorage to the neighbouring *partition*, since it is of thinner wood than a *door frame*. A facing of *plywood* or other *wallboard* to an interior wall is also a lining.

(2) [joi.] *Insulating fibre board*, woodwool slab, or corkboard, fixed to a *framed partition* under the surface covering to improve the sound insulation or *absorption* or the thermal insulation of a partition.

(3) [pai.] Continuous or discontinuous parallel troughs or ridges in the direction in which a *paint* or *varnish* has drained or been brushed. This defect shows that the media were not *compatible*. Silking is very fine-grained lining.

(4) [pai.] *See* **lining tool.**

lining paper Paper pasted on to *plaster* as a base on to which wallpaper is pasted.

lining plate In *flexible-metal roofing* a metal strip nailed to *eaves* or *verge* and hooked into the roofing sheet to hold it.

lining tool or **liner** [pai.] A small flat *fitch* with a slanting edge used for painting lines with the help of a *rule*. *See* **house painter** (illus.).

link dormer A large *dormer*, sometimes with lights at the sides. It may join one part of the roof to another part or incorporate a projection from the roof such as a chimney.

linked switch [elec.] Two or more switches joined by bars so as to open or close simultaneously or in sequence.

link fuse [elec.] A *fuse* which is not protected by a cover plate.

linoleum or **lino** A detachable floor covering built up from *linseed oil* and *hessian* canvas. Its thickness is measured in mm and varies from 2 to 6·7 mm, made in widths up to 2 m. If glued to the floor all over, it has longer life than if nailed. It must only be laid on a dry floor. Rubber floor finishes are an improvement on lino, but more costly and not removable.

linseed oil [pai.] The most valuable oil in the *paint* and *varnish* trades, obtained by crushing the seed of flax. When exposed to the air it darkens and thickens to a tough *film* by *oxidation*, called *drying*. It is also called Baltic, or Black Sea, or Plate oil, after the sea or river of despatch. *See* **boiled oil, raw oil.**

lintel or **lintol** A small beam over a door or window *head*, usually carrying wall load alone.

lip or **lipping** [joi.] A *banding* on a *flush door*.

lip union [plu.] A *union* with a ring-like inner projection which prevents the gasket from being forced into the pipe and partly blocking it.

liquidated damages *See* **damages.**

liquid driers [pai.] *Soluble driers.*

listing [tim.] Removing waney edge.

litharge or **lead monoxide** (PbO) [pai.] A *drier* and *pigment* of pale yellow to brown colour.

lithopone [pai.] A white very opaque *pigment*. It is a co-precipitated mixture of barium sulphate ($BaSO_4$) and zinc sulphide (ZnS), and is much used in interior paints, *water paints*, *distempers*, and in *enamels* with *synthetic resins*. It was developed first in about 1870 as a non-poisonous paint which was more opaque than zinc oxide.

little joiners [joi.] Small pieces of wood which are used to hide and fill holes in wood; for example, *pellets* and *inlays*.

live edge [pai.] An edge of paint is said to be live if it has been painted for some time and can be blended with newly applied paint without the *lap* showing.

live knot or **intergrown knot** [tim.] A *knot* whose fibres are intergrown with the wood. It is allowable in structural timber within certain size limits. *Compare* **dead knot.**

livering [pai.] *Feeding*.

live wire [elec.] A conductor with the electrical power switched on to it, and therefore dangerous to touch. It may be called alive (opposite, dead).

load [tim.] In *radio-frequency* heating, the *plies* and *glue* inserted between the electrodes.

loadbearing *See* (*C*), *also* BSCP 111.

loading coat or **loading slab** A concrete slab laid over asphalt *tanking*, to ensure that it is not pushed upwards by water pressure below it.

loading of pipe [plu.] Before bending a pipe, it is filled with a *bending spring*, or with molten lead, low melting-point alloy, pitch, *resin*, or compressed sand. This loading prevents the pipe distorting during bending.

Local Authority In Britain a City, Town, Borough, Rural District, Urban District, or County Council, or similar body, with power to control building and to make *zoning* regulations and *by-laws* or to waive them in case of need. The *Engineer* (*C*) or *Surveyor* (*C*) or *Architect* administers its by-laws.

location plan A plan which shows the dimensions and position of a building site, usually also the building proposed.

lock [joi.] A fastening for a door, operated by a key which shoots a bolt, the commonest types being *mortise*, *rim*, and *cylinder locks*. Compare *latch*.

lock block [joi.] A wood block in a *flush door* into which a *lock* can be fixed.

locking bar A development of the *hasp and staple* for fastening gates or barn doors. The hasp or bar pivots on a pin on the door and hooks over the staple, being secured with a padlock.

locking stile [joi.] A *lock stile*.

lock joint or **lock seam** A *seam* in *flexible-metal roofing*.

lock rail [joi.] The *rail* in a door which carries the *lock*.

lockshield valve or **balancing valve** [plu.] One of the two valves attached to a radiator, that enable the radiator to be removed for repair or replacement without stopping the heating system and necessitating the laborious work of draining the *circulation*. The other is the control valve, in daily use for turning the radiator on or off. The lockshield valve is only used once in the lifetime of a circulating system, on the occasion when the system is balanced to ensure that each radiator receives its adequate share of hot water.

lockspit A narrow V-shaped cut in the ground surface made by a *ganger* (*C*) to mark the line of a dig.

lock stile or **shutting stile** [joi.] The door *stile* at the edge which carries the *lock*.

loft (1) A storage space under a roof.

(2) (USA) May mean an entire upper floor of a commercial building, let to a tenant who subdivides it as he thinks best.

loft ladder or **disappearing stair** A folding ladder which is fixed on the top of a trap door into a loft or attic space and is invisible from below when the trap door is closed. The trap door is hinged to open downwards and is counterbalanced to carry the ladder by weights or springs. These ladders are indispensable for transforming an inaccessible loft into usable living space. They were made of wood but are now often of *light alloy*.

log [tim.] A length of trunk after felling, barking, and trimming.

London stocks *See* stock brick.

long dummy [plu.] A *plumber*'s tool for straightening kinks in lead pipe.

long float [pla.] A *float* which needs two men to handle it.

longitudinal bond A *bond* used in thick walls, in which occasional *courses* of bricks are laid all as *stretchers*.

long oil [pai.] A high ratio of oil to *resin* in a varnish. *See* oil length. *Compare* short oil.

long-oil alkyd [pai.] An *alkyd resin* containing more than 60% of oil as a modifying agent (BS 2015). *Compare* short-oil alkyd.

long-oil varnish [pai.] An *oleo-resinous varnish* other than an *alkyd*, containing not less than $2\frac{1}{2}$ parts by weight of oil to 1 part by weight of *resin* (BS 2015). *Compare* short-oil varnish.

longscrew [plu.] A *connector*.

lookout [carp.] (USA) A short wooden bracket (or projecting rafter or ceiling joist) which supports the overhanging roof.

looping in [elec.] A method of reducing the number of T-joints in house wiring conduits by keeping one cable permanently connected to the lampholder, the other passing through the switch. More wire is needed, as the conductor must double back on itself, but the cable is cut at fewer points and there are therefore fewer joints.

loop vent or **circuit vent** (USA) A *ventilation pipe* which is a continuation of a *soil pipe* above the last soil branch.

loose butt hinges [joi.] *See* lift-off butts.

loose-fill insulation Insulating materials such as granulated *cork*, loose *asbestos*, *expanded clay*, or other *lightweight aggregate*, *gypsum insulation*, or *mineral wool*. Loose fill is placed between or over ceiling rafters or in the gap of a cavity wall to increase the insulating value of a dry air space. Some of these fills, in particular mineral wool, are used to insulate, quickly and cheaply, houses already built, by blowing them through pipes with an air blast. *See* cavity filling.

loose knot [tim.] A *knot* which is not held tight and may drop out, and is therefore a *defect*.

loose-pin butt or **pin hinge** [joi.] A *butt hinge* with a withdrawable hinge pin which enables the door to be unhung merely by removing the pins from the hinges, without the labour of unscrewing the butts.

loose side or **slack side** [tim.] Of *veneer* cut with a knife, that side which was last in contact with the blade. It is bent outwards and slightly broken, being the inner side of the veneer when it was cut from the log. *Compare* **tight side.**

loose tongue [joi.] A *cross tongue*.

lost-head nail A round wire *nail* with a very small head.

lot (USA) A site in a town whether for building or not.

loudness Loudness is measured by phons. For instance, a sound judged by a normal observer to be 'x' *decibels* above the threshold of audibility has an intensity of x phons.

louvre or **louver** (1) A ventilator sometimes in a window, in which horizontal sloping slats allow ventilation and exclude rain. It was originally a turret to let out smoke from a medieval hall with no *chimneys*. *See* **lever boards.**

(2) A grille connected to a ventilating duct, usually one delivering air to the room (output louvre).

low-pressure system The normal way of heating water for washing purposes in small houses. The circulation of the water is natural, that is it is caused by the temperature difference between the *flow* and *return* pipes. The boiler must therefore be at the lowest possible point in the building to obtain the best circulation.

lucarne (USA) A *dormer window* in a roof or spire.

lug (1) A small projection from a frame or pipe for fixing purposes, for example, a door frame is fixed to a wall by building a steel lug into the wall.

(2) [elec.] A terminal on the end of a wire to make good electrical connection.

lug sill A *sill* with its ends built into the *jambs*. *See* **slip sill.**

lumber [tim.] (1) In USA the term for converted logs which have been sawn and sometimes *re-sawn*, the word *timber* being reserved mainly for round logs. Boards of lumber may be up to 7·3 m (24 ft) long while timbers may be 18 m (60 ft) long. In flat-sawn lumber the maximum width is the diameter, in quarter-sawn lumber it is half the log diameter. *Yard lumber* consists of *strips* and *boards. See* **dimension lumber.**

(2) Imported square-edged sawn hardwood of random width (BS 565).

lumber core [tim.] American term for *coreboard*.

luminaire [elec.] In USA a light fitting or lighting standard in a house or on a street.

lump hammer A *club hammer*.

lump lime *Best hand-picked lime*.

lump-sum contract A *contract* in which the contractor submits a price for construction (and maintenance for a short period) of the work shown on the contract drawings. A *bill of quantities* is sometimes drawn up by the engineer or *architect* to help the contractor, but not normally. This is a simple type of *fixed-price contract* which is suitable for small buildings but not for large work of any sort. It has the further disadvantage that *tender* prices submitted by different firms cannot be compared since they are not usually based on the same bill.

lune or **gore** A figure enclosed by two arcs of circles. A number of the figures, fitted together, form a balloon shape. Similarly, if cut in two and fitted together they form a hemisphere or dome. Half lunes are therefore used in cutting out metal, felt, or timber to cover a dome.

lying panel [joi.] A *lay panel*.

M

machine mason A *banker mason* who works a stone-working lathe or planing machine. *See* **mason.**

machinist A skilled *tradesman* or at least a semi-skilled man who sets and operates a wood- or metal-working machine. Since there are many woodworking machines and even more metal-working machines, the term is vague.

made ground or **made-up ground** Ground which has been raised by *fill* (*C*).

magazine boiler A coal- or coke-fired boiler for a hot-water or central-heating system, which has a bunker fitted to it, large enough to contain 24 hours' fuel. It thus needs attention only once a day.

magnesian lime or **magnesian quicklime** *Lime* with 5% to about 40% MgO. *Compare* **dolomitic lime**

magnesite flooring A hard *jointless floor*, made with the cement (oxychloride cement, Sorel's cement) formed by mixing magnesia (MgO) and magnesium chloride ($MgCl_2 \cdot 6H_2O$) with sawdust, sand, or similar fillers. When wet it corrodes metal and should not be used in kitchens or bathrooms for this reason.

magnesium alloy *See* **light alloy.**

magnesium hydroxide ($Mg(OH)_2$) the substance produced when water is added to magnesia (MgO). *Compare* **slaked lime.**

magnesium-oxychloride cement *See* **magnesite flooring.**

mahogany [tim.] Many hardwoods are called mahogany, for instance, *Cuban mahogany*, Honduras mahogany, *African mahogany*.

main The supply authority's pipe or cable. *See* **communication pipe**

main contractor or **general contractor** or **prime contractor** A *contractor* who is responsible for the bulk of the work on a site, including the work of *sub-contractors*.

main couple [carp.] A timber *truss* of *principal rafters*.

maintenance period or **retention period** The period, after completion of a *contract*, during which a *contractor* is required to make good at his own expense any work which needs repair. *See* **retention money.**

maisonette or (USA) **duplex apartment** A self-contained *flat* on two levels having its own internal stairs. Large blocks of flats are sometimes built in this way, with the intention of obtaining quietness for the occupants by planning, rather than by the costly methods of *discontinuous construction*.

make good To repair as new.

makore [tim.] (Mimusops heckelii) An *African mahogany*.

mall or **maul** A heavy *mallet* or beetle.

mallet A tool like a *hammer* with a wooden, rawhide, or rubber head. Wooden mallet heads should be of beech, hickory, or well-chosen applewood.

mallet-headed chisel A *mason*'s term for those of his *chisels* which have a rounded steel head and are designed to be struck with a *mallet*. *See* **hammer-headed chisel.**

management contract A contract in which the *main contractor*'s chief function is not as a builder but as a manager of sub-contractors and an adviser to the architect, structural engineer and other consultants. He is appointed at an early stage of the job, when his advice on construction methods can be most useful. Later his main functions are programming and running the job. Abroad, especially in the USA, the system is well known, but by 1973 fewer than 50 such contracts had been let in Britain. The system is most suitable for buildings with complicated electrical and mechanical equipment that is difficult to price. One advantage is that it allows procedures to be telescoped because of the intelligent co-operation that is possible from the start between consultant and contractor. Various parts of the job can thus be simultaneously at completely different stages, whether at design or detailing or tendering or construction. This is not possible with most ordinary contracts when the job must be completely billed, and partly designed and detailed beforehand for the contractor to price it.

mandrel [plu.] A cylindrical piece of *hardwood* which is pushed through a lead pipe to enlarge it or remove distortion at bends.

manganese drier [pai.] Manganese dioxide (MnO_2), and other organic and inorganic salts of manganese. *See* **drier.**

man-hour The amount of work done by one man in one hour.

manipulative joint [plu.] A *compression joint* in which the ends of the copper tubes are slightly opened out. One advantage of this arrangement is that a *gland* is not usually needed, the soft copper of the belled-out tube making a seal.

man-made fibres The first man-made fibre, artificial silk, was made in 1885 from nitrocellulose, viscose was first made in 1910, cellulose acetate in 1911, nylon in 1936 and terylene even more recently.

mansard roof (1) or **curb roof** A roof which is often gabled, and has on each side a relatively flat top slope and a steeper lower slope, usually containing *dormers* to use the attic space of the building. *See* **roofs** (illus.). In USA this sort of roof is called a *gambrel roof.*

(2) (USA) A roof without *gables*, therefore hipped, with all four sides of the roof mansarded, that is having two slopes. The lower roof space contains dormers.

manufactured gas Gas piped to customers after processing originally of coal (town gas), mainly now of oil. It will gradually be phased out and replaced by natural gas, involving the conversion of appliances by the gas authority, but in the meantime even natural gas is sometimes used to help make manufactured gas. Its calorific value is about half that of natural gas.

marble facing Marble about 19 mm ($\frac{3}{4}$ in.) thick with 13 mm ($\frac{1}{2}$ in.) air space between it and the wall. It is fixed to the wall by strong wires dowelled into the marble and sealed into *plaster dabs*.

marbling [pai.] Copying the appearance of marble or other stone with paint.

marezzo marble An artificial marble like *scagliola*, which differs from it mainly in having no chips of added coloured material. When precast, it is cast on a smooth sheet of plate glass or slate to give a polished surface.

margin (1) or **gauge** The exposed surface of a slate or tile. *See* **gauge**.

　　(2) [joi.] The projection of the *close string* of a stair above the line of *nosings*.

　　(3) [joi.] The exposed flat face of a *stile* or *rail*.

　　(4) [joi.] A border mitred round a hearth.

　　(5) *See* **drafted margin**.

margin draft *See* **drafted margin**.

margin light [joi.] A narrow *pane* of glass at the edge of a *sash*.

margin templet [joi.] A *pitch board* with an edge strip equal to the width of the *margin*.

margin trowel [pla.] A narrow rectangular trowel used for working in a narrow width (BS 4049).

marine glue [tim.] A *glue* which was used before the discovery of *synthetic resins*, containing 1 part of rubber, 2 parts of *shellac*, and 3 parts of pitch.

marked face [joi.] The *face side* of timber.

marking gauge or **butt gauge** [joi.] A beechwood bar with a steel point projecting at right angles from it near one end and a *hardwood* block sliding along it which can be locked at any point along the bar. It is used for marking lines parallel to that face of the wood which the block travels along. *See* **carpenter** (illus.), **cutting gauge, face plate, mortise gauge**.

marking knife [joi.] A steel bar with a cutting edge at one end, like a skewed chisel, and a point at the other end, used for marking wood before cutting or drilling.

marouflage [pai.] To *glue* a canvas to a wall which is to be covered by a mural painting. It is put on by rollers, glued with a paste (often of *gold size*), and forms a strong, matt *ground* for the painter.

marquetry [joi.] Wood surfaces decorated by cutting out hollows and then inlaying into them other contrasting wood, metal, mother of pearl, ivory, etc.

mash hammer or **mash** Scots term for a *club hammer* or *sledge hammer* (*C*).

masking [pai.] During painting, the edges of a painted surface are protected from paint by masking them with tape or paper stuck on, or by holding a paper, metal, or cardboard mask over them.

mason A stone worker or stone setter. In Scotland or USA a *bricklayer* is usually also a mason. He is generally in England either a *fixer* or a *walling mason*. *See* illustration overleaf.

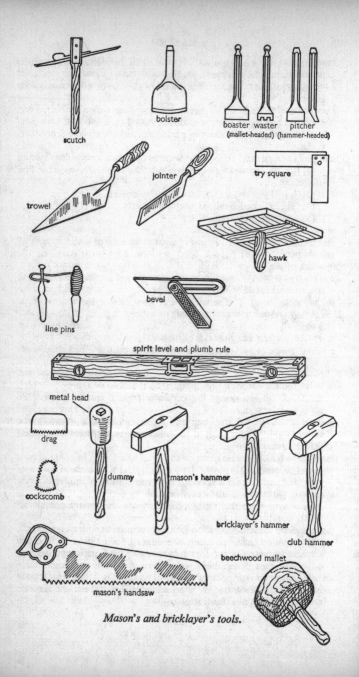

Mason's and bricklayer's tools.

masonry Stone and the craft of stone wall building, including the preparation and the fixing of the stones. In Scotland masonry includes brickwork and laying tiles. In USA it includes *concrete* block, *hollow block*, and sometimes also poured *concrete* or *gypsum* walls. *See* BSCP 121.

masonry cement A *cement* consisting mainly of ordinary *Portland cement* together with *lime* and *plasticizers*, or clay, *whiting*, paraffin, or secret materials to form a plastic, *water-retentive* mortar for building walls. It is sold ready mixed, and at the site merely needs to be mixed with clean sand and water.

masonry fixings *Cramps, anchor bolts* (*C*), *lewis bolts* (*C*) and plates for fixing stonework in place.

masonry nails Cadmium-plated *drive screws* which can be driven into ordinary bricks with a hammer or into predrilled holes in harder material. *Compare* **concrete nail.**

mason's joint A mortar *joint* consisting of a projecting triangle of mortar.

mason's labourer A skilled *labourer* who helps in the *mason*'s yard, lifting stones and cutting *lewis holes* with a compressed-air tool. On the site he helps the mason *fixer* generally, supplies him with *mortar* and grout and other materials, and cleans down stonework. *See* **rubbing-bed hand.**

mason's mitre or **mason's stop** An apparent *angle joint* formed by shaping the corner out of the solid stone. The actual joint is not a *mitre* but usually a butt joint away from the corner. It is also sometimes used in joinery.

mason's putty *Lime putty* mixed with stone dust and *Portland cement* for jointing *ashlar*, the mix being approximately 5:7:2.

mason's scaffold A *scaffold* which stands free of the wall, is supported on two rows of *standards*, and built of baulks or steel sections, well braced. Mason's scaffolds cannot (like bricklayer's scaffolds) be partly supported on the wall because the *putlog* holes would disfigure the stonework.

mason's stop A *mason's mitre.*

mastic (1) The resin of the pistachio or mastic tree, chewed in the East, hence the name. It is used as a *varnish* when dissolved in alcohol. The finest grade comes from Chios, Greece.

(2) A quick setting waterproof pointing or plastering material containing litharge and *linseed oil*, now replaced by *Portland cement*.

(3) A permanently plastic waterproof material, which hardens on the surface so that it can be painted, used for bedding roof lights, window frames, sealing joints in prefabricated buildings, sealing expansion joints, jointing pipes, *gutters*, *flashings*, and so on. If a truly satisfactory cheap mastic could be found, it would very greatly help *dry construction* and accelerate building. Gun mastics are applied to window glazing as *face putty* with a *pressure gun* which

lays it on very much more quickly and cleanly than the old hand method using a *putty knife*. A *sealer* may be needed on a wood surface under the mastic. Mastics may also be applied hot.

(4) The French word mastic also means glazier's putty, *iron cement*, and dental stopping. Some of these meanings (and other senses) are coming into English and complicating this unhappy word even further.

(5) *See* **mastic asphalt** (*C*), **pitch mastic**.

matchboard or **matched boards** or **match lining** [tim.] Boards laid side by side, and shaped with *mouldings* or *rebates* on each edge so that the *straight tongue* on one, fits (matches) the groove on the other. They are used for flooring, wall lining, and so on. *See* **vee-joint**.

matched floor [carp.] A floor made of *matchboard*.

matching [joi.] (1) *Matchboard*.

(2) [tim.] The arrangement of sheets of specially figured *veneer* in ways which bring out, by contrasts, the best in the colour and *figure* of the wood. Matching is practically confined to *sliced veneers*, since *rotary veneers* rarely have a figure which is interesting enough. *See* **book matching, four-piece butt matching**.

match planes [joi.] Pairs of *planes* which cut the tongue and the groove of *matchboard*. Sometimes the same plane will take both cutting irons (at different times). Most matchboard is now cut by machine. *See* **universal plane**.

mat sinking A depression inside a front door, designed for a doormat.

matt [pai.] A very low *gloss*.

mattock A tool like a *navvy pick*, with the difference that one end is broadened out like a hoe, the other end being pointed like a pick or shaped like a *chisel* or the blade of an axe. It is used for cutting tree roots and digging in stiff ground.

matt varnish [pai.] *Dammar* dissolved in turpentine with wax added will give a matt finish.

maturing (1) [pla.] *Fattening up*.

(2) [pai.] The *ageing* of *varnishes* to improve their properties.

maturing bin or **maturing pit** [pla.] A tank or wood-lined pit into which *milk of lime* is run after passing through a screen with 3 mm ($\frac{1}{8}$ in.) openings from the *slaking box*. The milk of lime is left two to four weeks for settling and *fattening up*, after which it is called *lime putty* and dug out. *See* **boiling hole**.

maul or **mall** or **beetle** A heavy wooden *mallet*.

maximum demand [elec.] The greatest instantaneous power demand from a consumer. It is important because some methods of charging for electricity are based on the maximum demand, but *see* **diversity**.

measure-and-value contract [q.s.] A *fixed-price contract* in which the *contractor* receives, with the tender drawings, a *bill of quantities*. When there is time to prepare the bill as well as the drawings, this sort of *contract* is preferred by consulting engineers or *architects* because the comparison of *tenders* from different contractors is simple. Contractors have precise information for tendering and less work to do to make up their prices than with unbilled contracts.

measurement [q.s.] The *quantity surveyor*'s duties of estimating from drawings the amount of work to be done and billed, and, later, his measuring on the site of the work done and to be paid for. On civil engineering *contracts* this work is often done by engineers in accordance with the simple and rational measuring methods of the Institution of Civil Engineers. *See* **standard method of measurement.**

measuring frame A *batch box.*

mechanic (1) [mech.] A *fitter* (*C*), a man skilled in mechanical engineering.

(2) A *tradesman* in any of the building trades.

mechanical core or **sanitary core** or **plumbing services** [plu.] Prefabricated pipes for hot and cold water supplies, wastes, drains, and gas, as well as prepared cable for electric services, all ready for installation with the minimum of site work in an *industrialized building method.*

mechanical engineer A person qualified in *mechanical engineering*, in Britain usually a member of the Institution of Mechanical Engineers, often a university graduate. The best theoretical background for a *heating and ventilation engineer* is a training as a mechanical engineer.

mechanical engineering The design and construction of engines and machines of every sort, particularly in metal. It merges into electrical engineering but does not include the design of electrical circuits.

mechanical plastering [pla.] Modern plasterers use mechanical mixers, grinding and polishing machines, sprays, vibrators, trowelling machines and pumps for raising mixed plaster to a multi-storey building. *See* **plastering machine.**

mechanical saw [tim.] The *circular saw, band saw*, and *jig saw* are the best known mechanical saws for wood.

mechanic's level A *level.*

medium [pai.] The liquid part of a *paint* or *enamel*, which becomes the *binder* of the film after it has hardened.

medullary ray or **pith ray** [tim.] Fine radial lines seen in the *silver grain* of quartered oak, but often too small to be noticed. Since the growing part of the tree is the outside part, not the medulla or pith, the term ray or *wood ray* is often preferred.

meeting rails or **check rails** [joi.] The *rails* of *sash windows* which touch when the window is closed.

meeting stile [joi.] A middle *stile* (or shutting stile or closing stile) of a *folding door* or casement.

melamine formaldehyde [tim.] A *synthetic resin* used for gluing or surfacing *veneers* or *laminated plastics*. It is not damaged by cigarette burn, is used for making stoved finishes, and keeps colour well.

221

melamine-surfaced chipboard [tim.] Resin-bonded *chipboard* with a decorative, smooth, *melamine formaldehyde* surface, is cheap, stable, fairly strong, and became available in Britain about 1963.

member A part of a *moulding*. *See also* C.

mending plate A flat steel plate drilled with holes for countersunk screws, used for repairing *carpentry* work by screwing it on to sound wood on each side of the break.

mensuration [q.s.] The measurement of lengths, and the calculation of lengths, areas, and volumes.

merchantable bole [tim.] The length of the tree trunk which is usable as timber.

merulius lacrymans [tim.] The fungus which usually causes *dry rot* in Britain.

metal coating A thin film or films of nickel, copper, cadmium, *chromium*, aluminium, or zinc laid over corrodible metal surfaces. The coating protects either by completely enveloping the surface or by *sacrificial protection* (*C*) when it is locally worn through. *Galvanizing* (*C*), *sherardizing*, and *chromium-*, *cadmium-*, or nickel-*plating* are usually applied before a part is built in. Zinc or cadmium coatings can, however, now be sprayed on after erection by a special oxy-acetylene blowpipe (metal spraying). The cost of such a metal coating on a structure is higher than the best painting, but metal coating is likely to last five times as long as paint, and because of this may allow the structural designer to use thinner metal in his structure. The metal can be applied thickly enough to be polished after grinding off the surface roughness. *See* **protective finishes**.

metal cramp A bent bar for fixing stones in the same course. *See* **cramp**.

metal lathing *Expanded metal* (*C*) not less than 0·56 mm thick weighing 1·6 kg/m² (3 lb/yd²) at least, used as a base for plaster. It should be galvanized or painted. It is stronger and more fire-resistant and vermin-resistant than *laths* and generally easier to fix. Expanded metal lathing may be with or without stiffening ribs. Other types are perforated steel sheet, *dovetailed lathing*, and *clay lath*.

metallizing [mech.] Applying a *metal coating*.

metal painting [pai.] Painting with *bronzing fluid* as a vehicle.

metal-sheathed mineral-insulated cable [elec.] Cable like *Pyrotenax*.

metal sheeting *See* flexible-metal roofing.

metal spraying *See* metal coating.

metal trim *Architraves*, *skirtings*, picture rails, and angle beads of sheet metal, which are fixed before plastering and incorporated in the plaster surface. They often have the advantage of being flush with the plaster and therefore not collecting dust. *See* **pressed steel**.

metal valley [plu.] A *valley gutter* lined with lead, zinc, copper, or aluminium.

methylated spirit [pai.] Industrial alcohol containing some wood alcohol (methyl alcohol) to make it doubly poisonous, used for making *knotting*, *French polish*, etc.

mica Minerals with such excellent cleavage that they can be used in thicknesses so small as to be transparent. They are also very good electrical insulators and are used as *extenders* to help the *leafing* of aluminium paint and thus to resist the penetration of moisture. *See* **heat-resisting glass.**

mica-flap valve A sheet of mica at a *fresh air inlet*, hinged to allow air to flow inwards only.

microbore [plu.] *See* **minibore.**

mid-feather (1) [joi.] A *parting slip*.

(2) A central *withe* in a chimney.

milkiness [pai.] The defect of a whitish or translucent appearance in a *varnish* film.

milk of lime [pla.] *Slaked lime* in water. *See* **maturing bin.**

milled lead Lead rolled into sheets from cast slabs.

mill-run mortar *Mortar* made in a *pug mill* or other mixer.

millwork [tim.] (USA) *Prefabricated* joinery (doors, windows, panels, stairs) made at the mill and partly assembled there. *See* **planing-mill products.**

mineral black [pai.] Black *pigments* which may be the carbonaceous clays of Devon or the crushed *slate* of the Continent of Europe. Graphite also is a black mineral pigment but is not called mineral black.

mineral-insulated cable [elec.] *See* **Pyrotenax.**

mineral streak [tim.] A greenish or brown discoloration of *hardwoods* which does not affect the strength.

mineral-surfaced bitumen felt A heavy *bitumen felt* whose upper surface is dressed with particles of *slate* or other stone, its lower surface with talc or sand. It can be used as a single layer on sloping roofs or as a top layer in multiple roofing. It is made 90 cm (3 ft) wide in rolls 11 m (36 ft) long and weighing 36 kg. (80 lb).

mineral wool or **rock wool** or **slag wool** A flexible, resilient, flocculent material made from slag or mineral fibres. It can be used loose or made up in 1 to 5 cm ($\frac{1}{2}$ to 2 in.) thick *blankets* in waterproof paper, *scrim* cloth, or aluminium foil as required. It is also made up in wire-netting reinforced mattresses 2 to 15 cm (1 to 6 in.) thick and in semi-rigid felted slabs 2 to 10 cm (1 to 4 in.) thick (*see* **bat**). In these states its thermal *conductivity* is from 0·033 to 0·037 W/m deg. C. (0·23 to 0·26 Btu in./ft²h deg. F.) and its weight from 60 to 225 kg/m³ (4 to 14 lb/ft³). *Compare* **glass silk, perlite, vermiculite.**

minibore or **microbore** [plu.] A description of a central heating circuit using circulation pipes smaller than those used for *small bore*, usually of 6, 8, 10 or 12 mm outside diameter. Since the advent of small-bore systems, low-cost circulating pumps have been developed that can adequately deal with the higher pressures involved with these smaller tubes. The tubes are easy to bend and install, less conspicuous than larger pipes and they lose less heat. Water distribution manifolds of suitable diameter, to which the main flow and return pipes are led, provide convenient connections for up to nine radiators.

mismatching [joi.] (1) Bad fit at a joint.

(2) A lack of fit in *figure* or *grain* of matched *veneers*.

mission tile (USA) *Spanish tiles*.

mist coat [pai.] (1) A very thin sprayed coat, usually of cellulose *lacquer*.

(2) A very thin coat of *emulsion paint*, placed over a wall covered with washable distemper as a *sealer* before painting it with emulsion.

mitre or (USA) **miter** [joi.] An *angle joint* between two members of similar cross section. Each is cut at the same bevel, 45° for a right-angle corner, so that the straight line of the joint is seen to bisect the angle. The term is also used in masonry, concrete, and welded steel. The strength of a simple glued and butted mitre may be greatly increased by inserting a glued *cross tongue* or by making of it a *mitre dovetail* or a plain *dovetail*. *See* **mason's mitre**.

mitre block or **mitre box** [joi.] A U-shaped or rebated block of wood with saw cuts in it at 45° to the axis of the block. The piece of wood to be cut is held firmly, or clamped to the block and can thus be sawn to an exact mitre for a 90° corner.

mitre board [joi.] A *mitre shoot*.

mitre brad [carp.] A *corrugated fastener*.

mitred-and-cut string [carp.] A *cut-and-mitred string*.

mitred border [carp.] A *margin* to a hearth.

mitred cap [joi.] A *newel cap* into which the *handrail* is mitred.

mitred closer A brick *closer* cut at an angle.

mitred knee [joi.] A mitred intersection between a horizontal part of a *handrail* and a steeply falling part, curved like a bent human *knee*.

mitre dovetail or **blind** or **secret dovetail** [joi.] A *dovetail* joint in which the pins cannot be seen and only the line of the *mitre* joint shows.

mitred valley or **hip** A *close-cut valley* or hip.

mitre joint [joi.] A *mitre*.

mitre machine [tim.] A *trimming machine*.

mitre saw [joi.] A *tenon saw*.

mitre shoot or **mitre board** [joi.] A frame for holding a *moulding* while the *mitre* is being planed.

mitre square [joi.] A *bevel* with the blade fixed at 45° to the stock.

mitre templet [joi.] A small rebated frame made to guide the *chisel* when *mitring small mouldings*.

mitring [joi.] Making a *mitre*.

mitring machine [tim.] A *trimming machine*.

mixed glue or **ready-mixed glue** [tim.] *Synthetic resin* glue with an *accelerator* mixed with it in the pot. *Compare* **separate application**.

mixed stuff *Best reed* containing some gladden (Iris pseudacorus or foetidissima) and lesser reed mace (Typha angustifolia).

mixing varnish [pai.] A varnish *medium* which can be mixed into an *oil paint* to give additional *gloss*.

model by-laws Draft building *by-laws* issued until 1965 by the British Ministry of Housing to guide Local Authorities in forming their own. *See* **Building Regulations.**

modelling [pla.] Shaping a plaster or clay surface.

modular masonry unit (USA) A *brick* or *building block* which measures, laid up with its mortar joint, a multiple of 4 in. in plan both ways. Vertically a whole number of courses (1, 2, 3, 4, or 5) lays up to 4, 8, 12, or 16 in. (10, 20, 30 or 40 cm). *See* **brick format.**

modular system The planning of buildings and their components to fit a *planning grid* related to a *module*. *See* **dimensional coordination.**

modulated control [mech.] Automatic control of central heating is usually by simple on-or-off switch from the room *thermostat* or water thermostat. For a hot water system the switch may open fully or close completely the boiler air supply. For a *warm-air system*, the switch will stop or start the fan. This method is widely used because it is cheap and sturdy, but it is abrupt and often noisy. A modulated control does not switch on or off abruptly, it reduces or increases the heat flow gently. It is therefore smoother, more responsive, and (most important) quieter. Most companies are developing modulated control.

module A unit of length by which the planning of buildings can be to some extent standardized. Before metrication some British *architects* used plan modules of 1·015 m (40 in.) based on the width of one person at shoulder height plus construction thickness and tolerances. When a module is used, few dimensions are put on the architect's drawing, all main dimensions being indicated by the grid lines. The module thus saves drawing-office work and *quantity surveyor*'s work. Japanese architecture was governed in its planning before any European influence came to Japan, by the module of the 3×6 ft (0.90×1.80 m) floor mat. *See* **brick, blockwork.**

moellon French term for *rubble walling* resembling *Kentish rag*.

moisture barrier A *vapour barrier* or a *damp-proof course*.

moisture content [tim.] The amount of water in wood expressed as a percentage of its *oven-dry* weight. The moisture content often exceeds 100% for freshly felled timber. At the time of erection the moisture content should not exceed 22% in carpentry or 17% in joinery. In dried-out buildings these figures are reduced to about 16% and 14% with a further 2% reduction if the building is continuously heated. The smallest moisture content at which *fungus* can grow in wood is 20%. The lowest moisture content demanded is 8% to 12% for wood-block floors in hospitals. The weights of *timbers* given in this book are at 12% moisture content. To obtain the corresponding weight at any moisture content between 5% and 25%, 0·5% of the weight should be added for each 1% increase in moisture. Below 25% moisture, the moisture content can be measured to an accuracy of 2% or slightly closer by electrical moisture meters, which measure the electrical resistance of the wood between two sharp electrodes driven into the wood with a special hammer. Reduction of moisture content from the green state to about 25% has little or no effect on the strength, but a reduction from 25% to 12% may increase compressive strength by 100%. *See* fibre saturation point.

moisture gradient [tim.] The variation of *moisture content* between the outer and the inner part of a piece of wood.

molding (USA) *Moulding.*

moler brick Brick made of *diatomite.*

monitor or **monitor roof** A continuous *lantern* light in a steel *truss* used for daylighting single-storey workshops, used in the large-span roofs of American factories.

monk bond or **Yorkshire bond** or **flying bond** *Flemish bond* modified to show on the face two *stretchers* and a *header* repeating in each course.

monkey tail [joi.] A downward scroll at the end of a *handrail.*

monkey-tail bolt [joi.] An *extension bolt.*

monumental mason A *mason* who carves stone, cuts lettering, creates *polished work*, etc., with or without machines.

mopboard A *skirting board.*

mopstick handrail or **mopstick** A *handrail* which is circular except for a small flat surface underneath.

mortar A mixture of *Portland cement*, *lime putty*, and sand in the proportions by volume of 1:1:6 or 1:2:9, etc., for laying bricks or stones. Until the use of Portland cement became general, lime–sand mortars were universal. Occasionally cement–sand mortar is used, but it is less easy to lay and more liable to crack than cement–lime–sand mortar. Mortars should be *water retentive* so that they do not stiffen in contact with absorbent brick; they should also be plastic (spread easily) and cling to the trowel. *See* fat mortar, masonry cement.

mortar board A *hawk.*

mortar-cube test A test for a cement performed by crushing a cube of it made in a standard way with standard sand and measuring the crushing strength. *See* Leighton Buzzard sand.

mortar mill A mixer for *mortar*, often a *pug mill.*

mortgage An agreement by which a house owner (the mortgagor) gives his property conditionally to the mortgagee, usually a *building society* or insurance company, as a security for the payment of a debt. Very often the mortgagee lends the money to buy the house, but the house does not become his unless the mortgagor defaults in payment.

mortise or **mortice** (1) [joi.] A rectangular slot cut in one member, in which usually a *tenon* from another member is glued or pinned, or a *lock* is fixed. The thickness of a mortise should not exceed one third that of the wood in which it is cut. *See* **open mortise, blind mortise.**

(2) A rectangular sinking cut in a stone to receive a *cramp* or other locking device or a *lewis*.

mortise-and-tenon joint [joi.] A joint usually between members at right-angles to each other, such as a door *rail* tenoned into its *stile*. *See* **mortise.**

mortise chisel [carp.] A *chisel* strong enough to be struck by the *mallet*, for cutting mortises; stiffer than the *firmer chisel*.

mortise gauge or **counter gauge** [joi.] A *marking gauge* with two marking points, of which one is movable. It is used for marking out *mortises* and *tenons*.

mortise joint [joi.] A *mortise-and-tenon joint*.

mortise lock [joi.] A lock set in a *mortise* within the door thickness. The lock is hidden and the joinery is thus of better quality than with the *rim lock*.

mortising machine [tim.] A power-operated machine which cuts *mortises* in timber. It may be one of two general types, the square chisel, or the chain mortiser, which has a projecting jib and chain with cutting teeth. The latter is used for cutting large mortises. The square chisel is an *auger bit* rotating in a square steel shell with holes cut into the sides through which the chippings are thrown out.

mosaic Floor, ceiling, or wall surfaces built up from small cubes of marble, glass, or pottery laid in cement to a pattern, a technique known to the ancient Romans. When the surface is abraded and thus smoothed after laying, it is called *terrazzo*.

mosaic cutter A *floor-and-wall tiler* who fixes *Roman mosaic*, cuts the cubes, and glues them to the paper as designed. If he makes his own designs he is known as a pattern maker.

motorized valve [plu.] A valve that closes or opens the heating *circulation*; it can easily be fitted into an automatically controlled central heating system because it is operated by an electric motor.

mottle [tim.] A *figure* which greatly increases the value of *veneer* for cabinet work, for example *fiddleback*.

mottler [pai.] A flat thick *brush* for *graining* and *marbling*. *Compare* **overgrainer.**

mottling [pai.] Uniform, rounded marks, a defect of sprayed coats.

mould [pla.] A zinc sheet cut to the profile of a *moulding* and fixed to a wooden stock cut to the same profile. *See* **horsed mould.** *See also C.*

moulded (1) A description of stone, plaster, or wood, on which a *moulding* has been cut or cast.

(2) Any product cast in a mould.

moulded bricks (1) *Bricks* of ordinary quality, generally for *facing*, which are neither *pressed* nor *wirecut*.

(2) Bricks moulded to shape in Tudor times so that when laid they formed an ornamental brick *corbel*, *chimney-stack*, arch, window, or

door *jamb*, *transom*, or *mullion*. All of these moulded bricks, except in chimney stacks or *finials*, were at times plastered to imitate stone. In about 1650 moulded bricks lost their vogue in favour of *carved brickwork*.

moulded insulation or **sectional insulation** *Insulating material*, shaped to fit round steam, hot water, or refrigerator pipes and *fittings*, a ready-made *lagging*.

moulded plywood or **ply plastics** [tim.] *Plywood* bent during gluing by pressure, usually from *flexible bags*, using heat and *thermo-setting* glues so that when it cools it sets in a curved shape. Aircraft wings and other surfaces curved in two planes are made in this way.

moulding (1) A continuous projection or groove used as decoration to throw shadow, sometimes also to throw water away from a wall. It may be in stone, brick, plaster, joinery, cast iron, aluminium, plastics, and so on. In joinery, mouldings are either formed on the solid (when they are called *stuck*) or applied by gluing or nailing (when they are called *planted*).

(2) [joi.] The operation of cutting mouldings with woodworking machinery or hand *planes*. The *spindle moulder* is the most versatile machine for cutting mouldings, but other machines are used.

moulding cutter [joi.] A *solid-moulding cutter*.

moulding machines Machines for cutting *mouldings* on wood or stone. *See* spindle moulder.

moulding plane [joi.] A hand *plane* used for cutting *mouldings*. *See* universal plane.

mouse or **duck** [joi.] A short, curved piece of lead tied to a string and slipped over a *sash pulley*. The other end of the string is tied to the *sash cord*, which can be pulled over the pulley by the string as the mouse is drawn through the *pocket*.

mouth [joi.] The slot in the *sole* of a *plane* through which the *cutting iron* projects and into which the shavings pass.

moving stair An *escalator*.

mudsill [carp.] A *sole plate*.

muffle [pla.] A layer of *gauged stuff* covering a *horsed mould* to the thickness of the *finishing coat*, used for roughing out the core of a *moulding* which is too large to be made without a core. Before the finishing coat is run, the muffle is chipped off the mould.

mullet [joi.] A piece of wood grooved to a certain width and slipped along the edge of a *panel* to ensure that it is of the correct thickness.

mullion or **munnion** A vertical dividing member of a frame between the *lights* of a door or window, each of which may be further subdivided into *panes* by *glazing bars*. *See* transom.

multi-ply [tim.] *Plywood* with more than three plies. *See* balanced construction.

multi-point water heater An instantaneous (non-storage) gas water heater that supplies hot water to several taps. So as to reduce its running cost it should be installed nearest to the tap most frequently used. A typical multi-point unit heats 7 litres (1½ gal.) of water per minute through 45° C. (81° F.) whereas a single-point heater (instantaneous sink heater) heats only 3 litres (5 pints) per minute through 45° C. (81° F.). The Building Regulations allow only *balanced flue* heaters in bathrooms.

multi-unit wall (USA) A wall built of two or more half-brick thicknesses, called *withes*.

munnion A *mullion*.

muntin [joi.] (1) A subsidiary vertical framing member in a door, framed into the *rails*, separating the *panels*, usually of the same width as the *stiles*.

(2) (USA) A *glazing bar* or a *mullion*.

N

nail The commonest nails are cold-forged from bright round steel wire of diameter between 1·2 mm (0·05 in.) and 6 mm (¼ in.). The largest nails (over 13 cm (5 in.)) are known as spikes; the smallest are pins, tacks, oval brads (not to be confused with flooring brads), and *glazing sprigs*. Clout nails are large-headed, galvanized *wire nails* for fixing *slates*, *roofing felt*, or *sash cords*. Those nails which are not made from wire are usually black and of rectangular cross section. They are either sheared from steel plate and called *cut nails* or have forged heads such as the rose-head nail, when they are called wrought nails. Nails of brass, aluminium alloy or copper are made for roofing, since they last longer than galvanized or *zinc* nails. *Sherardized* nails are made for nailing *gypsum plasterboard*. Nails are usually sold by weight. The lengths of nails in USA are described by the penny system, supposed to be derived from the old price per 1000 or per 100 nails, so that the highest penny number goes to the longest nail. The approximate lengths are: fourpenny (or 4d), 3·2 cm (1¼ in.); 6d, 5 cm (2 in.); 8d, 6 cm (2½ in.); 10d, 7·6 cm (3 in.); 20d, 10 cm (4 in.); 60d, 15 cm (6 in.). Intermediate sizes are roughly proportional.

nailable A material which holds *nails* and into which nails can be driven.

nail float [pla.] A *devil float* with nails projecting from it.

nailing block A *fixing brick*.

nailing ground A *common ground*.

nailing strip *See* no-fines concrete.

nail puller or **pry bar** [carp.] A tool used for drawing *nails*, more delicate than a *claw hammer*, shaped something like a screwdriver with a curved, forked blade.

nail punch or **nail set** or **(USA) brad setter** or **brad punch** [joi.] A short blunt steel *punch*, tapering at one end to the diameter of a small *nail*. It is struck by the *hammer* when a nail head is to be driven below the surface. Its point is concave to hold the nail head.

naked flooring [carp.] A wooden floor before the floorboards are laid.

naphtha [pai.] A *thinner* distilled from organic material at temperatures from 160° to 270° C., particularly from petroleum, mainly consisting of paraffins and olefines. They are used in painting but with caution owing to their strong smell. *See* white spirit.

narrow-ringed timber or **close-grown** or **close-grained** or **fine-grained** or **fine-grown wood** Wood which has grown slowly, has narrower, less conspicuous *annual rings* and is therefore stronger than wood which has grown quickly. *Compare* wide-ringed timber.

National Building Agency (NBA) An organization with architectural staff, directed by an eminent architect, and started with Government help in 1964 to encourage the use of industrialized building by local authorities so as to lower the cost, improve the quality and speed the construction of housing and related matters. It later undertook consulting work on a fee basis and was at one time expected to pay for itself. The NBA appraises complete building systems, now not only for housing but also for other purposes, doing on a large scale what the Agrément Board does on a small scale, appraising elements that can be used in any building, not necessarily a system.

National Building Specification (NBS) A four-volume book, first published in 1973 by NBS Ltd, a subsidiary of the RIBA. NBS is a 'library' of standard *specification* clauses from which architects and other consultants can selectively copy when building up specifications and *preambles* to bills of quantities. The clauses are simple, short and numerous, to give flexibility in use. Explanations appear alongside the clauses, with copious reference to *BS* and there are a number of worked examples.

natural bed A stone is said to be laid on its natural bed when its bedding planes are horizontal. Sometimes the natural bed must be marked on the stone at the quarry, since, particularly with igneous rocks, it is not always possible to see it. It is advisable to lay stones on their natural bed if they are to carry heavy load and be out of doors. *See* **face-bedded.**

natural cement A limestone containing clay or a clay containing *lime*, which, when burnt, makes without additions a *hydraulic* cement.

natural gas Any fuel gas that flows from the ground. Usually it is mostly methane (CH_4) with a small proportion of other paraffins and olefines. North Sea gas is natural gas from holes drilled under the North Sea.

natural seasoning [tim.] The drying of timber by stacking it so that it is exposed to the air all round but sheltered from sun and rain, a process which usually takes years and is therefore costly. Since this method is older than *kiln* seasoning, it is sometimes preferred.

natural stone Stone which has been quarried and cut, not *cast stone*.

navvy pick A heavy double-pointed *pick*, or a pick with a point at one end and a *chisel* edge at the other, generally with a *helve* 90 cm (36 in.) long.

neat A description of *cement* or *plaster* mixed with water only, without sand or lime.

neat size [carp.] The dimensions of a piece after cutting and planing.

neat work The brickwork above the footings, that is the visible brickwork.

needle (1) or **needle beam** A short, horizontal, wood or steel beam which passes through a shored wall, carries the wall, and transfers its weight to *dead shores* (illus.).

(2) In *flying* or *raking shores* (illus.) a short (about 45 cm (18 in.) long) horizontal, *hardwood* or steel piece which passes through the

vertical *wall piece* and the wall. It thus holds the wall piece in place and forms an abutment for the sloping shores.

needle bath [plu.] (USA) A shower bath with jets which strike the bather from every direction, mainly horizontally.

needle scaffold A *scaffold* hung on *needles* driven into the wall.

Neoprene Trade name for an oil-resistant American synthetic rubber (like PVC), which has other excellent properties of non-inflammability and light-resistance. Puttyless glazing, with weathertight Neoprene gaskets enclosing the glass on both inside and outside edges, is now often used on large buildings, partly because of the speed of glazing. *See* **elastomer.**

nest of saws [joi.] Several saw blades which can be used at different times in the same handle.

network analysis *See* **PERT.**

neutralizing [pai.] Preparation of *concrete* or cement *mortar* or *plaster* surfaces for painting, so that the free *lime* of the ground does not attack the paint. *See* **alkali resistance.**

newel (1) or **newel post** [joi.] A post in a *flight* of stairs carrying the ends of *outer string* and *handrail* and supporting them at a corner.

(2) A stone column carrying the inner ends of the *treads* of a spiral stone staircase. *See* **solid newel stair, open newel.**

newel cap [joi.] A wooden top to a *newel* post.

newel drop [joi.] A downward decorative projection of a newel post through a *soffit*.

newel joint [joi.] A joint between *newel* and *string* or *handrail*.

nib (1) or **cog** A downward-projecting lug at the upper end of a tile which hooks over the tiling *batten*.

(2) The part of the top edge of a vertical sheet of asphalt which fits into a chase in the wall which it protects.

(3) [pai.] A small solid particle which projects above the surface of a film, usually of *varnish*. A film with nibs is called *bitty*.

nib guide [pla.] or **nib rule** A 5×2 cm ($2 \times \frac{3}{4}$ in.) *straight edge* nailed on the floating coat of a ceiling on which a *cornice* mould is to be run. It holds the upper end of the *horsed mould* in position as the *running rule* holds the lower edge.

nicker (1) [carp.] *See* **center bit.**

(2) A mason's broad *chisel* for grooving stone before splitting it.

nidged or **nigged ashlar** Stone, particularly granite, dressed roughly with a pointed hammer. The hardest granite can be dressed only in this way.

night lock [joi.] A *cylinder lock*.

night vent or **ventlight** or **vent sash** or **ventilator** [joi.] A small *opening light* with horizontal *hinges* at the top of a *casement window*.

nippers or **crampon** or **stone tongs** Two curved levers hinged together near their middle and lifted by a crane at their upper ends. The pull of the crane at the upper end draws the lower ends of the nippers together and they are thus forced to nip the block of stone to be lifted.

Nippers (or crampon or stone tongs).

nipple (1) [plu.] A short pipe threaded outside at each end with a *taper thread*, used for joining two *couplings* or internally threaded pipes.

(2) A small valve at the high points of a hot-water system by which air can be released to prevent air locks.

(3) [mech.] A small brass tube screwed into a machine part for injecting grease into it through a grease gun.

Nissen hut An arch-shaped hut built of corrugated steel sheet, extremely cheap but draughty.

nitrocellulose or **cellulose nitrate** or **guncotton** An important constituent of most modern *lacquers* used for lacquering metal, wood, or textiles. It is also used for making *plastic wood* and some *glues*.

no-fines concrete Concrete made without sand. It therefore contains a volume of large pores, and for this reason provides no *capillary* passage for water. It has been used in Britain for building house walls (Crawley new town) in roughly the same thickness as the *cavity wall* which it replaces. A typical no-fines aggregate consists of 95% between 19 and 9·5 mm ($\frac{3}{4}$ and $\frac{3}{8}$ in.) and 5% smaller than 9·5 mm ($\frac{3}{8}$ in.). Since it is not *nailable*, *chases* are formed during casting and a *foamed slag* nailing strip is plastered into the chase after the shuttering has been stripped. This consists of 1 part dry slaked lime, 2 cement, 2 fine foamed slag, 2 coarse foamed slag. No-fines concrete must not be vibrated or rammed, but lightly punned. It is usually not reinforced except for a few diagonal bars across the corners of openings. It must be rendered outside to strengthen and protect it. It is often regarded as a *lightweight concrete* especially when made of *lightweight aggregate*.

nog (1) A *fixing brick*.

(2) *Nogging*.

nogging (1) or **nogging piece** or **nog** [carp.] Horizontal short timbers which stiffen the *studs* (verticals) of a *framed partition*.

(2) *See* bricknogging.

noise absorption The noise level within a room can be reduced before it is built, by *discontinuous construction*, or after it is built by noise *absorption*, that is by surrounding it or filling it with absorbent (non-echoing) material.

noise insulation The prevention of sound transmission by *discontinuous construction* of walls, floors, and ceilings.

nominal size [tim.] The dimension of timber after sawing but before planing or otherwise working it. It is usually about 6 to 9 mm ($\frac{1}{4}$ to $\frac{3}{8}$ in.) larger that the final size after planing (*dressed size*).

non-bearing wall A wall which carries its own weight and wind load only, as opposed to a *loadbearing wall* (*C*).

non-combustible A term which is now preferred by the Fire Protection Association and other fire authorities to the old word *incombustible*, meaning that which does not burn.

non-flammable A term now used by fire authorities to mean material which will not burn with a flame. *See* **flame spread.**

non-hydraulic lime [pla.] *High-calcium lime.*

non-inflammable That which does not burn with a flame. The term now preferred by British and US fire authorities is *non-flammable*.

non-manipulative joint [plu.] A *compression joint* which requires no work on the pipe other than cutting the ends square, and is therefore more used, at least for the smaller diameters of pipe, than the manipulative joint. *See* **gland.**

Norfolk latch [joi.] A *thumb latch.*

normal roll pantiles *Pantiles* which have rolls of the same width from head to tail.

Norman brick (USA) A *brick* measuring $30 \times 10 \times 6.7$ cm ($12 \times 4 \times 2\frac{2}{3}$ in.) including the mortar *joints*. Three courses lay up 20 cm (8 in.) in height.

north-light roof or **saw-tooth roof** A sloping roof having one steep and one gentle slope. In the northern hemisphere the steeper slope is glazed and faces north. It is built facing south in the southern hemisphere.

nose (1) Any blunt overhang, a *nosing.*

(2) The lower end of the *shutting stile* of a door or *casement.*

nosing A half-round, overhanging edge to a stair tread, flat roof, window sill, etc., in concrete, stone, or timber. *See* **flight** (illus.).

nosing line A line touching the edges of the nosings of a stair. The *margin* of a *close string* is measured from it.

notch or **gain** [carp.] A groove in a timber to receive another timber.

notch board [carp.] (1) A *cut string.*

(2) (USA) A *close string.*

notching [carp.] Joining two timbers by cutting a part out of one or both. *See also C.*

novelty siding [tim.] *German siding.*

nylon One of the stronger plastics, though it has only one tenth the strength of steel and 100 times its thermal expansion, nylon is so wear-resistant, corrosion-resistant and cheap that it is now used for making such hardware as barrel bolts for cupboard doors, wood screws up to 6·5 cm (2½ in.) long and even gear wheels.

O

oak [tim.] (Quercus robur, Quercus petraea, etc.) Hardwoods which grow in temperate climates throughout the world. English oak is difficult to work because of its hardness and twisted grain, but it has for long in England been a valued structural timber which is very decorative when *quarter sawn*. It is not generally so stiff nor so strong as either beech or ash. All imported oak is more straight-grained than English oak, and therefore easier to work. Slavonian oak from Austria or Yugoslavia is used as *wainscot* oak. Japanese oak is paler than English oak, and also used as veneer. Of the many North-American oaks, the white oaks most resemble English oak. *See* **deathwatch beetle, workability.**

oak shingles *Shingles* made from oak heartwood, generally $30 \times 9 \times 1 \cdot 6$ mm ($12 \times 3\frac{1}{2} \times \frac{5}{8}$ in.) at the butt to 13 mm ($\frac{1}{2}$ in.) thick at the tip.

oakum Untwisted rope or hemp, tarred or oiled and used sometimes as a caulking in prefabricated buildings, between precast concrete pieces and so on.

oak varnish [pai.] An oil *varnish* with a high proportion of oil to resin, $1\frac{1}{2}$ to 1 or less, normally used indoors. Elastic oak varnish can be used out of doors.

oblique butt joint [carp.] A *butt joint* at an angle other than 90° to the length of the piece.

oblique grain [tim.] *Diagonal grain.*

obscured glass or **vision-proof glass** or **translucent glass** *Glass* through which light can pass, although things cannot be seen through it. It is sand-blasted or moulded to make the surfaces irregular.

ochre [pai.] Hydrated iron oxide used as a *pigment.* It is yellow or yellowish brown, paler than *umber* or *sienna*, and may contain a little clay mixed with it.

offset (1) A ledge in a wall where the wall thickness changes.
    (2) A *swan neck.*

offset screwdriver [carp.] A screwdriver which turns screws at right-angles to its length.

offshoot A *water table.*

off-white [pai.] A *colour* which is not white but has not enough of one colour in it to be given the name of that colour.

ogee joint A type of *spigot-and-socket* joint in pipes.

oil-bound distemper [pai.] *Distemper* which contains *drying oil* in the *medium.* It is more often called oil-bound water paint. *See* **emulsion paint.**

oil-fired central heating [plu.] The most expensive and probably the most luxurious *central heating. See* **flue lining.**

oil gloss paint Paint made with *boiled oil* and some *raw linseed oil.* It is suitable for interior use. *Compare* **hard gloss paint.**

oil length [pai.] The ratio of oil to *resin* in a varnish *medium*. *See* **long oil, short oil.**

oil paint Paint with a binder of *drying oil* or oil *varnish* mixed with *thinner*, as opposed to *water paint*.

oil paste or **colours in oil** [pai.] A paste-like, concentrated mixture of *pigment* and oil. It can be used for *tinting* or for making paint by adding oil or *thinners* and *driers*.

oil slip [joi.] An *oilstone slip*.

oil stain [pai.] A thin *oil paint*, with very little *pigment*, used for staining wood floors. *Compare* **spirit stain, water stain.**

oilstone [joi.] A *hone*.

oilstone slip or **slip stone** or **gouge slip** [joi.] A small *hone* with a curved edge or edges for sharpening *gouges* and concave *cutting irons* of *planes*.

old woman's tooth [joi.] The original form of the plane now called a *plough*. It consists of a *chisel* like a *cutting iron* in a block of beech-wood, held by a wedge.

oleo-resin [pai.] A pine gum obtained either by distilling dead wood or by bleeding the living tree. *See* **resins, turpentine.**

oleo-resinous varnish [pai.] A *varnish* containing vegetable *drying oil* and hardening *resin*, natural or synthetic. *See* **gold size.**

oncosts *See* **overheads.**

one-pipe system (1) In house drainage, a soil and waste system that uses one pipe for waste and soil water together. The system may also need *ventilation pipes*. *See* **two-pipe, single-stack.**

(2) A heating circuit in which all the flow and return connections to the radiators come from the same pipe. The furthest radiator is therefore much cooler than the radiator which is nearest the boiler. *See* **two-pipe system.**

opacity or **opaqueness** [pai.] The *hiding power* of a paint, the opposite of transparency.

open assembly time [tim.] The time which elapses between the application of the *glue* to a *veneer* or joint and the assembly of the veneers or parts of a joint.

open cornice or (Britain) **open eaves** [carp.] In American timber house construction, an eaves overhang in which the rafter *soffits* and usually also the slates or roof *sheathing* can be seen. *Compare* **box cornice.**

open defect [tim.] Any hole or gap in timber or ply, such as *checks*, splits, knothole, wormhole, or open joints.

open floor [carp.] A floor, in which the *joists* are exposed beneath, since it is not ceiled.

open grain [tim.] *Coarse-textured.*

open-grained [tim.] *Wide-ringed.*

opening leaf [joi.] A leaf of a *folding door* which opens, as opposed to a *standing leaf*.

opening light [joi.] A light which opens, as opposed to a *dead light*.

open mortise or **slot mortise** or **slip mortise** [joi.] A *mortise* open on three edges. It is not a true mortise but is much used for table legs and can be cut with a circular saw. *See* **forked tenon**.

open-newel stair A *geometrical stair* (one without newels).

open planning [t.p.] Planning of tall buildings without *air shafts*. It may also mean the designing of a house with few fixed *partitions*.

open roof A roof in which the *principals* can be seen from below, since it has no ceiling.

open slating or **spaced slating** *Slates* laid with a gap between those in the same *course*. The gap should not be more than the slate width minus 20 cm (8 in.) (this allows for 10 cm (4 in.) side lap). This roof is sometimes used for cowsheds, etc. The same technique can be used with tiles.

open stair (USA) A *stair* which is open on one or both sides. *Compare* **closed stair**.

open string [joi.] A *cut string*.

open tendering [q.s.] *See* **tendering**.

open valley A *valley* in which the *slates*, *tiles*, or *shingles* are so cut and laid that the lead sheet or other waterproof material under them in the *valley* is exposed. *Compare* **secret gutter**.

open vented system [plu.] The usual water heating circuit, with an *expansion pipe* open at the top, not a *sealed system*.

open-well stair A *stair* with a generous *well*.

open wiring [elec.] Electrical wiring held on porcelain insulators away from the wall.

orange peeling [pai.] Small circular craters in a *finish* applied by a spray gun; a *flow* failure caused by incorrect air pressure or badly chosen *solvents* or the use of a *lacquer* which is cooler than the room temperature. *See* **pinholing**.

ordinary Portland cement A *hydraulic cement* made by heating to clinker in a kiln a slurry of clay and limestone. This is the cheapest and most widely used *cement* in building in Britain. *See also* **C**.

oriel window An upper-storey, overhanging window. Unlike the *bow window*, it is carried on corbels.

O-ring joint [plu.] A joint widely used between plastics pipes, in which watertightness is ensured by a rubber ring between the spigot and the socket. It has the advantage that it is not a completely rigid joint.

orpiment [pai.] Arsenic sulphide (As_2S_3), a lemon-yellow mineral *pigment*.

outband A stretcher stone visible in a *reveal*.

outer lining [joi.] An *outside lining*.

outer string or (USA) **face string** [joi.] The *string* of a wooden *stair* furthest from a wall, as opposed to the *wall string*.

outlet (1) A vent, in particular an opening in a *parapet wall* through which rainwater discharges into a rainwater hopper.

(2) [elec.] A *socket outlet.*

out-of-wind Description of a wood or masonry surface which is plane, at right angles to the neighbouring surfaces, and therefore not winding.

outrigger A beam projecting from a building and wedged against a ceiling within it. It carries a *fan* or *flying scaffold. See also C.*

outside casing or **outside facing** [joi.] An *outside lining.*

outside glazing *External glazing* placed from outside the frame.

outside lining [joi.] or (USA) **outside casing** The boards forming the outside of a *cased frame.*

out-to-out Description of an *overall* measurement.

oval-wire brad A *wire nail* formed from oval wire in lengths from 19 mm ($\frac{3}{4}$ in.) to 10 cm (4 in.). It is supplanting *wrought* or *cut nails.* Not to be confused with *brads* for flooring.

oven-dry timber Timber which loses no moisture in a ventilated oven at $103° \pm 2°$ C. (BS 565).

overall An adjective describing wall measurements made from outside to outside of walls.

overcloak In *flexible-metal roofing,* that part of the upper sheet which laps over the lower at a *drip* or *roll* or *seam. See* **undercloak.**

overgrainer [pai.] A brush like a *mottler,* but thinner and longer, used also for *graining* and *marbling.*

overhand work *Facing bricks* laid from inside a building by men standing on a floor or on a *scaffold.*

overhang A projection of a roof, floor, or other horizontal part beyond the wall which carries it.

overhanging eaves *Rafters,* tiles, slates, etc. projecting over the *wall plate* and clear of the wall.

overhead door [joi.] A door which opens by being lifted up and slipped into a horizontal position at the door *head.* It may be in one or two leaves.

overheads or **overhead expenses** or **establishment charges** [q.s.] The costs of electric light, roads, supervision, accounting, director's fees, etc., which cannot fairly be charged to one job and must therefore be distributed over all the *items* on a *contract.*

oversailing course A brick or stone *string course* or corbelling.

oversite concrete A layer of about 15 cm (6 in.) of concrete under the ground floor of a house, whether this is of wood or of other flooring material.

over-tile The imbrex of *Italian* or *Spanish tiling.*

overtime Payment for time worked above the normal weekly number of hours at a fraction over the normal hourly rate. Sundays are paid at twice the normal rate, *double time. See also* **time-and-a-half, time-and-a-quarter.**

oxidation [pai.] The hardening of *drying oils* in air by their absorption of oxygen to form a durable *film.*

oxter piece [carp.] An ashlar piece, the vertical in *ashlering*.

oxy-acetylene flame [plu.] A flame obtained at a gas jet supplied from large steel cylinders of oxygen and of acetylene, through high-pressure hoses. The most powerful flame in common use, it is hot enough to weld steel or to cut it by burning, or to make brazed joints or for *lead burning*. For most plumbing purposes the modern *propane blowlamp* is adequate, needs no hoses, and is much safer and more portable.

oxychloride cement *See* **magnesite flooring.**

P

package deal An arrangement whereby a *contractor* agrees with his *client* to take full responsibility for design and construction. This eliminates some of the delays involved in co-operation with consultants but it increases the risk to the client of unsuitable construction. Package deals, only conceivable with a sophisticated, conscientious contractor, are less secure for the client than a *management contract*.

packing Small stones used for filling the gaps in *rubble walls*. *See also C*.

pad (1) or **padstone** or **template** A stone or precast concrete block placed under a heavy load such as the end of a girder to spread the load in a *loadbearing wall* (*C*).

   (2) A *tool pad*.

paddle mixer or (USA) **twin pug** A *mortar* mixer with two horizontal shafts rotating in opposite directions. *See* **pug mill**.

padsaw A *saw* blade which fits into a *tool pad*.

padstone A stone or concrete *pad* in a wall.

paint A liquid applied to building materials in *coats* which dry hard within about forty-eight hours. It protects them from corrosion, and is usually decorative. Paint consists of *pigment* and *medium*. *See below*; also **dispersion** (*C*), **emulsion**, **fire-resisting finishes**, **oil paint**, **plastic paint**, also BSCP 231, Painting of Buildings.

painter *See* **house painter**.

painter's labourer A skilled labourer who helps a *house painter* by preparing surfaces, stripping old wallpaper, washing ceilings, etc. He may apply the first coat of *whitewash* or *distemper*, or creosote fences or rough timber, in which case he is called a brush hand.

painter's putty *Glazier's putty*, used as a *filler*.

paint harling Throwing paint-coated stone chips on to a sticky paint film to make a rough surface.

painting The use of *paints*, *varnishes*, and *stains* for protection or decoration of materials. *See* **house painter** (illus.).

paint kettle A cylindrical steel pail used by painters for holding paint and dipping the brush into.

paint remover A liquid which softens a *paint* or *varnish* film, so that it can be easily scraped or brushed off. They may be of two types: (1) Those containing soda which may damage the surface and always injure the hands. (2) Those containing organic *solvents* which cannot damage either the surface or the hands. These can easily be distinguished from soda-based liquids, as they do not mix with water.

paint system A succession of *coats* designed to protect a surface and give a decorative finish. The first coat is a *sealer* on wood or plaster, or a *priming coat* on corrodible metal. The next is usually an *under-*

coat with good *hiding power*, which is followed by a high *gloss* coat and sometimes by a *varnish*.

pair Two oppositely *handed* but otherwise similar objects.

pale or **paling** An upright board or narrow stake in a *palisade*.

pale boiled oil [pai.] A *linseed oil* through which a little air has been blown at about 150° C. It contains a small amount of *driers* and is used for making oil and *hard gloss paints*.

palette board or **palette** [pai.] A small thin board with a handle or a hole through it for the thumb. It is used for mixing colours, mainly by artists.

palette knife [pai.] A knife with a long, narrow, symmetrical, springy blade rounded at the point. It is used for handling colours on a palette board. *See* **house painter** (illus.).

palisade An enclosure of pointed wooden or metal uprights (*pales*).

pallet or **pallet slip** A *fixing fillet*. *See also* C.

pallet brick A brick rebated at one edge to receive a *fixing fillet*.

pan-and-roll roofing tiles (USA) *Italian tiling*.

pan breeze The mixture of small coke (coke breeze) and furnace clinker from the pan beneath a furnace burning coke breeze. It was used as an aggregate for making concrete blocks (*breeze blocks*).

pane (1) or **square** A sheet of glass cut to size to fill part of a *light* between *glazing bars*.

(2) A panel.

(3) [mech.] The *peen* of a hammer.

panel (1) [joi.] An infilling of glass, wood, or other material let into grooves or rebates in *panelled framing*, leaving the panel free to move relatively to the *frame*.

(2) The brick infilling to the steel or concrete frame of a structure.

(3) A single span or *bay* of a continuous concrete slab.

panel box or **panel board** [elec.] A *cutout box*.

panel heating [elec.] Electrical resistance heating by flat panels with a surface temperature of about 38° C. (100° F.) flush with the wall, or by panels about 5 cm (2 in.) clear of the wall at a higher temperature. The term has been used for coils of hot-water pipes hidden in walls or floors, but these are now usually called concealed heating or *coil heating*. *See also* **electric panel heater**.

panelled door [joi.] A door built of a framed surround with the spaces between the framing members (*stiles*, *rails*, *muntins*) filled with *panels* of thinner material. These were the best doors in existence before *plywood* was made, since they are so built as to have no excessive *moisture movement* (C). However, plywood flush doors, even of the cheapest sort, have less moisture movement than *panelled framing*.

panelled framing [joi.] Framing consisting of *stiles* (vertical) with *rails* (horizontal) tenoned into them and, in wide frames, *muntins* (vertical) tenoned into the rails. The frames contain panels.

panel mould [pla.] A mould in which plaster panels are cast.

panel pin [joi.] A round wire *nail* between 1 and 1·6 mm thick and from 1 to 5 cm ($\frac{1}{2}$ to 2 in.) long. It is very slender, has a small head, and is used for *joinery*. It can be nearly invisible when driven below the surface.

panel saw [joi.] A *cross-cut saw* about 56 cm (22 in.) long with closely set teeth. It has about 4 points per cm (10 per in.).

panel wall Brickwork or other walling in *skeleton construction*. It is restrained by the building frame, generally on at least three of its four edges, carries only its own weight, and is carried at each floor by the building frame.

panic bolt [joi.] A door bolt often used at the double exit doors of theatres. It is opened by pressure from inside on to a horizontal bar within the door at waist height.

pan mixer A machine for mixing *mortar*.

pantile A *single-lap tile* shaped like an S laid horizontally – thus ∽. The British standard size is about $34 \times 24 \times 1$ cm ($13\frac{1}{2} \times 9\frac{1}{2} \times 1$ in.). Each tile overlaps its neighbour on the right (looking at the roof from the ground in front of the house) and is overlapped by its neighbour on the left by about 5 cm (2 in.). About 205 tiles are needed per 10 m² (107·6 ft²) and the weight of the pantiles alone is about 49 kg/m² (10 lb/ft²). Pantiles generally have a nail hole and a *nib* at the head. Larger, heavier pantiles exist, up to 42×33 cm but they weigh less per 10 m² than the small ones.

pap The vertical outlet from an *eaves gutter* downwards.

paperhanger A *tradesman* who hangs paper and prepares surfaces for it by stopping cracks and sizing. He usually strips old paper. *See* house painter and decorator.

parallel coping A coping, not *weathered* but of uniform thickness, for covering a sloping surface such as a *gable*.

parallel gutter or **parallel parapet gutter** A *box gutter*.

parallel thread [mech.] A *screw thread* of uniform diameter used on mechanical connections such as *bolts* (*C*) but not in pipe fitting except on *connectors*. *Compare* **taper thread**.

parapet A low wall guarding the edge of a roof, bridge, balcony, etc; that part of a house wall which passes above the roof. Since it is exposed on its face, back and top to the weather, it needs detailing with more care than other walls.

parapet gutter *See* box gutter.

parapet wall A *parapet*.

parenchyma [tim.] The wood tissue of the *medullary rays*, seen in the *silver grain* of oak, also the main part of the *heart centre*, soft and weak.

parge [pla.] The mixture used in *pargetting* (1) either cement mortar or *coarse stuff* with *hair* and cowdung.

pargetting or **parging** or **pargeting** [pla.] (1) Rendering the inside of a brick *flue* with *parge*, superseded by modern *flue blocks* and *linings*.

(2) Decorative plastering to the outside walls of Elizabethan houses with lime plaster. Repetitive patterns, sometimes very beautiful, were modelled in the plaster before it hardened. Examples can be seen in Suffolk and Essex.

(3) (USA) Rendering the inner face of an outer leaf of *cavity* brickwork; sometimes called back mortaring.

paring chisel [joi.] A long, thin, bevel-edged *chisel* which is never struck with the *mallet*.

paring gouge [joi.] A thin, long *gouge* sharpened on the inside.

Parkerizing *Phosphating*.

parliament hinge [joi.] An *H-hinge*.

parpend A *perpend stone*.

parquet floor A wooden floor covering of *hardwood* blocks in geometrical patterns glued to the floor and polished. It can now be obtained fixed to *plywood* so that large areas can be quickly laid (plywood parquet).

parquet-floor layer or **wood-block floor layer** or **hardwood-strip flooring layer.** A *tradesman* who lays prepared *parquet floor* on a wood *subfloor* by gluing or pinning. As a wood-block floor layer he lays *hardwood* blocks in adhesive on concrete. As a hardwood strip-flooring layer he lays prepared *parquet strip* on a wood sub-floor covered by three-ply. He works to a drawing if need be, finishes the floor with a hand scraper or *sanding machine*, and sometimes stains and polishes it. *See* **wood-block paving.**

parquetry The inlaid patterns of a *parquet floor*.

parquet strip or **strip** or **overlay flooring** [tim.] A floor consisting of tongued and grooved *hardwood* boards 1 cm ($\frac{3}{8}$ in.) actual thickness which are *secret nailed* and glued to a wooden *sub-floor*. *See* **strip.**

particle board [tim.] *See* **chipboard.**

parting bead or **parting strip** [joi.] A narrow vertical strip of wood fixed to the *pulley stiles* of *cased frames* of *sash windows* to separate the upper sash from the lower sash when they are being opened or closed and therefore sliding past each other.

parting slip or **mid-feather** or **wagtail** or **pendulum** [joi.] A long narrow vertical slip of wood, which hangs from the pulley level to the bottom of the *cased frame* of a *sash window*, and keeps the sash weights from colliding when the window is being opened or closed.

parting tool or **vee tool** [joi.] A V-shaped *gouge* used by wood carvers and wood turners.

parting wall or **party wall** A wall between two buildings and used by them in common.

partition A wall separating adjacent rooms, generally non-loadbearing and only one *storey* high. Partitions may be *framed partitions* or built up like a brick wall from *building blocks* or bricks or *gypsum plaster* or glass. *Compressed straw* or *woodwool* slabs, built like *brick*, are made in sizes about 2·4 × 0·6 m (8 × 2 ft) usually 5 cm

(2 in.) thick. Slabs of *corkboard* can be built like brick but are more usually fixed to a partition as an additional insulating cover. Metal removable partitions are used in office buildings. *See* BSCP 122.

partition coverings *See* **coverings.**

partition infillings *See* **infillings.**

partition plate or **partition cap** or **partition head** [carp.] The uppermost timber of a *trussed partition*, on which the *joists* rest.

party fence or **wall** or **parting fence or wall** or (USA) **common wall** A fence or wall separating two properties and shared by them.

passings or **laps** [plu.] The distance by which one lead sheet overlaps the next in *flashings*, ridge coverings, *gutters*, etc.

paste drier or **patent drier** [pai.] A stiff paste consisting of *drier* mixed with *drying oil* and an *extender*. *Soluble driers* are now more popular.

paste filler [pai.] A *filler*.

patch or **shim** [tim.] An *insert* of veneer in *plywood*.

patent glazing Any system of dry (puttyless) *glazing*. The *glazing bars* usually span 2·1 m (7 ft), but spans of 3·3 m (11 ft) are known. They may be made of bare bronze or aluminium, steel covered with a *lead sheath*, or concrete, galvanized steel, or wood. The last three are protected by *cappings* of zinc, copper, lead or plastics. The glass is bedded on a *cushion* of oiled asbestos, etc. Patent glazing is used in *monitors*, *north lights* and other industrial roofing, whether vertical or horizontal. It can be double glazed. *See also* **condensation groove, draught fillet, shoe, storm clip,** BSCP 145, Patent Glazing.

patent plaster [pla.] *Hard plaster* like *Keene's cement*, based on *gypsum* with an *admixture*.

patent plate *See* **plate glass.**

patent-roofing glazier A *glazier* who fits glass in roof lights with wood or metal frames, fixes assembled *roof lights*, and makes them weatherproof. He is often specialized either in assembling the frame or in the fixing of it in the roof or in the *glazing*. *See* **patent glazing.**

patent stone *See* **cast stone.**

patina A thin, stable, protective film of oxide which forms on metals exposed to air, particularly the green coating on copper or its alloys (verdigris) which usually takes many years to form, but has recently been made artificially by chemical means in much less time.

pattern staining [pla.] The discoloration of plasterwork caused by the different conductances of backings. Where *gypsum plasterboard* is fixed over steel *joists*, the part in contact with the joists becomes darker than the part with no backing. This part has a bigger temperature difference compared to the air in the room. The air therefore circulates over it more freely and drops more dust on to it.

pavement light or **vault light** A *light* formed of solid glass blocks cast into concrete or set in a cast-iron frame over a basement so as to let in daylight.

pavement prism A *glass block* fitted in a *pavement light*.

pavilion roof or **polygonal roof** or **pyramid roof** A sloping roof with equal hipped areas all round it.

peacock's-eye veneer [tim.] *Veneer* figured like *birdseye*.

pean [mech.] The *peen* of a *hammer*.

pearl essence [pai.] Mother of pearl paint made by dissolving fish scales in *lacquer*. Sometimes solid fish scales are laid on by a lacquer.

pearlite *See* **perlite**.

pebble dash An external plaster which has been surfaced with clean pebbles or chips thrown on to the second coat of plaster while plastic. *Compare* **rough cast**.

pebble walling A wall built of rounded pebbles. It can be as beautiful as *flint walling* or more so, and is built in the same way.

pecking *See* **picking**.

pecky timber Wood showing signs, occasionally decorative, of *decay*.

peeler log [tim.] A log chosen for *rotary cutting*, particularly a *Douglas fir* log.

peeling (1) [tim.] *Rotary cutting* of veneers.

(2) The dislodgement of plaster or paint from its backing.

peen [mech.] or **pein** or **pean** or **pene** or **pane**, etc. The blunt, wedge- or ball-shaped end of a *hammer* head, opposite the striking *face*. *See* **ball peen**, **cross peen**, **straight peen**.

peen hammer A *mason*'s hammer with no flat striking face but two cutting *peens*.

peg (1) An *oak* or galvanized steel rod 6 mm ($\frac{1}{4}$ in.) dia. passed through a *roofing tile* to hold it in place.

(2) [joi.] A wood *dowel*.

(3) [joi.] A metal pin which secures glass to a metal window frame.

peggies Small random *slates* from 25 to 35 cm (10 to 14 in.) long, sold by weight.

peg stay A casement stay which holds a window in place by a peg through one of its holes.

pein [mech.] *See* **peen**.

pellet [joi.] A small circular piece of wood, cut to match the grain of wood into which it is inlaid to cover the head of a countersunk screw.

pelmet [joi.] A built-in head to a window for hiding the curtain rail, blind fittings, and so on. It may also be a very short curtain which has the same purpose.

pencilling Painting mortar joints of brickwork with white paint to bring out the contrast between the joints and the brickwork.

pendulum saw or **swing saw** or **swinging crosscut** [tim.] A *circular saw* which hangs on a frame pivoted at ceiling level and is brought down to a log to *cross-cut* it.

penny system *See* **nails**.

penthouse (1) A small house or *apartment* built on the flat roof of a building with walking space round it. *See* **bulkhead**, **penthouse roof**.

(2) A projecting hood over a window or door or wall to protect it from rain, sometimes spelt pentice.

penthouse roof or **pen roof** A roof sloping in one direction only. It differs from a *lean-to roof* in that it covers the higher wall.

perfections *Western red cedar* shingles 46 cm long and 1·4 cm (18 in. and $\frac{9}{16}$ in.) thick at the *butt*.

perforated brick A *brick* with vertical perforations through the *frog*. They reduce the weight, allow the brick to dry quickly, and thus lower the cost of burning. They are much used on the Continent of Europe particularly in Germany. *See* **V-brick.**

perforated gypsum plasterboard A *gypsum plasterboard* which has uniformly spaced circular holes to give it a high sound *absorption*.

pergeting *Pargetting.*

perimeter diffuser In *warm-air heating*, a diffuser placed at the edge of a room, for example under a window, to neutralize its cold effect.

perlite or **pearlite** A volcanic glass found in USA. When heated, it expands to lightweight, glassy, spherical particles about the size of small peas. It is much used in USA as an insulating *aggregate* for precast concrete and is believed to absorb less water than exfoliated *vermiculite. See* **mineral wool.**

perlite plaster [pla.] Gypsum plaster which contains only *perlite* aggregate, and no sand. It is a good insulator, and is easy for the plasterer and his labourer to work with because it is light in weight.

perpends (1) or **face joints** The visible part of *cross joints* in brickwork or masonry. *See* **keeping the perpends.**

(2) The corners of a brick wall erected first and carefully plumbed to serve as a guide for the wall between.

(3) The sloping (so-called vertical), joints between adjacent slates or tiles.

(4) (USA) *Perpend stones.*

perpend stone or **parpend** A *bond stone* which passes through a wall and is seen on both faces.

Perspex or **Plexiglas** Transparent acrylic resin which can be obtained tinted or colourless, in thicknesses from 1 to 25 mm ($\frac{1}{32}$ to 1 in.), flat or corrugated, domed or moulded. It is stronger and lighter than glass, admits more ultra-violet light and ordinary light, and does not splinter into sharp pieces. It can be (and is) made to the shape of slates, tiles, or asbestos-cement sheets and used with them to light an attic. Perspex sheets can be joined together by gluing or by cementing, that is by coating the edges with an organic solvent and pressing them together. Tinted Perspex is used for interior panelling or *partitions* to give a tinted *borrowed light*. Perspex is inferior to glass in fire resistance since it is *thermoplastic* and is not allowed for *glazing* in London for *fire resisting* purposes.

PERT (**project evaluation and review technique**) or **project network analysis** or **network analysis** A general form of *critical path* scheduling in which any of the various limited resources can be chosen for analysis. In critical path the main emphasis is on determining the shortest time to completion, and expenditure of other resources is secondary to expenditure of time.

pet cock [mech.] A small *drain cock* or small valve which is opened to release air from the upper part of a water pipe.

petrifying liquid A thin preliminary coat applied to make a concrete or masonry or plaster surface suitable to receive paint. It is of different materials according to the type of paint to be applied, and may be a *sealer* (1) like *clearcole*, or a thinner for *distemper*.

Petrograd standard or **Petersburg standard** [tim.] The unit of timber measurement formerly used in Britain, 165 cu. ft (4·67 m³).

petrol intercepting chamber or **petrol trap** A *trap* into which the waste from petrol (gasoline) filling stations or washdown yards for cars flows before entering the *sewer* (*C*). It consists of three separate chambers with scumboards across the top to keep the floating petrol from flowing out. The chambers are ventilated so that the *volatile* and explosive part of the petrol evaporates and is removed. *Compare* **intercepting trap**.

phenol formaldehyde resin [tim.] A *synthetic resin* which forms a *glue* more moisture-resistant than any known glue and which is even boil-proof. It is also immune to attack by bacteria. It hardens with heat (i.e. is thermo-setting), so that *plywood* must be glued with it in a *hot press*. No technique of *separate application* is yet possible. Phenol glues began to be used commercially about 1930. *See* **film glue**.

Phillips recessed-head screw [joi.] *See* **recessed head screw**.

phon The unit of loudness of sound. Painfully loud sound is at 130 phons, a pneumatic drill at 10 ft 90 phons, soft speech 40 phons, a very quiet room 30 phons. *See* **decibel**.

phosphating Protection of a metal surface by hot phosphoric acid. Like *pickling* (*C*) it is a pretreatment and *inhibiting* coat rather than a finished surface. Phosphating should therefore be followed by surfacing with oil, wax, paint, or *lacquer*, and many proprietary processes make use of it, including Coslettizing, Bonderizing, Jenolizing, Parkerizing, and Walterizing. Phosphating can now also be done on the building site.

phosphorescent paint A *paint* which emits visible light for some minutes or hours after visible or ultra-violet light has fallen on it. There are several types of phosphorescence, shown respectively by zinc sulphide, strontium sulphide, and calcium sulphide. *Fluorescent* material emits visible light immediately, not after the radiation has fallen on it.

pick A digging tool like an axe, but with two long sharp points, used for breaking loose rock or digging stiff clay or gravel. *See* **navvy pick**.

pick and dip or **New England method** (USA) Bricklaying by picking up a *brick* with one hand and simultaneously lifting enough *mortar* on the trowel with the other hand to butter the end of the brick. This is also a common method of bricklaying in England, but has no specific name.

pick axe [c.e.] A *navvy pick*.

picket A *pale* in a fence or driven into the ground.

pick hammer A *slater*'s tool for holing *slates* and drawing or driving *nails*. *See* **plasterer** (illus.).

picking or **wasting** or **stugging** (or **clouring** in Scotland) Surfacing a stone in *rubble walling* with a steel point struck at right angles to the surface to make many small, closely spaced pits.

picking up [pai.] (1) *Pulling up*.

(2) Joining *live edges*.

pieced timber A timber from which a damaged part has been cut out in a *dovetailed* shape. A good piece of timber which matches the grain is cut and fitted to the dovetail.

piece moulding [pla.] A moulding process used for fibrous plaster when a single reverse mould is not possible. The mould is made in several pieces held together by a case made of fibrous plaster.

piecing or **dry piecing** [pla.] A line where plastering finished one day and was started again next day or later. Piecings are often a disfigurement but they can be made less conspicuous by locating them carefully, for example at a corner or behind a rainwater pipe. If the whole of a wall can be plastered in a day, there will be no piecing.

piend Scots for a *hip*.

pier (1) The loadbearing brickwork in a building between openings.

(2) Short *buttresses* (*C*) on one or both sides of a wall, bonded to it to increase its stability. *See* **pilaster**.

pigeon-holed wall A *honeycomb wall*.

pig lug (Scotland) In *flexible-metal roofing* a *dog ear*.

pigment [pai.] An insoluble, finely ground, usually opaque powder. Its solubility in water is generally below $\frac{1}{2}\%$, and its fineness is such that only $\frac{1}{2}\%$ is allowed to be larger than 0·06 mm for most British standard pigments. This fineness makes a *paint* with high *spreading rate* and smoothness. Pigments are generally ground in a *vehicle*, often oil. Most pigments are either minerals or *lakes*. White pigments are not considered to have *colour*, black pigments are. Pigments are therefore classified as white or coloured. *See* **extender, stainer**.

pilaster A rectangular *pier*, sometimes fluted, projecting from the face of a wall, having a cap, shaft, and base. It buttresses the wall.

pile *See C*.

piling (1) [pai.] The behaviour of a very quick drying *paint* which during application by *brush* becomes so sticky that the resulting *film* is thick and uneven (BS 2015).

(2) *See also* **pile** (*C*).

pillar A *column* or *pier*, the term is ordinarily now confined to stone and cast iron.

pillar tap [plu.] A water tap used on washbasins and baths, having a long vertical screw thread at the bottom, passing through the edge of the basin or bath, and thus connecting to the supply pipe.

pilot hole [carp.] A guiding hole, usually of smaller diameter than the main hole.

pilotis A reinforced concrete column projecting through an open ground-floor space to carry the structure above.

pilot light (1) A small flame on gas appliances which is left always burning so as to ignite the main gas burners immediately heat is required.

(2) [elec.] *See* (*C*).

pilot nail [carp.] A temporary *nail*, driven to hold timber while the main nails are being driven.

pin (1) [joi.] A slender *wire nail* such as a *panel pin*.

(2) [carp.] A *treenail* or *dowel*.

(3) [joi.] The *dovetail* tenon inserted into a dovetail joint.

(4) To wedge brickwork up to a wall or floor above it with slates bedded in *mortar* (*underpin* (*C*)).

(5) A *pintle*.

pinch bar or **jemmy** or **claw bar** or **wrecking bar** or **case opener** A bent piece of hexagonal or round steel 13 to 19 mm ($\frac{1}{2}$ to $\frac{3}{4}$ in.) dia., about 45 cm (18 in.) long, with a claw bent to a U at one end and a rough *chisel* point at the other. When longer than 90 cm (36 in.) it is called a *crowbar* (*C*) and the U-shaped claw becomes straight.

pinch rod A rod used like a *storey rod* for checking the width of a gap such as the dimension between floor slabs, or a door or window opening.

pine [tim.] (*Pinus*) Many *softwoods* are called pines but the only one mentioned in this book is *redwood*.

pine oil [pai.] A strong *solvent* made from the *oleo-resin* of pine trees or synthetically. It is an *anti-skinning agent* and has been used to give good *flow* properties to paints.

pine shingles European *shingles* made from pinewood.

pin hinge [joi.] A *loose-pin butt*.

pinhole [tim.] *Worm hole* not larger than 1·5 mm ($\frac{1}{16}$ in.), without bore dust and usually dark stained. Pinhole borers do severe damage but only to green timber.

pinholing [pai.] Tiny holes in a *varnish* film, usually sprayed. It may be caused by moisture or other foreign matter in the spray tubes, insufficient breaking up of the varnish in the *spray gun*, too thick application, or too low a pressure. *See* **orange peeling**.

pink primer [pai.] A *priming coat* for wood, originally containing mainly white and red lead *pigments* but now of very vague composition.

pin knot [tim.] In Britain a *knot* smaller than 6 mm dia. ($\frac{1}{4}$ in.). In USA it may be a knot smaller than $\frac{1}{2}$ in. (13 mm) and called a *cat eye*.

pinnings (Scotland) Stones of different colour or texture set in a *rubble wall* to give a chequered effect.

pinoleum A material of which sun blinds are made. It consists of narrow wood slats, bound with strips of strong canvas into a continuous sheet. It admits a little light and air.

pin rail [joi.] A wooden rail carrying wood pegs on which coats can be hung.

pintle [joi.] The pin of a hinge for a door, lock gate, etc. If fixed to the gate it projects downwards. If fixed to the post it projects upwards.

pipe Pipes for most purposes may be of plastics, copper, cast iron, steel, or asbestos cement. Water supply pipes are often of galvanized *dead-mild steel* (*C*) (called wrought iron because they used to be made of wrought iron). Plastics pipe (polythene or rigid P V C) is also used for cold water, but not yet for hot water supply or circulation because it expands about ten times as much as metal for the same temperature range, and is therefore liable to buckle. It is used for waste pipes because the lengths are usually short and one end is free, allowing expansion. *See* **drain pipes**.

pipe cutter [plu.] A tool for cutting steel pipes. It carries hard-metal cutting discs which bite into the metal as the tool is twisted round the pipe. Some of the metal is forced inwards as the pipe is cut, reducing the pipe dia. The pipe bore is recovered with a *burring reamer*.

pipe drill A tubular *plugging chisel* which cuts round holes in brickwork for wood *plugs*.

pipe duct A *duct*.

pipe fitter [plu.] A *tradesman* who instals pipes for water, steam, gas, oil, or chemical plant. The pipe may be screwed, flanged, loose-flanged, Victaulic, or gas- or arc-welded and the fitter may specialize with any of these sorts of pipes or others.

pipe fittings [plu.] *See* **fittings, copper fittings**.

pipe hook A spiked *fastener* driven into a wall *joint* or a timber. It has a curved end for holding a pipe.

pipe-jointing clip [plu.] An asbestos and metal ring that envelops a pipe joint which is to be filled with molten lead.

pipe layer or **drain layer** A *skilled man* who joints, in the trench, pipes of glazed stoneware, concrete, iron, steel, or asbestos cement, laying them to correct levels. A service layer can also cut threads in the ends of metal pipes.

pipe sleeve *See* **expansion sleeve**.

pipe stopper A *screw plug* for drains.

pipe tongs [plu.] *Footprints*.

pipe wrench or **cylinder wrench** or **Stillson** [plu.] A heavy wrench with serrated jaws for gripping, screwing, or unscrewing steel pipe. *See* **plumber's tools** (illus.).

pisé de terre Walling made of *cob*.

pitch (1) The ratio of the height to the span of a *stair* or roof, or its angle of inclination to the horizontal.

(2) The distance between parallel objects at uniform spacing, such

as nails in wood, reinforcing bars in concrete, rivets in steel, the threads of a *screw*, etc.

(3) [carp.] The distance from one *nosing* to the next of a stair. *See* pitch board.

(4) or **rake** [carp.] The slope of the *face* of a *saw-tooth* measured from the perpendicular to the line of the points.

pitch board or **step mould** or **gauge board** [carp.] A triangular *templet* for a stair, cut with one side equal to the rise, the second to the *going*, and the third, the hypotenuse, equal to the distance from one *nosing* to the next (pitch). It is used for setting out the lines to which the *strings* should be cut or *housed*. A *margin templet* gives the correct distance from the top of the string to the *nosing line*. Additional templets for treads and risers are used for marking out their thicknesses on the string.

pitched roof The commonest type of roof, usually one with two slopes at more than 20° to the horizontal, meeting in a central *ridge*. *See* flat roof.

pitcher tee, pitcher cross, etc. [plu.] A *malleable iron* (*C*) tee or cross on which the *branch* is cast with a gently curved turn instead of the sharp *elbow* which is usual with ordinary *tees* and *crosses*.

pitch-faced stone or **pitched-face stone** or **rock-face stone** Stone which has been worked at the quarry with the *pitching tool*. It may be *rubble* or *ashlar* but is always rough.

pitch fibre pipes Black pipes made 25% of wood or asbestos fibre, and 75% of refined coal tar pitch. They are usually buried, and are suitable for soil and rainwater drainage, but not for continuously flowing hot liquid, nor for pitch solvents such as any petroleum spirit. In first cost they are competitive with concrete or *stoneware*, they are also more flexible and lighter in weight. They are very quickly laid (120 m (400 ft) per hour) for two reasons: the average pipe length is 2·4 m (8 ft) not 90 cm (3 ft) and the jointing method is much quicker. No mortar nor other jointing fluid is used, merely light taps with a hammer. Being smooth the pipes can be laid at flat gradients, e.g. 10 cm (4 in.) pipes at 1 in 85.

pitch-impregnated fibre pipe *Pitch fibre pipe.*

pitching piece or **apron piece** [carp.] A horizontal timber fixed into the wall near the *landing* of a *stair* to bear the *carriages*, strings, and landing-floor *joists*.

pitching tool or **pitcher** A *hammer-headed chisel* about 23 cm (9 in.) long with a thick broad edge (about 6 mm (¼ in.) thick). *See* pitch-faced; *also* mason (illus.).

pitch mastic A mixture of aggregate and coal-tar pitch which is fluid when hot and can be spread as a topping for *jointless floors*. Like asphalt it is laid 16 to 19 mm (⅝ to ¾ in.) thick. Not to be confused with *mastic*.

pitch pocket or **resin pocket** or **pitch streak** [tim.] An opening between the *growth rings* containing *resin* in certain *softwoods*. A 'large' pitch pocket is one larger than 3×76 mm ($\frac{1}{8} \times 3$ in.).

pith [tim.] The *heart centre* of a log.

pith fleck [tim.] A pith-like discoloration caused by insect attack on living timber.

pith knot [tim.] (USA) A minor defect, a *knot* whose only fault is a pith hole smaller than 6 mm ($\frac{1}{4}$ in.).

pith ray [tim.] *Wood ray*.

pit-run gravel Untreated gravel from a pit.

pit sawing [tim.] The method used for *ripping* timber before mechanical saws were used. One man stands in a pit and another man stands at the top, each pulling at the same saw.

pitting [pla.] The *blowing* of plaster.

plafond French for a ceiling, hence a *soffit*, particularly of a *cornice*.

plain ashlars Surfaced stones, smoothed with a *drag* or other smoothing tool.

plain-sawn timber *Flat-sawn timber* (illus.).

plain tile or **plane tile** The common, flat *roofing tile* of concrete or burnt clay, which in reality has a slight spherical camber, convex above. Its size is standardized in Britain at $26 \cdot 5 \times 16 \cdot 5$ cm by $1 \cdot 0$ to $1 \cdot 5$ cm thick ($10\frac{1}{2} \times 6\frac{1}{2}$ in. by $\frac{3}{8}$ to $\frac{5}{8}$ in.) with at least two *nibs* at the head and two nail holes nearly 6 mm ($\frac{1}{4}$ in.) dia. Each tile overlaps two *courses* below it, and for this reason plain tiling is heavy, about 590 being needed to cover 10 m^2 ($107 \cdot 6$ ft^2) of roof at 10 cm (4 in.) gauge, and weighing 78 kg/m^2 (16 lb/ft^2). *Single-lap tiles* are not so heavy; *see* **pantiles, tile-and-a-half tile, under-ridge tile.**

plan The layout of a building drawn on a horizontal plane. Much of the *architect*'s work consists in designing layouts; that is planning in this sense. *See* **open planning, planning, town planning,** *also* **C.**

planceer piece or **soffit board** A horizontal timber forming the wooden or plaster *soffit* to an *overhanging eave*.

plane [carp.] A tool for smoothing and shaping wood, of which very many different sorts exist. Many of the *moulding planes* have been superseded by machinery which can cut mouldings much more quickly than they can be cut by hand. The *bench planes* are still much used. *See also* **back iron, badger, bead plane, cutting iron, hollow plane, match plane, plough, round, sole, spokeshave, universal plane.**

plane iron [carp.] The *cutting iron* of a *plane* as opposed to its *back iron*.

planer [mech.] A *planing machine* for surfacing metal or stone or timber.

plane stock [carp.] The body of a *plane*, holding the *cutting iron* and *back iron*.

planing machine [tim.] A machine which smoothes wood surfaces by the adzing action of a *cutter block*. It will also, with care, remove

253

twist from a surface, do bevelling, chamfering, moulding, and tongueing and reduce different pieces of wood to the same thickness. Many types exist. *See* **surface planer.**

planing machinist [tim.] A *woodworking machinist* who sets and operates a *planing machine* to give true face and edge to timber. Working as a thicknessing machinist, he sets the *thicknessing machine* to give precise thickness.

planing mill [tim.] (USA) A sawmill where timber is also planed and made into match or floorboarding.

planing mill products [tim.] (USA) Floorboards, ceiling boards, and *weather-boarding*, as opposed to *millwork*.

plank [tim.] (1) In *softwoods*, square-sawn timber 48 to 102 mm ($1\frac{7}{8}$ to 4 in.) thick and 279 mm (11 in.) or more wide (BS 565). *Compare* **deal.** In *hardwoods* the meaning varies too much for any definition to be useful, *board* and plank being understood in opposite senses in different places.

(2) (USA) A piece of lumber thicker than 1 in. (25 mm) laid with its face horizontal like a floorboard. *See* **joist.**

plank-on-edge floor or (Britain) **solid-wood floor** [carp.] A floor used in USA for its fire resistance and solidity. The floor *joists* are laid touching each other; no rough floor is laid, the finish floor being fixed to the edges of the joists.

planning In house or building design, the planning is the organization of the details of the layout, and is best done by an *architect*, though many owners prefer to do it themselves. *Town planning* is a cooperative process in which architects, economists, engineers, landscape architects, doctors, sociologists, surveyors, and even lawyers have each the right to state their opinion. Municipal or civil engineers with an architectural bias, or architects with an engineering bias, have the best qualifications for leading a *town-planning* team.

planning grid A network of perpendicular and parallel lines usually one *module* apart. It is used by *architects* to help them arrive at a building layout and is not a *grid plan*.

planted [joi., pla.] A description of *fibrous plaster* or of a *moulding* or strip fixed by nailing, screwing, or gluing on to the piece which it decorates, and not cut or moulded in the solid. *See* **sticking, stuck moulding.**

plaster [pla.] A substance which later hardens, but is applied to walls and ceilings while it is *plastic*. Plasters may be of *Portland cement* or *gypsum plaster* or *lime putty* with sand. *See* **coarse stuff, finishing coat, plasterer, stucco,** *also* BSCP 211, Internal Plastering.

plaster base [pla.] A ground for plaster. It may be wood, *metal lathing*, brickwork, *masonry, insulating board*, or *gypsum lath*.

plaster bead [pla.] An *angle bead*.

plasterboard The usual name for *gypsum plasterboard*.

plasterboard nail A galvanized or *sherardized* nail for fixing *plasterboard*.

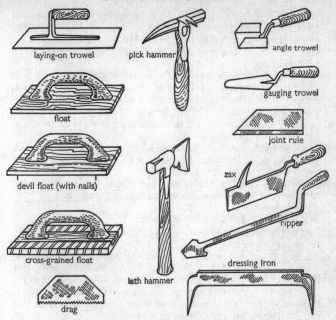

Tools of the plasterer (left) *and the slater* (bottom right).

plaster dab A small lump of *gypsum plaster* which sticks to brickwork or *lathing* and is used as a fixing for *wall tiles*, *marble facing*, or *joinery*.

plasterer A *tradesman* who may be a *fibrous plasterer* or a plasterer in solid work. The latter lays successive coats of plaster or rendering and fixes *fibrous plaster* such as mould *cornices* and wall patterns. He can use a *horsed mould*, erect *lathing* for plaster, and apply stucco. Plasterers in some districts also lay *granolithic floors* (*C*) and fix *wall tiles*. *See* above.

plasterer's float [pla.] *See* float.

plasterer's labourer A helper who mixes plaster or brings it from the mixer in a hod, wheelbarrow, etc. He may also slake *lime*, sieve materials, and beat *hair* to separate the lumps, fix *laths*, work a *gin block*, raise *scaffolding*, and prepare a wall surface for plaster.

plasterer's lath hammer A *lath hammer*.

plasterer's putty *Lime putty*.

plastering machine A hopper-fed machine that applies *pre-mixed*

255

plaster to a wall or ceiling, using a metal applicator rotor applying the full thickness of plaster in one coat through a flexible hose. The plaster is then smoothed by hand, the men using a metal *feather-edge rule*. The final process is electrically powered trowelling followed by hand finishing.

plaster of Paris or **hemihydrate plaster** or **casting plaster** [pla.] *Gypsum* which has been heated to drive off some of its water and has become $CaSO_4 \cdot \frac{1}{2}H_2O$. When mixed with water it warms up and sets in about ten minutes, expanding slightly; because of this expansion it is an excellent casting plaster, but it is little used in building except as retarded *hemihydrate plaster*.

plaster slab Precast, solid, or hollow blocks of *gypsum plaster* and sand used for making *partitions*. They may be laid with *cement mortar* or with *gypsum plaster* as *mortar* or grout.

plastic (1) [pla.] Said of plaster which is easy to trowel and spread. *Lime putty* of *high-calcium lime* is generally the most plastic.

(2) *See* **plastics**.

plastic emulsion [pai.] *See* **latex emulsion**.

plastic or **plastics glues** *Synthetic resin* glues for timber, *epoxide resins* for gluing *light alloys*, etc.

plasticity [pla.] or **workability** or **fatness** The property in a plaster or mortar, of sticking to the trowel, and of being smooth to work with the trowel, obtained with cement mortars by adding an *air-entraining agent* (*C*) or matured slaked lime, or by using a masonry cement. *Gypsum plasters*, being much more finely ground than sands, are usually much more plastic than cement mortars.

plasticizer (1) An *admixture* in mortar or concrete, which can increase the *workability* (*C*) of a mix so much that the water content can be extremely low, and the mortar or concrete strength can be high. However, plasticizers do exist which have the effect of reducing concrete strengths. Calcium chloride ($CaCl_2$) is often added to such plasticizers because it is a useful *accelerator*, and the net result may be a good 'frostproofing' additive. *See* BS 4887, Mortar Plasticizers.

(2) [pai.] A non-volatile substance mixed with the *medium* of a paint, varnish, or lacquer to improve the flexibility of the hardened *film*.

(3) In plastics, plasticizers are added during manufacture to make them soft, flexible and easy to process.

plastic paint or **texture paint** A paint which can be manipulated after application to give a patterned or *textured finish*. On conscientiously scrimmed plasterboard, there is no need for even a skim coat of plaster under this paint.

plastics Some of the most important plastics are *polyvinyl chloride* (PVC) which has been in production since the early 1930s; polythene (since the early 1940s), polypropylene, polymethyl methacrylate (Perspex), and *glassfibre-reinforced resins*. Not all plastics are *synthetic resins*. Plastics are either *thermo-setting* (those which harden once for all when heated), or *thermoplastic* (those which soften whenever they are heated). (The word plastics is used in the singular as a noun to distinguish it from the adjective plastic.) *See* **plastic glues, resin.**

plastics coatings [pai.] Enamels, lacquers, varnishes, paints and even *jointless floorings* made of plastics are becoming more and more usual, particularly *alkyds*, amides, polyamides related to *nylon,* phenolic resins, silicones, polyvinyl resins and *chlorinated rubber*.

plastic steel This useful proprietary repairing material is believed to be a synthetic resin like that in *glassfibre reinforced resin*, with a filler of iron filings. Also known as plastic padding.

plastic wood [tim.] A paste of *nitrocellulose*, wood flour, *plasticizers*, *resins*, and other materials dispersed in *volatile* solvents. It is used for repairing wood, filling holes, etc., and its surface can be painted over in about one hour.

plate [carp.] (1) A horizontal timber about 5 × 10 cm (2 × 4 in.) supported throughout its length, particularly one along the top of a wall on which the ends of roof timbers are laid (*wall plate*). In *frame construction* it is the corresponding member capping the studs.

(2) (Scotland) A broad thin board, planed on one or both sides.

(3) [mech.] *See C*.

plate cut [carp.] A *foot cut*.

plate glass or **polished plate glass** *Glass* of better quality than *sheet glass*, which has been cast (originally on an iron plate) and polished, and therefore has two smooth faces. *See* **float glass.**

platen [tim.] The hot steel plates used in *hot pressing* for making *plywood* with *thermo-setting* glues. *See* **phenol formaldehyde resin.** *See also C*.

platform frame or **Western frame** [carp.] A timber *frame house* in which wall, floor, and roof frames are independently built. The floor platforms are carried over the full thickness of the wall frames which are built in separate *storeys*. Diagonal boarding braces the floor, all external walls and the roof; additional stiffer braces are usually provided in the wall frames.

platform roof Scots for a *flat roof*.

plenum chamber An air chamber kept at a pressure slightly above atmospheric.

plenum system A method of *air conditioning* a factory or large building by keeping the pressure in it above atmospheric pressure. Clean air is blown into rooms near the ceiling level and foul air is withdrawn near floor level at the same side of the room, or allowed to escape through cracks in doors and windows. *Unit heaters* which draw fresh air directly from outside through a hole in the wall are often now preferred.

Plexiglas *See* **Perspex.**

pliers A pair of pliers is a gripping tool, pivoted like a pair of scissors, usually with blades for cutting thin wire, built into its jaws. *See* **seamer.**

plies [tim.] Plural of *ply*, a sheet of *veneer*, *bitumen felt*, etc.

plinth A slight widening and thickening at the base of a *column*, wall or pedestal.

plinth course The top course of a brick *plinth*.

plot An area of land usually for building, called a lot in USA.

plough (1) or **router** or (USA) **plow** [joi.] A *plane* which makes grooves, usually along the *grain* of wood. By varying the position of the *fence*, the distance of the groove from the edge of the wood can be adjusted. The plough was invented in about the sixteenth century together with the *panelled* construction for which it is an essential tool.

(2) *See C.*

ploughed and tongued joint [joi.] A *feather joint*.

plough strip [joi.] A strip of wood which has been ploughed, used for fixing the edge of a drawer bottom.

plow (USA) A *plough*.

plug (1) A small pointed wooden peg pushed into a hole in a wall where a *screw*, *nail*, or other *fastening* is to be driven. Metal and fibre plugs (Rawlplugs) also exist, but wood is cheap and very effective for small loads. *See* **plugging.**

(2) A *bag plug* or *screw plug*.

(3) An electrical connection to fit into a *socket outlet*.

(4) [plu.] A pipe *fitting*, threaded outside, which closes an end of pipe by screwing into it.

plug-centre bit [carp.] A *bit* used for widening holes. The central part is a plug inserted in the hole already drilled.

plug cock or **plug tap** [mech.] A simple valve, in which the fluid passes through a hole in a tapered plug. The valve is closed by turning the plug through 90°.

plug-driving gun *See* **stud gun.**

plugging Drilling a hole in *masonry*, which is to be filled with a *plug* of wood or fibre or metal as a fixing for a nail, *drive screw*, wood *screw*, etc.

plugging chisel or **drill** or (USA) **star drill** A short steel bar held in the hand, and struck with the *hammer* to make holes from about 3 mm ($\frac{1}{8}$ in) to about 13 mm ($\frac{1}{2}$ in.) dia. in brick, concrete or masonry. It is used like the *jumper* (*C*). *See* **pipe drill.**

plug-in connector or **flex cock** or **plug-and-socket gas connector** [plu.] A method of connecting a short, flexible gas pipe to a wall plug as conveniently as electrical flex. The plug is pushed into a bayonet socket and, when turned, allows gas to pass to the appliance. Removal of the plug automatically closes the gasway.

plugmold [elec.] (USA) A hollow, pressed steel or plastics *moulding* fitted to the wall as a *chair rail* or *skirting board*, containing electric cables with power plugs as required. *See* **raceway.**

plug tenon or **spur tenon** [carp.] A *stub tenon* shouldered on all four faces as for an angle post (BS 565).

plumb Vertical.

plumb bob or **plummet** A weight hanging on a string (the *plumb line*) to show the direction of the vertical.

plumb cut [carp.] (USA) The vertical cut in the *birdsmouth* at the foot of a *rafter* where it fits over the wall plate; also the vertical cut at the *ridge* (top cut). *Compare* **foot cut.**

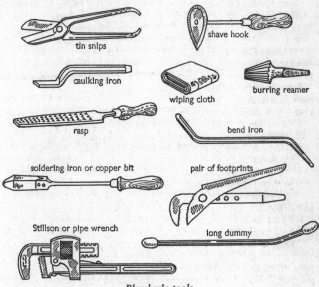

Plumber's tools.

plumber A *tradesman* who shapes and fixes *flexible-metal roofing*, cuts, bends, and joins water pipes, instals *soil pipes*, waste pipes, and water systems, sometimes gas pipes also. The usual metals he works in are copper, lead, iron, zinc, and mild steel, though the term plumber originally meant lead worker. In the north of England, plumbers are usually *glaziers*. In Britain a plumber may hold the technical certificate of the City and Guilds of London or the highest qualification, that of Registered Plumber (R.P.). *See* **lead burning, chemical plumber.**

plumber's labourer or **plumber's mate** A skilled *labourer* who helps the plumber by carrying his tools and bringing his materials to him, and may help to make *wiped joints*, etc.

plumber's solder or **coarse solder** An alloy of lead and tin varying from 1:1 to 3:1, used for joining lead pipes, since its melting point (about 185° C.) is below that of lead. *See* **solder, fine solder.**

plumber's union *See* **union.**

plumbing (1) *See* **plumber.**

(2) Fitting out the sanitation, hot- and cold-water services, and heating of a building.

(3) The pipework and sanitation of a building.

(4) Transferring a point at one level to a point vertically below or above it, usually with a *plumb bob* or *plumb rule*. *See C.*

plumb level or **plumb and level** A spirit *level* fitted with a small bubble at right angles to the main bubble to show the plumb direction. The small bubble is central when the main bubble is held vertical.

plumb line [sur.] A string on which a weight is hung to stretch it in a vertical direction. The string should be braided like fishing line to avoid spinning of the weight.

plumb rule A wood *straight edge* from which a *plumb line* is hung at the top and a hole cut out near the bottom to allow the *plumb bob* to swing freely. It is used for plumbing walls. *Compare* **plumb level;** *see* **mason.**

plume [tim.] The figure formed by *crotch* in mahogany.

plummet A *plumb bob*.

ply (1) [tim.] A thin sheet of wood usually called a *veneer*, used for making *plywood* or *laminated wood*.

(2) One sheet of *roofing felt* or of other material built up in several layers.

plymetal *Plywood* faced on one or both sides with a sheet of metal which may be galvanized steel, aluminium, Monel metal, or many other metals.

plywood [tim.] Structural board, stronger and more dimensionally stable than wood, because it is glued from an odd number of sheets of *veneer* with the *grain* of adjacent sheets at right angles to each other. *Three-ply* is the commonest and cheapest but *multi-ply* is also much used. Plywood was used by the joiners of Europe, particularly France, in furniture making some centuries ago, and it is also believed to have been used by the ancient Romans and Egyptians. Plywood, as we know it now, was used by the American piano industry in 1830 for the planks which held the pins to which piano cords are attached. It was then made of *sawn veneer*. At this time, obviously, piano makers understood its superiority over wood in strength and stability in varying conditions of dampness. *Blockboard* desk tops were made in 1883 and plywood *panels* for doors in 1890, flush doors following much later. Plywood was known in USA as 'veneered stock' until 1919 when, to avoid the ignorant but widespread prejudice against veneer, the old Veneer Association changed its name to the Plywood Manufacturers' Association of USA. It differs from *laminated wood*, in which the plies have parallel grain. *See also* **balanced construction, composite board, cross band, moulded plywood, rotary cutting, sliced veneer, synthetic resin.**

plywood clips [tim.] Specially shaped L, Z, and other intricate shapes of *plywood*, made for fixing plywood sheets to frames of wood or metal. They have the advantage over metal clips that they are easily glued.

plywood parquet or **plywood squares** A form of *parquet* consisting of tiles formed of *plywood* with a top *veneer* 3 mm ($\frac{1}{8}$ in.) thick of *oak*, *birch*, or *ash*. The tiles are 23 to 90 cm (9 to 36 in.) square and pinned to the wood *sub-floor* (and glued in the best work). The nail holes are filled and the floor *glasspapered* after laying. If no wood sub-floor exists a 1 to 2 cm ($\frac{3}{8}$ to $\frac{3}{4}$ in.) thick plywood layer must be put down as a base for the parquet. If the sub-floor is rough, it should be made smooth by a plywood underlay about 5 mm thick all over it.

plywood sawyer [tim.] A *tradesman* who sets and works power saws for cutting *plywood* (*circular saws, straight-line edger, band saws*, and so on).

pneumatic architecture *See* **air house.**

pneumatic water supply A water-supply system used in isolated houses in North America. The *cistern*, in the basement, is a closed tank from which the water is forced into the house by compressed air.

pocket (1) An opening in a wall in which a beam is to be inserted.

(2) or **weight pocket** [joi.] The hole in which the *pocket piece* fits. The sash weight is passed through it when it is fixed to the *sash cord*.

pocket chisel or **sash chisel** [joi.] A *chisel* for cutting the pocket in the *pulley stile* of a window frame. It has a wide, thin blade, honed on both sides.

pocket piece [joi.] The piece of the *pulley stile* at the foot of a window frame which can be removed to insert the sash weights or to string new *sash cords* on them.

pocket rot [tim.] Decay of timber in small lens-shaped areas, which eventually become round holes.

pock marking [pai.] *Orange peeling* or other unsightly depressions formed in the *drying* of a *paint* or *varnish*.

pod A bathroom pod is a prefabricated bathroom that can be added to a house that does not have a bath. It can be placed in position on its prepared slab within 20 minutes of its arrival on the site, but connecting to the water supplies, wastes and gas or electricity will need a few hours more. It is then ready for use. Because of the cost of hiring the crane needed to lift the pod over the roof of the house, it is usually not economic to install fewer than six on any particular site. The roof is often flat, and made of *glassfibre reinforced plastics*.

pod auger [carp.] A very old type of *auger*, of which the upper part is shaped like a bean pod to contain the chippings.

point (1) [carp.] The sharp end of a *saw-tooth*. The number of teeth per inch of a saw is one fewer than the number of points per inch.

(2) [elec.] A lampholder, *socket outlet*, or other terminal from which power can be taken from the fixed wiring of a house. A power point is one which is designed for circuits more powerful than lighting circuits, but both consume power.

(3) Any outlet of a gas system, connected or ready for connection.

pointing (1) Raking out mortar *joints* 2 cm (¾ in.) deep and pressing into them a surface mortar. The bedding mortar is disturbed and may not bond with the surface mortar. *Jointing* is therefore more durable, but pointing provides the opportunity of using a mortar of different colour. *See* **tooling**.

(2) The completion of joints between ridge or *hip tiles* or tiles on walls with mortar. *See* **torching**.

pole plate [carp.] A horizontal beam resting on, and perpendicular to, the tie-beams or *principal rafters* of a wooden roof *truss*, supporting the feet of the common *rafters* and the inner edge of the *box gutter* if there is one.

polished plate *Plate glass*.

polished work or **glassed surface** or **polished face** Building stone of crystalline texture (limestone, marble, or granite) which has been smoothed with *abrasives* and buffing to form a glass-like surface.

polishing varnish or **rubbing varnish** [pai.] A *short-oil varnish* used for fine joinery, generally one which is so hard and dry that it can be polished by *abrasive* and mineral oil without dissolving the resin.

poll (1) [mech.] A striking *face* of a *hammer*.

(2) To split (knap) flints.

pollarding [tim.] Annual lopping of the shoots on the poll (the top) of a tree. It forms a valuable figure, such as the *burr* in *veneers*.

polychromatic finish [pai.] A *finish* of many *colours*.

(2) A finish with a metallic lustre and an iridescent scintillating effect from different angles. The effect is produced by *lacquers* or *enamels* containing metal flake powders as well as transparent colouring matter.

polyester resin Synthetic resins used in *glassfibre-reinforced resin*.

polymers Organic compounds, including many *synthetic resins*. They have large molecules containing many hundreds of smaller molecules of the same compound linked together. Polymers are usually solid, and are formed when synthetic resins harden, whether as *glue* or casting resin. *Polyvinyl chloride* is an example. Not all polymers are synthetic resins, for example *casein* is a polymer but not a *resin*. *See* **cure, elastomer**.

polymerization or **condensation** [tim.] The *cure* of a *synthetic resin*. *See* **polymer**.

polystyrene A material for making *thermoplastic* wall tile of extremely light weight. It cannot be scrubbed with a gritty cleaning powder, softens at 60° C. (140° F.), and is marked by splashes of boiling water or turpentine or paint thinner. In spite of these disadvantages it has been used in block form to make transparent partitions, like a *glass-block* wall. It becomes yellow on exposure to ultra-violet light. It has also been used for cold-water pipe. *Expanded polystyrene* is an *insulating material*.

polytetrafluorethylene or **ptfe** [plu.] A plastics which, in the form of a substitute for *jointing compounds*, came into the British building industry in about 1962, and has a number of uses, especially in plumbing.

polythene or **alkathene** Abbreviation for **polyethylene**. A chemically inert, electrically insulating, synthetic rubber of which sheets are made for covering the floors of chemical plants (alkathene sheets). The foil is 0·25 mm thick, laid on a very smooth screeded floor, and

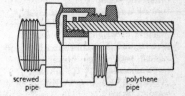

Polythene pipe. Brass or gun metal fitting for joining polythene coldwater pipe to screwed water pipe. Note the threaded metal sleeve inserted within the polythene tube, and screwed into it to ensure a tight joint.

screwed pipe

polythene pipe

covered with a second screed, which is covered usually with floor tiles. Polythene is also used for making pipes for chemicals or cold water and for making translucent bottles which bend but do not break.

polyurethane A plastics which as foam has great possibilities. It began to be used commercially in sandwich form in the USA in the early 1960s and developed rapidly as a *sandwich* core. Like *expanded polystyrene* it can be foamed to a density as low as 30 kg/m³ (2 lb/ft³). Both have high ratios of strength/weight, low thermal and acoustic conductivities, controllable density, no smell, low transmission of water vapour, low absorption of water, and high dimensional stability.

polyvinyl chloride or **PVC** A vinyl resin, a *polymer*, a rubbery almost non-combustible material used for making pipes, for insulating electrical cables, as a waterproof membrane to form an expansion joint in a wall, in vandal-resistant glazing, and as a flooring material. Like *linoleum* it must be laid on a dry floor. It can be used for surfacing floors, walls or ceilings, being impervious to water, oils, petrol and many chemicals. At a thickness of 2·5 mm it is obtainable in widths up to 1·8 m (6 ft) and lengths up to 27 m (90 ft). In Germany, sales of window frames made of hollow PVC sections had attained 10 per cent of the market in 1972, from nothing in 1966. The frames have low leakage, are self-extinguishing, and are made to measure, by fusion welding of the sections, to tolerances as close as 4 microns.

pommel (1) A knob such as a ball *finial* to a roof.

(2) A *punner* (*C*), with an iron foot for ramming earth.

Poole's tiles Flat, clay *interlocking tiles* standardized in Britain at $41 \times 34 \times 1$ cm ($16 \times 13\frac{1}{2} \times \frac{1}{2}$ in.). They have 7·6 cm (3 in.) side lap and differ from the *double Roman tile* in that the central ridge extends only half way up. Like the double Roman tile it has two waterways, two nail holes, and no *nibs*.

poor lime [pla.] *Lime* containing a high proportion of material which is insoluble in acids (and is therefore not pure lime, CaO). The impurities can vary from 6% to 30%.

poplar Poplar trees should not be allowed nearer than 18 m (60 ft) from buildings founded on clay unless the foundations have gone deeper than the poplar roots. Poplars grow very quickly and cause clay to shrink rapidly in summer because of the water which they draw out of it. They thus damage foundations by causing the clay to shrink.

popping [pla.] *Blowing* of plaster.

pores [tim.] The small round holes, seen on the end grain of *hardwood*, which lead the sap upwards.

porous woods [tim.] *Hardwoods*.

portable belt conveyor A rubber or steel *belt conveyor* (*C*) 7 to 11 m (25 to 35 ft) long with a belt from 35 to 50 cm (14 to 20 in.) wide driven by a motor from 2 to 4 hp, frequently used for loading lorries. The belt is usually troughed and provided with short steel angle cleats riveted on, to prevent material slipping back down steep slopes. It may be used therefore for delivering bricks to second floor level. These conveyors are carried on a frame resting on two wheels and can be tilted to any angle. An economical belt speed which gives little belt wear is 30 to 45 m/min. (100 to 150 fpm). Some types have a power drive to one road wheel. The brick elevator is a similar machine.

portable electric tool A hand-held tool driven, like an electric drill, by an electric motor or vibrating armature in the body.

Portland cement So called since about 1820 when it was first patented in England, because when hard it resembled the stone from Portland in Dorset. *Roman cement* was its forerunner. *See* **concrete, ordinary Portland cement,** *and C.*

Portland stone An oolitic limestone from Portland on the south coast of England, much used for facing important buildings in London. It weathers well in London and forms strong contrasts between the parts which get wet and stay white and those which are sheltered and blacken.

post [carp.] A main vertical support of a building frame or *partition* or sub-frame, thicker than a *stud*.

postal knocker [joi.] A *letter plate* which includes a door knocker.

posts and timbers [tim.] (USA) Square timbers 13×13 cm (5×5 in.) or larger, *stress-graded* for use as struts.

pot floor A *hollow-tile floor* (C).

pot life or **spreadable life** or **working life** or **usable life** The maximum time for which a *mixed glue* (or paint) remains usable in the pot, as opposed to its *shelf life*.

pot type boiler A coal- or coke-fired boiler with a plain cylindrical water jacket. It is efficient, but is not aimed to heat the room in which it is placed, since most of its heat passes into the water.

pounce [pai.] A pattern transferred on to a surface below a piece of paper by pricking holes in it and rubbing pumice powder (pounce) through it on to the surface below.

pouring rope [plu.] A *joint runner*.

powder-post borings [tim.] *Worm holes*, 1·5 to 3 mm ($\frac{1}{16}$ to $\frac{1}{8}$ in.) dia., with a fine, dust-like flour from the borings. The holes are slightly larger than *shot hole* and are often found in *hardwoods*.

power panel [elec.] A *cutout box*, for power circuits rather than for lighting.

preamble [q.s.] The introduction to each trade in a *bill of quantities*.

preboring for nails [carp.] Holes are prebored to a dia. equal to $0.8 \times$ the nail dia. when driving a *nail* might split the wood or when the nail is required to have a high *withdrawal load*.

precast stone *Cast stone*.

precure or **precuring** [tim.] The setting of *glue* in a joint before it has been clamped. *See* **cure**.

prefabricated building All buildings are to some extent prefabricated, since many building components are brought on to the site completed (doors, roof trusses, etc.). The word prefabricated is therefore usually reserved for those buildings of which walls, roofs, or floors are completed either off the site or in a site factory. *See* **industrialized building**.

prefabricated tie (USA) A *wall tie* consisting of one wire of about 3 mm ($\frac{1}{8}$ in.) dia. in each leaf of a *cavity wall* joined by similar wires at right angles, welded to them at intervals, like fabric reinforcement.

preliminaries [q.s.] The introduction to a bill of quantities, *compare* **preamble**.

preliminary works [q.s.] Demolition, diversions, and so on.

pre-mixed plasters [pla.] Bagged plasters usually containing *perlite* or *vermiculite*. The low density of these aggregates makes them easy for the plasterer and labourer to apply or mix. (Water only needs to be added to them.) Their high thermal insulation values make them warm in cold weather, and also valuable for the *fire protection* of steelwork.

preservatives (1) Preservatives for steel or cast iron are *metal coating*, *phosphating*, painting, tarring, *Bower-Barffing* (C), airless spraying and so on. *See* **protective finishes**.

(2) For stone, preservative treatment should include washing with liberal amounts of clean, warm water every three years to remove the corrosive salts which collect on the surface. It is generally best to use neither chemicals nor soap. Many proprietary treatments exist. Steam cleaning without chemicals may also be harmless and effective even if the stone has been allowed to accumulate dirt for many years. *See* **silicone.**

(3) [tim.] *See C.*

press [tim.] An arrangement of *platens* (steel plates) between which sheets of *veneer* are glued to form *plywood* under pressure. *Hot pressing* is generally worked by hydraulic pressure. Screw presses are only used for cold gluing.

pressed brick Bricks with sharp *arrises* and smooth surfaces formed by moulding under high pressure. The stronger qualities are called *engineering* or *semi-engineering bricks.*

pressed glass Bricks, *pavement lights*, and so on, of glass pressed to shape.

pressed steel Sheet steel, hot pressed into various shapes as *trim* in joinery or as window *sills*, *sub-sills* and *sub-frames*. For these uses it is generally 1·6 to 2·5 mm thick. It was first used on a large scale about 1890 for building rail wagons.

pressure [tim.] Timber may be glued either in presses which have no follow-up pressure (screw presses), or there may be a follow-up pressure, as in a hydraulic press which allows for the yielding of the wood. A third type of pressure is fluid pressure, exerted by steam, compressed air, hot water, etc., often through flexible rubber bags, sometimes called omni-directional pressure. *See* **flexible-bag moulding.**

pressure gun or **caulking gun** A tool like a grease gun, used for applying joint sealing material to a joint, *mastic* as a *face putty* to window bars and so on.

priced bill [q.s.] A *bill of quantities* which has been sent out to a *contractor* for him to *tender*. The contractor has entered his price opposite each item, *extended* the prices, and summed the totals. The priced bill with certain formal documents constitutes the contractor's offer or tender to do the work for the price he has stated.

pricking up [pla.] Applying the first coat of *plaster* on *lathing.*

pricking-up coat [pla.] The first coat of plaster on *lathing*. *See* **rendering coat.**

prick post [carp.] (1) A *queen post.*

(2) Any intermediate post in a frame.

prick punch A tool like a *nail punch* with a point on it, used for pricking holes through sheet metal, for making a small hole to start a *nail* and so on.

primary flow and return pipes [plu.] The pipes in which water circulates between boiler and *cylinder* in a water-heating system, the flow pipe being that by which water leaves the boiler, returning to it by the return pipe.

primary gluing [tim.] The gluing in *plywood* manufacture and other veneering work, as opposed to *assembly gluing*.

prime cost sum [q.s.] A sum entered in a *bill of quantities* by the *consultant*. The sum is provided to pay for the cost to the *contractor* of a specific article after deducting all discounts except the $2\frac{1}{2}\%$ discount for monthly settlement of accounts. Its purpose is to specify the quality required in the article.

primer (1) A bituminous adhesive coating for sticking a *roofing felt* on to a roof boarding.

(2) [pai.] A *priming coat. See also C.*

priming coat or **primer** or **priming** [pai.] A first coat on new metal or wood or on an old surface from which all old paint has been removed. It is put on before the *undercoat*. It ensures that the paint will stick to the surface, and with wood, plaster, or masonry, may partly seal the pores but should penetrate the surface deeply. A priming coat for steel contains *inhibiting pigments. See* **sanding sealer, surfacer, filler.**

princess posts or **side posts** [carp.] Subsidiary vertical timbers between the *queen posts* and the wall, to stiffen a *queen-post truss*.

principal (1) A *principal rafter.*

(2) A roof *truss.*

principal post [carp.] A *door post* in a *framed partition.*

principal rafters [carp.] or **principals** The main *rafters*, those in the roof *truss* which carry the *purlins* on which the *common rafters* are laid.

prismatic glass A window glass with small parallel triangular prisms rolled into the outer surface. If these are placed horizontally in the window, more light will reach the remote parts of the room than through ordinary glass.

processed shakes Common, sawn, *western red cedar* shingles which have been surface-textured on one face to look like split *shingles. See* **shakes.**

profile (1) A British term for *batterboard.*

(2) [pla.] A *templet* for shaping a mould.

progress certificate *See* **certificate.**

progress chart A wall chart showing the various operations in the construction of a building, such as clearing the site, excavation, concreting foundations, erecting steelwork, pouring ground floor, pouring first floor, pouring roof, building outer walls, partitions, plastering, painting, and so on. Each operation is shown with proposed starting and finishing dates. Immediately, it can be seen at what state each *trade* should be on a certain date, and whether it is delayed. *See* **critical path schedule** *and* illus. overleaf.

progress chaser or **chaser** or (USA) **expediter** A clerk employed by a *contractor* on a building site to verify the arrival on time of building

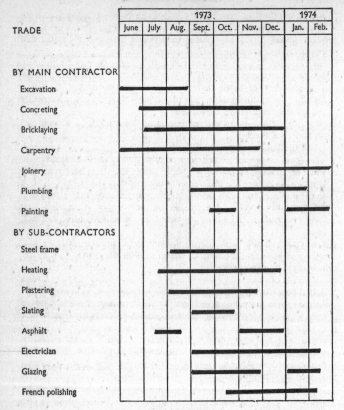

Progress chart

materials and plant. He is also expected to know the reasons why any section of work is delayed. The person (and terms) are borrowed from factory production.

progress report A weekly report by a *clerk of works* or *resident architect* on the work done during the week and, more especially, on the work which should have been, but was not, done.

projecting scaffold A working platform such as a *bracket scaffold*, built out from an upper storey, not reaching to the ground.

project network analysis *See* PERT.

propane blowlamp [plu.] A lamp supplied from a *bottled gas*. The lamp

268

has been available in Britain since the 1950s. The gas is obtainable in small steel bottles containing as little as 0·34 kg (12 oz) of the gas. Because of this and the high calorific value of the gas, a bottle with an appropriate burner screwed on to it can be used instead of the *oxy-acetylene* flame or the paraffin blow lamp. It has a directional flame which is reliable and quick to ignite, and it is fast displacing the paraffin blow lamp although it is twice as expensive in fuel refills.

protected metal sheeting A roofing and wall *cladding*, consisting of steel sheet, coated with bitumen and asbestos or other materials.

protected opening An opening in an internal, *fire-resisting floor* or wall which can be closed by a shutter or door of appropriate *fire grading*.

protective finishes to metal [joi.] Surface coatings which wholly or partly protect metal from rust are very numerous. The most important purely protective finishes are *hot-dip* (*C*) or electro-tinning or *galvanizing* (*C*), *sherardizing*, *cadmium* or *copper plating*, and *phosphating*. All these are applied to mild steel wood *screws*. Finishes which are also decorative are *chromium* or nickel plating. For large metal structures, *cathodic protection* (*C*), or *metal coating* are possible.

provisional sum [q.s.] A sum set aside in the *bill of quantities* by the *consultant* to provide for work whose scope is not clearly foreseen. It may be *sub-contractor*'s work.

Prussian blue or **ferrocyanide blue** or **Chinese blue,** etc. [pai.] A synthetic *pigment*, first made commercially in about 1770, consisting mainly of $Fe(CN)_6$ with some water, potassium, and additional iron. Some varieties are valued for their metallic lustre, a sheen like bronze when looked at obliquely.

pry bar [joi.] A *nail puller*.

ptfe [plu.] *Polytetrafluorethylene*.

P-trap [plu.] A trap with an ordinary U-shaped seal and a final horizontal outlet. *See* **S-trap.**

puff pipe An *anti-siphon pipe*.

pugging or **pug** or **deafening** or **deadening** *Soundproofing* floors with material inserted between the ceiling and floorboards of a wooden floor. Sand, slag wool, and *coarse stuff* have all been used. *See* **sound boarding.**

pug mill A machine for mixing (and breaking up lumps of) clay, *mortar*, or *plaster*. It consists of an open pan in which two rollers (or knives) on opposite ends of the same horizontal shaft are pushed round, and crush and mix the materials. The pug mill is also used for blending paints, a process called pugging.

pull [joi.] A handle for opening a door, drawer, etc., called a door or drawer pull or *sash lift* as the case may be.

pull box [elec.] A box placed in a length of *conduit*, at which the cables can be pulled. If pull boxes are placed at short intervals, the work of drawing the cables through the conduit is easier.

pulley head or (USA) **yoke** [joi.] The horizontal board in the *cased frame* of a *sash window* against which the sashes abut when pushed to the top.

pulley stile or **sash run** or (USA) **window stile** [joi.] The vertical board at each side of the *cased frame* to a *sash window*. The pulleys over which the *sash cords* pass are fixed in them.

pulling [pai.] *Drag.*

pulling over [pai.] Levelling a film of cellulose *lacquer*, usually on wood, by rubbing it in one direction only with a soft cloth or leather pad soaked in a mixture of organic *solvents* which only partly dissolve the lacquer.

pulling up or **picking up** [pai.] The softening of a dry *coat* of *paint* or *varnish* when another coat is put on. Brushing is difficult and there may be intermingling of the coats. *See* **lifting**.

pull-out strength of nails [carp.] *See* **withdrawal load**.

pumice A highly cellular rock quarried in Germany and elsewhere for use as a *lightweight concrete* aggregate or as an *abrasive* in painting. It floats in water. *See* **pounce**.

pumice concrete *Lightweight concrete* made from pumice. It is *fire resisting* and has, like *foamed slag*, a low thermal *conductivity* (K-value) from 0·21 to 0·43 W/m deg. C. (1·5 to 3·0 Btu in./ft²h deg. F.) for concrete densities of 720 to 1500 kg/m³ (45 to 95 lb/ft³), with corresponding crushing strengths of 1·4 to 3·5 N/mm² (200 to 500 psi). The highest strength is comparable with *clinker blocks*.

pumice stone [pai.] *Pumice* cut with plane faces to use as an *abrasive* for smoothing paint before repainting.

punch (1) [joi.] A *nail punch* or *centre punch* (*C*) or *handrail punch*, *solid punch*, or *prick punch*.

(2) A *mason*'s chisel with a cutting edge 10 cm (4 in.) wide. *See also C.*

punched work or **broached work** *Ashlar* faced with a *punch*, with rough, diagonal strokes across the face.

puncheon (1) [carp.] A short post in the middle of a *truss*, particularly a *trussed partition* or a *queen-post truss*.

(2) (USA) A roughly dressed, sometimes adzed *slab* of wood used for flooring. *See also C.*

purlin A horizontal beam in a roof, at right angles to the *principal rafters* or *trusses*, and carried on them. It carries the *common rafters* if there are any, or the *corrugated sheet*.

purlin roof A roof for small houses, in which the purlins are carried on cross walls instead of on *trusses*.

purpose-made brick Bricks *moulded* to a special shape, such as *compass bricks*. *Compare* **gauged bricks**.

push plate [joi.] A metal plate, on the *lock stile* of a door at the level of the hand, to protect the door from damage by people's hands.

push-pull rule A steel tape 12 mm (½ in.) wide, usually 2 m or 6 ft long, which can be rolled into a pocket box of about 5 cm (2 in.) dia. It is replacing the old *fourfold rule* since it is so convenient to carry about.

putlogs Short horizontal bearers, which carry *scaffold boards* in a *bricklayer's scaffold*. The putlog rests in a small hole left in the brickwork at one end, and on the *ledger* at the other end.

putlog holes Holes left in brickwork for *putlogs*.

putty *See* **glazier's putty, lime putty, mason's putty, white lead putty.**

putty and plaster [pla.] *Gauged stuff*.

putty knife A *glazier's* knife for applying *face putty* to a *glazing bar*.

PVA [pai.] *See* **emulsion paint.**

PVC *See* **polyvinyl chloride.**

pyramid For many hundreds or thousands of years the Great Pyramid was man's most massive structure. It contains 20 million m^3 of stone, but this figure is now surpassed by the Fort Peck and probably other dams. *See* **hydraulic-fill dam** (*C*).

pyramidal light A *roof light* in which the *glazing* slopes up to a point from a base shaped like a regular polygon.

pyramid roof A *pavilion roof*.

Pyrotenax or **copper-sheathed** or **mineral-insulated cable** or **earthed concentric wiring** [elec.] An electric cable consisting of copper wire or wires contained in a copper tube filled with magnesia, MgO. It is unaffected by fire, can be heavily overloaded without endangering the building, and is so reliable that it is often used in power stations. However, each run of cable must be sealed against moisture by welding or other special treatment of the ends. One development is casting Pyrotenax cables into concrete slabs as *coil heating*.

Q

quadrant [joi.] A metal, curved *casement stay*. *See also* **quarter-round.**

quadrant dividers [carp.] A pair of dividers in which one limb slides on an arc fitted to the other limb and may be temporarily clamped to it by a screw.

quaggy timber [tim.] Wood with many *shakes* at the centre.

quantities [q.s.] The amounts of building materials and of work put into a building, translated into the terms and units of the *bill of quantities* (in mm, m, m², m³, kg, tonnes, etc.).

quantity surveying [q.s.] The drawing up of *bills of quantities*, their settlement between *building owner* and *contractor*, and the *arbitration* of disputed points after the completion of the work. *See* **quantity surveyor.**

quantity surveyor [q.s.] A person with somewhat similar professional qualifications to a *building surveyor*, who looks after the technical accountancy of building *contracts*. He measures the work shown on the drawings, and writes the *quantities* down, after he has compiled the *bill of quantities*, for use in *tendering*. He measures the work done by the contractor on completion (also every month for the *certificates*) and advises the *client* on the correct sum to be paid to the *contractor* at any time. In USA and many other countries the work of the quantity surveyor is always done by the *architect* or civil engineer. In Britain, for civil engineering work the quantity surveyor is not always needed, but British building quantities are very much more complicated and usually require one. In the nineteenth century, groups of competing contractors would collaborate during tendering to employ one man to draw up a bill of quantities so that the cost of this labour was shared between them. This was the origin of the profession. This also explains why those countries which do not use competitive tendering do not have quantity surveyors.

quarrel A *pane* of glass in *leaded lights*.

quarry A *quarry tile*.

quarry face or **quarry pitched** or **quarry dressed** A description of stone as it comes from the quarry, that is, squared at the joints with a rough face.

quarry sap The moisture in stone freshly cut from the quarry. When this has dried out, the stone is harder to work.

quarry stone bond Any *bond* which exists in rubble walls.

quarry tiles or **quarries** Burnt clay, paving or wall-facing tiles of black, buff, or red colour. They are from 23×23 to 10×10 cm (9×9 to 4×4 in.) in size, unglazed, but not porous.

quarter (1) [joi.] A square panel.

    (2) [tim.] *Quarterings*.

272

quarter bend [plu.] A 90° bend in a pipe. Other bends are proportional to this, a one-eighth bend being 45°.

quartered [tim.] *Quarter-sawn*.

quartered log [tim.] A log cut into four quarters for *conversion* by *quarter-sawing*.

quartered partition or **quarter partition** [carp.] A *partition* built of *quarterings*.

quarter-girth rule [tim.] A method of computing the volume of the timber in a round log. It is assumed to be equal in cross sectional area to a square of side equal to the quarter girth of the log at the middle of its length, that is $0.616D^2$ which is equal to $\left(\dfrac{\pi}{4}D\right)^2$

quartering [tim.] *Quarter-sawing*.

quarter-round or **quadrant** A convex *moulding* like a quarter circle.

quarter-sawing or **quartering** or **rift sawing** [tim.] Sawing wood as nearly radially as possible, with no growth ring at an angle of less than 80° to the surface in fully-quartered timber, 45° being usually acceptable for floorboards. It is used for the best floorboards and for oak to give the fine figure of *edge grain*. It is so called because the log is usually first cut into quarters, after which each quarter is cut radially, wasting a considerable amount of timber in wedge shapes. *See* **flat sawing.**

quarter-sawn timber [tim.] Timber converted by *quarter-sawing*.

quarter-space landing A small square *landing* which is of the width and length of a *tread*. *Compare* **half-space landing.**

queen bolt A steel bolt used in place of a *queen post* in the *queen-post truss*.

queen closer A *brick* cut in half along its length to keep the *bond* with a 5 cm (2 in.) face width. Normally in England it measures about $23 \times 5.7 \times 6.7$ cm ($9 \times 2\frac{1}{4} \times 2\frac{5}{8}$ in.) deep. *See* **closer.**

queen posts [carp.] The two posts nearest the midspan in a *queen-post truss*.

queen-post truss [carp.] A *truss* which was used for spans from 10 to 20 m (35 to 70 ft) but is now little used for large spans in Britain where steel trusses are cheaper. It differs from the *king-post truss* in having no central post but two queen posts each side of the centre. It is suitable for larger spans also. *See* **Pratt truss** (*C*).

quetta bond A bond like *rat-trap bond* in which gaps are formed in the middle of the wall. The bricks are laid on bed, not on edge, and the usual wall is 34 cm ($13\frac{1}{2}$ in.) thick. The cavities in the middle are filled with grout as the wall rises and contain vertical steel. Each course of both faces is laid in *Flemish bond*. *See* **reinforced brickwork** (*C*).

quick-hardening lime [pla.] *Hydraulic* lime.

quicklime [pla.] Lime (CaO) which has not been slaked. *See* **lime putty.**

quick sweep [carp.] A description of circular work of small radius.

quilt *See* **blanket.**

quilted figure or **blister figure** [tim.] An elaborate *figure* consisting of apparent knolls in *birch* or maple due to uneven *annual rings*.

quirk [joi.] A furrow parallel to a *bead* and terminating it.

quirk bead [joi.] A semicircular *bead* with a *quirk* at one side to mark an edge of a board. A double quirk bead has quirks each side.

quirk router [joi.] A *plane* for cutting *quirks*.

quoin or **coin** (1) An outer corner of a wall.

(2) A brick or large stone set in a salient corner of a wall.

quoin header A corner *header* in the face wall which is a *stretcher* in the side wall.

R

rabbet or **rebate** or (Scotland) **check** [joi.] A long step-shaped rectangular recess cut in the edge of a timber, such as the part cut from each side of a *rail* to form a *tenon* on the end of it.

rabbet plane or **rebate plane** [joi.] A *plane* with a *cutting iron* and mouth reaching to the edge of the *sole*, so that *rabbets* can be cut with it.

raceway [elec.] (USA) A rectangular *duct* in which cables or bus bars travel through a building. It may be a *plugmold*.

racking back or **raking back** The normal way of building a *brick* wall consists of first building the corners or ends very carefully in steps rising one course at a time from the middle part of the wall. The gradual increases of height to the corner are called racking back.

radial brick or **radius brick** A *compass brick*.

radial road [t.p.] A road which leads directly out from the centre of a town.

radial shrinkage [tim.] The drying *shrinkage* of timber at right angles to the *growth rings*. It is about half the tangential shrinkage. *See* flatsawn (illus.).

radial step A *winder*.

radiant heating (1) (USA) *Coil heating*.

(2) [pai.] *Stoving* a finish by radiation from a hot surface.

radiator [plu.] A gilled container, usually for water, often part of a *central-heating* system. Being at a low temperature it loses less heat by radiation than by *convection*, and in theory it should therefore be called a convector.

radio-frequency heating or **high-frequency** or **dielectric capacity** or **electrostatic heating** [tim.] A method of rapidly heating thick *plywood* assemblies for gluing. Electrodes connected to a high frequency power source are placed in the assembly. *Compare* **strip heating**.

radius rod or **gig stick** [pla.] A strip of wood about 5×5 cm (2×2 in.) and slightly longer than the radius of an arch to be moulded. The *mould* is fixed firmly to one end of the rod at the correct radius. The gig stick is then pivoted at the centre of the arch by nailing it to a board. The moulding is formed as for a *horsed mould*.

radius shoe [pla.] A piece of zinc plate screwed to one side of a plasterer's *radius rod* over its centre point. It is drilled so that the centre pin or nail can pass through the radius rod.

rafter [carp.] A sloping timber extending from the *eave* to the *ridge* of a roof. It may be a *common rafter* or a *principal rafter*.

rafter filling or **beam filling** or **wind filling** Brick infilling between rafters at wall plate level.

rag felt (USA) *Bitumen felt*.

raglet or **raggle** or **raglin** A thin groove, in stone often *dovetailed*, cut in stone or in a mortar joint of brickwork to receive the end of a lead *flashing*, which is fixed by *burning in* or wedging.

ragstone Coarse-grained sandstone such as *Kentish rag*.

rag work Thin flat stones built into a *rubble wall*.

rail (1) [joi.] A horizontal secondary member with a *tenon* cut on it, framed into vertical *stiles*, the lowest in a door being the kicking rail.

    (2) [joi.] The upper, continuous part of a *balustrade*, a handrail.

    (3) Any longitudinal member in fencing.

rail bolt [joi.] A *handrail bolt*.

railing (1) An open fence made of posts and *rails*.

    (2) [joi.] A *banding*.

raindrop figure [tim.] A *figure* which may be *mottle* alternating with *ribbon grain*.

rain leader (pronounced leeder) or **rain conductor** (USA) *Downpipe*.

rainwater head or (USA) **conductor head** or **leader head** or **cistern head** The enlarged entrance at the head of a *downpipe*. It collects the water from the *eaves gutters*.

rainwater hopper A hopper-shaped rainwater head, sometimes also used in the middle of a long downpipe.

rainwater pipe A *downpipe*.

raised fibres [pai.] The fibres of wood rise from the surface when it dries naturally or after the application of a coat of paint or varnish. To achieve a glossy paint or varnish surface, the fibres should be sanded off with fine glass paper after every coat but the last.

raised panel or **fielded panel** [joi.] A *panel* which is thicker at the centre than at the edges.

raising plate [carp.] A *pole plate* or *wall plate*.

rake (1) or **batter** An angle of inclination to the vertical.

    (2) [carp.] *See* **pitch** (4).

    (3) [pla.] A pronged tool with a long handle for mixing *haired mortar*.

raked joint A mortar *joint* which has been cleaned of mortar for about 2 cm (¾ in.) back from the face. This is done before *pointing* or plastering.

rake moulding (USA) The sloping *moulding* at the top of a *barge board*, just below the *shingles* or *tiles* on a *gable* end.

raking back *Racking back*.

raking bond *Diagonal bond*.

raking cornice or **coping** A *cornice* or *coping* on a slope, e.g. over a *gable*.

raking cutting Cutting not at right angles.

raking flashing A *cover flashing* used e.g. between a stone *chimney* and a sloping roof. It is parallel to the roof slope and is let into a sloping *raglet*. A *stepped flashing* is not practicable with stone, as the joints are too far apart.

raking out Cleaning out mortar *joints* before *pointing*.

raking rise A *riser* which is not vertical and overhangs the *tread* below, to give more foothold. The tread is bigger than its *going*. *See* **flight** (illus.).

raking shore or **shoring** or **raker** A long *baulk* or several baulks erected to hold up temporarily a wall of a building. It is carried on a wooden *sole plate* at ground level and designed so that one baulk abuts against the building on a *needle* inserted into the wall at each ceiling level. *See* **rider shore**, **wall plate**.

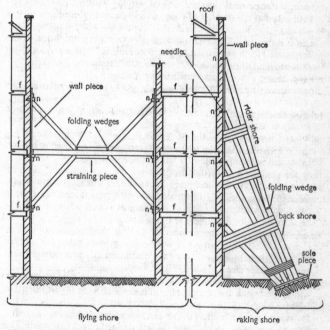

Flying shore and raking shore. Both types of shore bear against vertical wall pieces held in place by needles of timber or steel into the wall f = floors, n = needles.

rammed-earth construction The *cob* construction of the semi-arid parts of USA.

ramp (1) [joi.] A bend in a *handrail* or coping which is concave on the upper side. *Compare* **knee**.

(2) [plu.] A short, steep length of drain pipe.

ramp and twist Description of a shape of a stone surface which is twisted.

ram's horn figure [tim.] A ripple like *fiddleback*.

random ashlar American term for *coursed squared rubble*.

random courses *Courses* of varying depths.

random rubble *Rubble walls* built of stones which are of irregular shape and size and not coursed. See **coursed random rubble**.

random shingles *Shingles* varying from 6 to 30 cm (2½ to 12 in.) or more wide but of uniform length.

random slates or **rustic slates** *Slates* of varying width, sold by the ton with a maker's statement of their covering capacity in square metres per ton. They may be laid in *diminishing courses* or at random. Their rustic appearance is often more attractive than the smoothness of the dark-blue North Wales slates. The best known English randoms are *Delabole* and *Westmorland*. Compare **sized slates**.

random tooled ashlar Stone finished by *batting*.

range masonry or **coursed ashlar** or **range work** (USA) *Regular-coursed rubble*.

ranging line A string stretched between *batterboards* to mark a face of a wall or other line.

rank set [carp.] A set of the *back iron* of a *plane* which leaves a big space between it and the edge of the *cutting iron*. It is used for rough work and is the opposite of a fine set.

rasp See **plumber's tools** (illus.).

ratchet brace or **ratchet drill** [carp.] A carpenter's *brace* for use where, in confined spaces, a full turn cannot be given to the brace. It is fitted with a ratchet and pawl mechanism which allows the bit to be turned clockwise while it is in the hole.

rate of growth [tim.] The number of *growth rings* per cm of timber measured radially. *Softwoods* with many growth rings per cm are stronger than those with few, an indication of rapid growth and weakness. See **stress graded**.

rat-trap bond or **all-rowlock wall** or **silver-lock bond** A brick *bond* in which the bricks are laid on edge in *courses* 11·5 cm (4½ in.) deep, to build a 23 cm (9 in.) thick wall. It consists of two leaves 7·6 cm (3 in.) thick in which headers and stretchers alternate. A cavity, 23×7·6 cm (9×3 in.) in plan, is left opposite each stretcher. The wall is cheap and fairly strong. Compare **Quetta bond**.

raw linseed oil [pai.] Refined *linseed oil* which has not been 'boiled'. Unrefined linseed oil is never used in painting.

Rawlplug A small drilled plug, made usually of wood fibre, occasionally of soft metal (see **Rawlbolt** (C)) inserted into a hole in a wall, as a fixing for a nail or screw. Some metal Rawlplugs are made with a wide collar at the outer end to prevent them passing through a bottomless hole (screw anchor).

ray [tim.] *Wood ray*.

ready-mixed concrete *See C.*

real estate (USA) Land and buildings.

rebate *See* **rabbet**.

rebated weather-boarding [carp.] *Weather-boarding* of wedge-shaped cross section with a *rebate* along the inner face of its lower edge so that the top (thin) edge of the lower board fits into the rebate in the lower (thick) edge of the upper board. *German siding* and *shiplap siding* are also both rebated but not specially so called. *See* **siding**.

receptacle [elec.] (USA) *Socket outlet.*

receptor [plu.] (USA) The shallow sink of a shower bath. *See* **lead safe**.

recessed head screw [joi.] A *screw* with a head not slotted across its full width like the common wood screw but with a cross-shaped recess into which a cross-shaped screwdriver blade fits. The grip between screwdriver and screw is better, and the head is stronger and less likely to break off than with the slotted head. One disadvantage is that certain *protective finishes* are unsuitable. *Galvanizing* (*C*), *sherardizing*, hot-dip tinning, *japanning*, and Berlin black cannot be used, as they spoil the fit between the screwdriver and the recessed head. Other finishes such as *phosphating* and copper plating are suitable.

recessed pointing or **recessed joint** The mortar *joint* in *pointing*, is set back about 6 mm ($\frac{1}{4}$ in.) from the face of the wall. It therefore is strongly shadowed. There is less danger of mortar peeling than with joints which are near the face.

reciprocating drill [joi.] A small drill with a steeply threaded shaft which carries the drill *chuck* (*C*) at its lower end. A sleeve which fits this thread is pushed down the shaft and so rotates it. When the sleeve is drawn back up the drill, the grip within the sleeve does not hold the thread nor turn the shaft. It is used for drilling holes less than about 3 mm ($\frac{1}{8}$ in.) dia.

reconditioning [tim.] High temperature steam treatment of *hardwood* which has suffered *collapse*. It also reduces *warp*. The process was developed by Australians for their hardwoods.

reconstructed stone *Cast stone.*

rectangular tie (USA) A *wall tie* made of a bent rectangle of heavy wire about 5×15 cm (2×6 in.).

red lead [pai.] A red *pigment* consisting of mixed oxides of lead, mainly Pb_3O_4. It is an *inhibiting pigment*, is used on wood as well as on steel, but like most *lead paints* is poisonous.

red oxide [pai.] Red iron oxide, a *pigment* which does not inhibit corrosion.

reducer (1) [pai.] A *thinner*.

(2) [plu.] A taper pipe reducing in diameter in the direction of flow.

reducing power [pai.] The strength of a *white pigment*, that is the paleness of tint produced when a standard amount of it is mixed with a standard amount of a coloured *pigment*. The paler the tint the greater is the reducing power of the white. *Compare* **staining power**.

redwood (Pinus sylvestris) or **red deal** or **yellow deal** or **northern pine** or **Scots pine** or **fir**, etc. [tim.] A *softwood* which grows throughout northern Europe and Asia, used for *carpentry* and *joinery*. It is easy to work, durable, machinable, takes *nails* and *screws*, and also *glues*, *paints* and *stains* well. The word redwood is usually confined to timber imported to Britain, often from the Baltic.

reed In *thatching*, the best reed is arundo phragmites, but in the west of England reed means wheaten straw. *See also* C.

re-entrant corner An internal angle or corner, the opposite of a (projecting) *salient* corner.

refractory mortar *Mortar* suitable for boiler settings, or other furnaces. One suitable mix is 3 parts of *grog* to 1 of *high-alumina cement* (C). Another is 2 of *fireclay* to 3 of *grog* with sodium silicate solution.

refrigerant The working fluid in a *refrigerator*, which alternately vaporizes to cool the refrigerator and is compressed to liquid again by a compressor (*Freon*, sulphur dioxide, etc.).

refrigerator [mach.] The domestic refrigerator consists of a small compressor for the cooling fluid, often *Freon* or sulphur dioxide (SO_2), which is compressed to about 400 kN/m² (60 psi), cooled in a finned tank outside the refrigerator and then expanded to gas inside the tubes of the cold cupboard. This expansion and volatilization of gas takes up heat from the cold cupboard and gives it out in the cooling tank outside.

register (1) A *damper* to vary a chimney draught.

(2) An outlet into a room from a ventilating duct, provided with a *damper* to regulate the volume of warm air discharged into it. *Compare* **grille**.

reglet A *raglet*.

regrate To remove the outer surface of *hewn stone* so that it looks new.

regular-coursed rubble or (USA) **range masonry** or **coursed ashlar** or **range work** Coursed *rubble walling* in courses of uniform height, generally 7 to 15 cm (3 to 6 in.) but each different.

regulus metal or **antimonial lead** [plu.] An alloy of lead which is very much harder than lead. It contains about 10% antimony and is used as a wall *cladding* in sheet form.

reinforced bitumen felt A light *bitumen felt* made of fibre saturated with bitumen. A layer of loose jute hessian is embedded in the bitumen on one side. It is used as *sarking felt* on unboarded roofs and weighs about 1·2 kg/m² (¼ lb/ft²).

reinforced cames Steel cored *cames* for *leaded lights*, with a *lead sheath*.

reinforced concrete *See* C.

reinforced masonry *Masonry* of stone or *building blocks* reinforced in the bed or vertical joints like *reinforced brickwork* (*C*).

reinforced woodwool Building slabs of *woodwool* stiffened lengthwise with wood *battens*, or pressed steel U-sections.

rejointing *Pointing*.

relieving arch or **rough arch** or **discharging arch** An arch built over and clear of, a wooden *lintel* or other weak support, to carry load. It is generally of rectangular bricks with wedge-shaped mortar *joints*.

render and set [pla.] Two-coat plaster on walls, *rendering* covered by a *finishing coat*. It is used on *gypsum plasterboard*, *insulating board*, etc.

render, float, and set [pla.] *Three-coat plaster* on walls.

rendering (1) Applying *coarse stuff* to a wall or cement *mortar* to the inner face of a manhole or *stucco* to an outside wall. This is often done by bricklayers, but plasterers usually do any rendering which is to be covered by a *finishing coat* indoors or out of doors. *See* BSCP 221, External rendered finishes.

(2) Bedding slates in haired mortar (*torching*).

rendering coat [pla.] A first coat of *plaster* on a wall, usually about 1 cm (⅜ in.) max. thickness. A first coat on *lathing* is called a pricking-up coat. *See* **coarse stuff, stucco**.

re-saw [tim.] (1) To rip sawn timber into smaller sizes.

(2) A mechanical saw which re-saws.

resident architect An *architect* at a site who watches the interests of the *building owner* during construction, working under the *consultant*.

residual tack [pai.] A fault of finishes which do not harden, caused by vegetable oils such as dehydrated castor oil. It may last indefinitely in damp air.

resin bonded [tim.] Glued with synthetic *resin*, therefore usually moisture-resistant.

resin chipboard [tim.] *See* **chipboard**.

resin-impregnated wood [tim.] A *high-density plywood*.

resin pocket or **streak** [tim.] *Pitch pocket*.

resins (1) The natural resins are obtained from the sap of plants or pine trees or are found in the ground near where the trees have been. They are of three sorts, the gum resins (asafoetida, frankincense, and myrrh), the hard resins of which all varnishes contain some, and the *oleo-resins*. Examples of the hard resins are *copal*, *rosin*, *mastic*, and *amber*. Examples of the oleo-resins are Canada balsam, *dragon's blood* (from a palm tree of the Molucca Isles), and true *lacquer* from a sumach tree (used in China for about 1300 years). All hard resins are insoluble in water but may be dissolved in organic *solvents* or vegetable oils. Some gums are soluble in water, particularly those used as *glues*.

(2) *Synthetic resins* are organic compounds of which the first discovered members resembled the natural resins. They are well known in the timber trade for their outstanding qualities as *glues*,

also in the paint industry. Examples of popular resins are *urea-formaldehyde* (UF), *phenol formaldehyde* (PF), resorcinol formaldehyde (RF), *melamine formaldehyde* (MF), and *alkyd* resins. Some can be used as casting resins or for gluing metal (epoxide resins). *See* **polymer.**

resistance *See* **surface resistance** The thermal resistance is the reciprocal of the *conductance* and in this the electrical resistance is analogous.

resorcinol formaldehyde resin [tim.] A synthetic *resin.*

rest bend A *duck-foot bend.*

retarded hemihydrate plaster [pla.] *See* **hemihydrate plaster.**

retarder [pla.] An *admixture* such as keratin added to *plaster of Paris* to reduce its hardening rate enough for it to be used as a wall or ceiling plaster. *See also* C.

re-tempering [pla.] The practice of remixing *mortar* or plaster which has begun to set. It should be, and usually is, forbidden.

retention money [q.s.] A percentage of the money due to a *contractor* who has completed his contract. This money is not paid to him until the end of the *maintenance period* when the building owner and his consultants are satisfied that the contractor has fulfilled his obligations.

return (1) A change of direction of a wall or other member, usually at right angles.

(2) A *return pipe.*

return bead [joi.] A *bead* with a *quirk* each side of it on a salient corner.

returned end An end of a *moulding* shaped to the profile of the moulding.

return pipe [plu.] The pipe by which the water leaves the hot-water *cylinder* and returns to the boiler in a water-heating system. *See* **primary flow and return.**

return wall or **returned corner** or **return end** A short length of wall perpendicular to an end of a longer wall.

reveal The outer part of a *jamb* visible in a door or window opening and not covered by the frame. It is the visible side to a window or other opening.

reveal lining [joi.] The finish over a *reveal.*

reveal pin or **reveal tie** A screw inside a window opening used for clamping *tubular scaffolding* to the opening.

reverberation period or **reverberation time of an enclosure** The period of time, in seconds, required for sound of a certain frequency to decrease, after the source is silenced, to one millionth of its initial value, that is by 60 *decibels.* It depends mainly on the volume and on the *absorption* of the room. The higher the absorption, the lower is the reverberation period.

reverse [pla.] A *templet* cut to the reversed shape of a *moulding* and placed on it to check its accuracy.

reversible lock [joi.] A lock in which the *latch* can be taken out and reversed. This enables it to be used on a door of opposite *hand*, or on the opposite *stile* of the same door. In other words, the hand of the lock is reversible.

revolving door or **swing door** A door consisting of four leaves, ordinarily at right angles to each other, pivoted on a central post. They may also be locked parallel to each other in the shut or open position. It is used in public buildings, and acts as an air lock with the leaves at right angles to each other turning in contact with a circular vestibule (tambour).

rhom brick An experimental brick shaped like a parallelogram. The face resembles ordinary brickwork but the irregularity of shape makes the *bonding* easier, since it is very simple to prevent vertical joints coming parallel on top of each other. Special bricks with right-angle corners are needed for the ends of walls.

rhone or **rone** (Scotland) An *eaves gutter*.

R I B A The Royal Institute of British Architects.

riband [carp.] (1) A timber which runs along the head of the struts under a reinforced concrete beam, to fix the *formwork* (*C*) to the beam sides and to carry the weight during casting (*ribbon board*).

(2) A *rail* in a palisade.

ribbing [tim.] A corrugated surface of timber due to the different drying *shrinkages* of *springwood* and *summerwood*.

ribbing up [joi.] Building up circular *joinery* like *laminated wood* by gluing *veneers* together with parallel grain.

ribbon (1) An occasional course of ornamental slates or tiles.

(2) A *ribbon board*.

ribbon board or **ribbon strip** or **ribbon** or **girt strip** or **ledger board** [carp.] A *joist housed* into the *studs* at ceiling level, supporting the floor joists in *balloon framing*.

ribbon courses Alternate courses of slates or tiles, laid to shorter or longer *gauge*, thus giving alternately long and short exposed depths.

ribbon grain or **ribbon stripe** or **stripe figure** or **roe figure** [tim.] Alternating light and dark strips 3 to 13 mm ($\frac{1}{8}$ to $\frac{1}{2}$ in.) wide in *quartered* timber formed by the different reflections of light in *interlocked grain*.

ribbon rail [joi.] A metal *rail* connecting the tops of metal *balusters*, usually covered by a wood handrail.

ribbon saw [tim.] A narrow *band saw* about 5 cm (2 in.) wide only.

ribbon strip [carp.] A *ribbon board*.

ribbon stripe [tim.] *Ribbon grain*.

rich lime [pla.] *High-calcium lime*.

rich mix A mix of *concrete* or *mortar* which contains more cement than usual, or a plaster mix with less sand and more *gypsum plaster* or *cement* than usual.

riddle A coarse sieve.

ride [joi.] (1) A door which touches the floor when it opens is said to ride. This can sometimes be prevented by *cocking* the hinges.

(2) A joint which meets in the middle and is open at the ends is said to ride.

(3) A pin in a *strap hinge*.

rider shore [c.e.] A short, topmost *raking shore* which rests not on the ground but on a *back shore* on top of the highest full-length raking shore. Riders are not always needed but may be used for tall buildings.

ridge (1) The apex of a roof, usually a horizontal line.

(2) or **ridge board** or **ridge pole** or **ridge piece** [carp.] The horizontal board set on edge, at which the *rafters* meet. It is about 2 to 4 cm by 23 cm (1 to 1½ by 9 in.) in cross section.

ridge capping or **ridge covering** A covering over a *ridge*. It may be of *roofing felt*, purpose-made asbestos-cement sheet, clay roofing tile, sawn slate, sheet metal, or steel angle, depending on the roof construction.

ridge course The topmost course of slates or tiles, next the ridge, cut (or purpose-made) to the required length.

ridge pole [carp.] A *ridge*.

ridge roll [carp.] A *hip roll*.

ridge stop At an intersection between a ridge and a wall rising above it, a piece of *flexible metal* flashing dressed over the ridge and up the wall.

ridge tile A concrete or burnt clay *tile* for covering a *ridge*, made in four typical shapes in Britain: (1) Half-round, that is a semicircle of 20 cm (7¾ in.) internal dia., (2) hogback, that is half-round tiles which are cocked up in the middle and are therefore slightly sharper at the summit, (3) segmental, that is, flattened half-round tiles, (4) angle tiles which are made with a sharp right-angle bend and flat surfaces on each side of it. All these tiles are 30 or 46 cm (12 or 18 in.) long, usually bedded in mortar, and are often the same as *hip tiles*.

ridging Covering the *ridge* with tiles or other capping.

riffler A rasp bent for abrading concave surfaces, a *fillet rasp*.

rift sawn [tim.] *Quarter-sawn*.

rigger [pai.] A long-haired *brush* with a flat end for painting lines or bands of different thicknesses. *See also C*.

rim latch [joi.] A metal box screwed on to the inner face of the *shutting stile* of a door. It contains a *latch* for fastening the door, the latch being turned by door knobs. This is cheaper than a *lock*.

rim lock [joi.] A rim *latch* which can also be locked.

rindgall [tim.] A *callus*.

ring course The course nearest the *extrados* in an arch several *courses* deep.

ring main [elec.] A convenient method for wiring power circuits in houses, standardized in Britain about 1950, and in wide use in 1973. A ring of about ten power points is allowed by most authorities. Each *socket outlet* is connected to the main through two lengths of cable which together form the ring. Each *plug* contains its own fuse which protects its own appliance and, with the other plug fuses, the ring main. *See* **spur**.

ring-porous wood [tim.] *Hardwood* containing more and larger pores in the *springwood* than elsewhere. These pores show up as a ring. The contrary of *diffuse porous wood*.

ring shake or **cup shake** or **wind shake** [tim.] *Shake* along one or more *growth rings*, caused by wind or during seasoning. *See* **shell shake**.

rip [tim.] To saw parallel to the grain, also called *flat cutting*. *See* **cross-cut**.

ripper (1) A slater's long thin cranked blade which is inserted under the *slates* and pulled back to cut the *slate nails* when a roof is being repaired. *See* **plasterer's tools** (illus.).

(2) [tim.] A *rip saw*.

ripple [tim.] A beautiful *figure* caused by the buckling of fibres during growth. It is often seen in sycamore. *See* **fiddleback**.

ripple finish [pai.] An intentional uniformly wrinkled finish which is obtained usually by *stoving*.

rip saw or **ripping saw** [carp.] A *handsaw* made to cut parallel with the grain. It can have as few as 1·5 points per cm (4 per in.). *See* **setting**.

rise (1) The vertical height of an arch measured from *springing line* to the highest point of the *intrados*.

(2) The vertical height from the supports to the ridge of a roof.

rise-and-fall table [tim.] A circular-*saw bench*, which can be raised or lowered relatively to the saw.

rise and run [carp.] The *pitch* (1) of a member expressed as a certain vertical height for so much horizontal run.

riser (1) The upright face of a step; *see* **flight**.

(2) In *snecked rubble* a deep stone which is thicker than one *course* and may also be a *bond stone*.

rising-and-falling table [tim.] A *rise-and-fall table*.

rising and lateral conductors [elec.] Power or lighting cables in a *branch circuit*.

rising-butt hinges or **rising butts** [joi.] *Hinges* which cause a door to rise about 1 cm ($\frac{1}{2}$ in.) when opening. They are made with a helical bearing surface between the two leaves. The door therefore tends to close automatically as well as to clear a carpet when opening. They have the further advantage over ordinary *butts* that the door can be lifted off its hinges without unscrewing them.

rising main An electrical power supply cable or main gas or water supply pipe which passes up through one or more storeys of a building. (For water, BS 4118:1967 prefers 'rising pipe').

rive [tim.] To split *shingles*, *laths*, etc., making riven shingles or laths.

rivelling [pai.] *See* **wrinkle finish.**

riving knife [tim.] A steel blade which projects up out of a saw-bench. It protects the back edge of the *circular saw* blade and its thickness ensures that the wood shall not bind on the saw.

rocket tester or **smoke rocket** [plu.] A rocket which gives off a dense, lasting smoke which is directed into a drain under test. The drain is plugged at both ends and subjected to a slight excess internal pressure of 3·8 cm (1·5 in.) water gauge.

rock face The surface texture of *pitch-faced* stone.

rock wool *See* **mineral wool.**

rocking frame An oscillating frame on which moulds are set during the placing of *concrete*. The oscillation helps to compact the concrete.

rod (1) [joi.] A board on which the dimensions of a piece of *joinery*, etc., are set out full size, for instance, a *storey rod* for setting out *stairs*.

(2) *See below.*

rodding (1) Cleaning out drains with *drain rods* (*C*).

(2) (USA) Levelling plaster with a *floating rule*.

rodding eye An *access eye*.

roll (1) or **wood roll** A piece of wood over which *flexible-metal roofing* sheets are lapped and folded.

(2) *See* **double roll, double Roman, hollow roll, solid roll,** *and below.*

roll-capped ridge tiles *Ridge tiles* with a cylindrical projection above them.

rolled-strip roofing *Roll roofing.*

roller coating [pai.] (1) Applying *paint* by stationary plant consisting of rollers such as are used for roller-coating enamel.

(2) Applying paint or *enamel* or *distemper* by hand-operated roller.

roller-coating enamel [pai.] *Enamel* which is applied to steel or other metal sheet by the rollers of a roller-coating machine.

rolling shutters Shutters used for large doors or windows which are taken up or let down from a roller carried by a *lintel* over the opening. They may be of wood or metal. Metal rolling shutters give good fire protection.

roll roofing or **rolled-strip roofing** US terms for any roofing material sold in rolls like *roofing felt*.

rolok *See* **rowlock.**

Roman bricks (USA) Bricks which measure 30 to 40×10×5 cm (12 to 16×4×2 in.) including the mortar *joints*. They therefore lay up six courses per ft (30 cm) height.

Roman cement or **Parker's cement** Properly speaking *pozzolanas* (*C*), but the term has come to mean, in England, a cement made in the nineteenth century by calcining hard nodules of marl found in the London clay. It was the forerunner of *Portland cement*.

Roman mosaic or **tessellated pavement** [pla.] A *terrazzo* laid uniformly with pieces of marble about 1 cm (½ in.) square (tesserae) placed by hand. Geometrical patterns can be made by gluing the stone on to strong paper, then inverting the paper over the floor covered with wet *mortar*, when the stones and paper are pushed into the mortar. The paper is later removed and the stones are not abraded after laying. *See* **mosaic cutter.**

Roman tile A British, clay *single-lap roofing tile*. The single Roman tile is not standard and has one waterway. The *double Roman tile* is standardized and has two waterways divided by a central roll.

rone or **rhone** (Scotland) An *eaves gutter*.

roof A covering to throw off rain, etc., from a building. For domestic roofs and spans up to about 4·5 m (15 ft) *common rafters* are usual. Other timber roofs are the *king-post, queen-post, Belfast,* and *bow-*

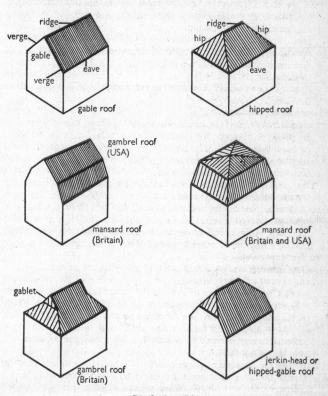

Pitched roofs.

string (*C*) trusses. Glued *laminated-wood* arches are also suitable for spans of the length of the bowstring truss (about 30 m (100 ft)). Generally in Britain now for ordinary buildings, timber roof *trusses* of larger than 6 m (20 ft) span are uncommon, because steel is cheaper, though sometimes less beautiful, and concrete is more *fire resisting* than either. *See* **concrete roofs** (*C*).

roof boards or **sheathing** Boards laid touching each other, usually with tongued and grooved joints, nailed to the *common rafters* as a base for asphalt, or *flexible-metal roofing*, or *roofing felt* under slates or tiles.

roof cladding Slates, tiles, *asbestos cement* sheet, *flexible metal*, corrugated steel sheet, *protected metal*, etc. *See* **cladding**, BSCP 143, 144.

roof covering *Roof cladding*.

roof decking Lightweight panels made by many different roofing specialists. They span up to 4 m (13 ft) and usually combine a structural material (aluminium, steel, or timber) with an *insulator* (woodwool, strawboard) and are usually covered with roofing felt or *flexible-metal sheet*. The weight is about 30 kg/m² (6 lb/ft²). This decking is popular when a building must be roofed quickly. Panels measure about 2 m² (21 ft²) and a roof is therefore quickly covered. *See* **built-up roofing, insulated metal roofing**, BSCP 199, Roof deckings.

roof guards *Snow boards* (2).

roofing felt or (USA) **rag felt** Sheets of matted fibres, treated with coaltar pitch, or bitumen, for waterproofing, and laid either under slates or in *built-up roofing* or as an *underlay* or for other purposes. Many sorts exist, mainly *bitumen felts*, and *impregnated felts*.

roofing-felt fixer or **roofing-felt layer** A *skilled man* who cuts to shape, lays, and fixes roofing felt. He joints the felt by lapping it, or mopping it with *sealing* or *bonding compound*, or by heating it with a *blow lamp*.

roofing nails The slating and tiling code, BSCP 142:1971, states that plain steel or galvanized nails should not be used. The recommended choice is between aluminium, copper, stainless steel and silicon bronze nails, the last being probably the most durable.

roofing paper *Building paper*.

roofing square [carp.] A *steel square*.

roofing tiles *Concrete*, burnt-clay, or *asbestos-cement* tiles for covering roofs. Clay tiles are the oldest and the most pleasant in appearance, for which reason the other types copy some of the shapes and colours of clay tiles. Clay tiles are of three general types: (*a*) *plain tiles*, (*b*) *single-lap* tiles, (*c*) *Italian tiling*, or *Spanish tiling*, expensive and heavy, but very decorative. *See* BSCP 142, Slating and Tiling.

roof ladder A *cat ladder*.

roof light or **skylight** A *dome, lantern, monitor, north light, patent glazing*, or a *saucer dome* in a roof. It may be an *opening light* or a *dead light*. *See* BSCP 153, Windows and Rooflights.

roof light sheet (1) An *asbestos-cement* roofing sheet with an opening, generally an upstand, in the middle for *glazing*.

(2) A corrugated sheet of plastic transparent material (*Perspex* or glass) which may be used as glazing for an opening light.

roof screen or **roof curtain** A vertical screen fitted into the roof space of a building so as to divide it into bays and to confine smoke and hot gases to the bay of origin.

roof space Unused space between the roof and the *ceiling* of the highest *storey*.

roof terminal (USA) The open end of a *ventilation pipe* at the roof.

roof tree [carp.] An old term for a *rafter* or the *ridge* of a roof.

roof truss A wood or metal frame which carries a roof. See **truss**.

roomheater A solid-fuel-burning appliance, usually now with fire doors of *heat-resisting glass*, formerly known as a slow combustion stove. Many now have *back boilers* that can heat six radiators and domestic hot water. BS 1846 mentions five types: free-standing, stand-in with either surround seal or chimney seal, built-in, and hopper feed.

room-sealed appliance [plu.] A gas appliance having the air inlet and the flue outlet isolated from the room where the appliance is installed. See **balanced flue**.

root [carp.] The part of a *tenon* which widens out at the shoulders.

ropiness [pai.] A surface in which *brush* marks have not flowed out is called ropy. This is caused by bad *flow* of the *paint* or *varnish* or by brushing after it has begun to harden.

rose (1) A decorative plate through which a door handle passes.

(2) A decorative plate or boss through which an electric light flex hangs from a ceiling (ceiling rose).

rose bit [carp.] A *countersink* bit for wood.

rose nail [carp.] A *wrought nail*.

rosin or **colophony** A natural *resin*, the solid residue from the distillation of *turpentine*, used for making *varnishes*, *size*, and as a soldering *flux*.

rot [tim.] *Decay* of timber.

rotary cutting or **peeling** [tim.] A *veneer* cutting method in which the hot soaked log is taken from the *cooking vat*, mounted in a powerful lathe and turned at a speed of 30 to 60 rpm (depending on the log dia.) against a long knife. The knife cuts off a continuous sheet of veneer, which is thus produced at a speed of about 1 m/sec (200 fpm); 90% of veneer is cut in this way, including most of the veneer from which *plywood* is made, but most *figured* veneers are made by *slicing* or as *half-round veneer*. See **birdseye**.

rotary veneer [tim.] *Veneer* made by *rotary cutting*.

rotten knot [tim.] An *unsound knot*.

rough arch A *relieving arch*.

rough ashlar A block of stone as brought from the quarry.

rough axed *See* **axed bricks**.

rough back An end of a stone hidden in a wall. *See* **clean back**.

rough bracket [carp.] A *bracket* under a stair.

rough carriage [carp.] A *carriage* under a stair.

rough cast or (Scotland) **harling** [pla.] Originally (Tudor times) consisted of a *rendering* of *coarse stuff*, rendered a second time and roughened. The rough cast, that is fine shingle mixed with hot *hydraulic* lime, was then thrown on. Nowadays the coarse stuff has a different composition, usually having some cement, and often including colouring matter (red iron oxide, yellow ochre, crimson lake) but always with pebbles or chips. *See* **pebble dash**.

rough coat [pla.] The *rendering coat* on a wall.

rough cutting [q.s.] Cutting of common brickwork. It is measured as an area, and if faced is presumed to have a 11·5 cm (4½ in.) thick facing which is measured elsewhere as *fair cutting*. A 34 cm (13½ in.) wall with facing would have 23 cm (9 in.) of rough cutting and a length of 4 m would be measured as $4 \times 0.23 = 0.92$ m². The fair cutting for the same job would be 4 linear m.

rough floor [carp.] (USA) Rough floor boards about 23 cm (9 in.) wide, usually square-edged, on which the finished floor is laid. In the best work a layer of building paper separates the two floors. *See* **subfloor**.

rough grounds *Common grounds*.

rough hardware (USA) Bolts, *nails*, *spikes*, and other metal fittings which are not seen and therefore are not *finish hardware*.

roughing-in Doing the first rough work of any trade such as plastering or plumbing. In *plumbing*, installing all water pipes as far as the points where they must be joined to their fixtures.

roughing-out (1) [carp.] Roughly shaping a piece before the final accurate work is done on it.

(2) [pla.] Laying a *core* of plaster of a large *moulding* for which the horsed mould is covered with a *muffle*.

rough plate Heavy *sheet glass*.

rough string [carp.] A *carriage* under a *stair*.

rough work Brickwork, which will eventually be hidden by plaster, *facing bricks*, *joinery*, etc.

round (1) A *rung* of a *ladder*.

(2) [joi.] A *plane* which cuts a groove. *Compare* **hollow**.

(3) Viscous, of paint, varnish etc. A round coat is a thick coat.

rounded step or **round-end step** A *bullnose tread*.

round knot [tim.] A *knot* cut roughly at right angles to its length. *Compare* **splay knot**.

round-log construction [carp.] Building houses with round logs, once practised in the forest clearings of North America.

round timber [tim.] Felled trees before they are *converted*.

round-topped roll In *flexible-metal roofing*, a joint formed over a *wood roll* with vertical sides and rounded top.

rout [joi.] To cut and smooth wood in a groove. *See* **router**.

router [joi.] (1) A plane developed from the *old woman's tooth*, often called a *plough*.

(2) The side wing in a *centre bit* which ploughs out the wood.

(3) A hand-held machine with a *cutter block* driven by an electric motor. Specialized variations of this machine up to 3 hp are used for cutting out for hinge *butts*, for *weather strips*, and even for lock *mortises* 11 cm ($4\frac{1}{2}$ in.) deep.

rove A washer, used with copper boat nails, of a truncated cone shape that flattens and widens to make a watertight joint when driven home by the nail head.

rowlock (1) or **rolock** A brick-on-edge course. *See* **rat-trap bond**.

(2) Decorative brickwork paving consisting of a ring of *headers* enclosing a panel of *stretchers* in a radiating pattern, the whole being set within a chessboard pattern of headers.

rowlock arch (USA) A brick *relieving arch* of concentric courses of *headers* set on edge.

rowlock-back wall A wall faced with bricks laid on bed and backed with bricks laid on edge.

rowlock cavity wall or **all-rowlock wall** A wall built in *rat-trap bond*.

royals *Western red cedar* shingles 60 cm (2 ft) long and 1·3 cm ($\frac{1}{2}$ in.) thick at the butt.

r.s.j. A *rolled-steel joist* (C).

rubbed brick or **rubber** A soft smooth brick without *frog*, suitable for rubbing to shape for *gauged brickwork*.

rubbed finish (1) A finish on concrete obtained by rubbing down with a carborundum stone or similar *abrasive*.

(2) [pai.] *See* **flatting down, polishing varnish**.

rubbed joint [joi.] A glued joint formed between two narrow boards to make of them one wider board. Both boards are planed smooth with a *jointer*, coated with *glue*, and then rubbed together until no more glue and air can be expelled from the joint. No clamp is needed and the joint is very strong.

rubber (1) or **polisher** or **terrazzo polisher** or **floor rubber** A skilled *labourer* who smoothes or polishes *terrazzo* surfaces with carborundum stone or by machine. He may correct irregularities by adding a little grout before the final smoothing.

(2) A *rubbed brick*.

rubber-bag moulding [tim.] *Flexible-bag moulding*.

rubber latex floor *See* **cement-rubber latex**.

rubber sheet A good but expensive floor covering, 3 to 13 mm ($\frac{1}{8}$ to $\frac{1}{2}$ in.) thick, obtainable in rolls up to 1·8 m (6 ft) wide and 30 m (100 ft) long. Tiles can also be obtained. The sheet or tiles can be glued to any dry floor. *See* **cement-rubber latex, sponge-backed rubber, polythene**.

rubbing (1) [pai.] *Flatting*.

(2) Rubbing with a carborundum or other stone.

rubbing-bed hand or **stone rubber** or **rubber down** or **floatsman** A *mason's labourer* in the stonemason's yard who helps the masons rubbing the stones to their final surface.

rubbing stone A grit stone on which *bricks* are rubbed to smooth them after they have been *axed* roughly to shape, for a *gauged arch*, etc.

rubbing varnish [pai.] A *flatting varnish*.

rubbish pulley A *gin block*.

rubble (1) Broken bricks, old plaster, and similar material.

(2) Walling stones which are not smoothed to give fine joints like ashlars but are sometimes squared and coursed. *See* **rubble walls**.

rubble ashlar An *ashlar*-faced wall, backed with rubble.

rubble walls Stone walls which differ from *ashlar* walls in the thickness of the mortar *joints*, which may be as much as 2·5 cm (1 in.) thick, *Random rubble* and *snecked rubble* are the main types. They are either uncoursed or occasionally coursed, but *squared rubble* is usually coursed, since uncoursed squared rubble would be the same as snecked rubble. *See* **chequer work**, *also* BSCP 121.

rule (1) A *straight edge* of any length or construction, for measuring, or for drawing straight lines or setting them out on the site.

(2) [pla.] A straight edge for working plaster or *dots* to a plane surface or for other purposes. They are of several types: *floating rules, joint rules, levelling rules, running rules*.

run (1) A *barrow run*.

(2) The general layout of the pipes or cables in a building.

(3) [plu.] The part of a pipe or *fitting* which is in the main line of flow. *See* **tee**.

(4) [pla.] To pass plaster or *lime putty* through a sieve.

(5) [carp.] The distances between the supports of a *stair* string or of a *rafter*.

(6) [pai.] A narrow ridge of *paint* or *varnish* which has flowed down, advancing from a small bulb or teardrop at the lower end of the painted area. The run continues after the paint has begun to set. A tear is a run like a teardrop.

rung or **round** [carp.] A horizontal bar used as a step in a *ladder*.

run line [pai.] A straight line painted by using a *lining tool* and straight edge, or by stencilling.

runner (1) [joi.] The guide in front of a *plough*.

(2) [carp.] A horizontal timber which carries the *joists* of the formwork under a concrete slab, or the folding wedges of an arch.

(3) A long strip of *withies* laid with others in horizontal bands on *thatch* and held down by *spars*. Straw bands are sometimes used instead.

running [pla.] (1) Slaking *lime* and pouring the *milk of lime* thus formed through a sieve into a *maturing pit*.

(2) Forming a *moulding* such as a *cornice*, in place with a *horsed mould*. *See* **running mould, running rule.**

running bond *Stretcher bond.*

running mould or **horse** [pla.] A *horsed mould.*

running-off [pla.] Applying the finishing coat of plaster to a *moulding*.

running plank or **gang plank** A plank in a *barrow run*.

running rule or **slipper guide** [pla.] A 5×2 cm ($2 \times \frac{3}{4}$ in.) *straight edge* nailed to the *floating coat* below a *cornice* moulding which is to be run. The *horsed mould* rests on it and slides along it, its upper end being held by the *nib guide*. *See* **running screed.**

running screed [pla.] A narrow band of plaster used instead of a *running rule* for running a *moulding*.

running shoes [pla.] Metal pieces on a *horsed mould* where it touches the *nib guide* and *running rule* to enable the *mould* to slide easily and prevent it wearing.

rusticated ashlar *Ashlar* on which the face is left rough and stands out from the *joints*, the stones being cut back at the edges by bevelling or rebating (*see* **drafted margin**). In other types of rustication (rustic quoins), alternate stones project about $2 \cdot 5$ cm (1 in.) beyond the others.

rustic brick A *brick* with a surface which has been roughened by covering it with sand, by impressing it with a pattern, or by other means. Rustic bricks are often of variegated colours and are then prized as *facing bricks*.

rustic joint The *joints* sunk back from the surface of stone, seen in *rusticated ashlar.*

rustics *Rustic bricks, rustic slates*, etc.

rustic siding [tim.] *Drop siding.*

rustic slates *Random slates.*

rust joint A joint between cast-iron pipes, *eaves gutters*, etc., made with *iron cement*, little used now. *See* **glassfibre reinforced resins.**

rust pocket An *access eye* at the foot of a ventilating pipe where rust collects and from which it can be removed.

rybate or **rybat** (Scotland) A rebate or *jamb stone* forming a rebate, that is a reveal stone.

S

sabin The unit of sound *absorption*, equivalent to the absorption of 1 ft² (0·092 m²) of open window (after Dr Sabine, pioneer of acoustics).

sable pencils [pai.] Pointed *brushes* of sable hair, set in quills varying in size from lark's through crow's to duck's and goose's, or set in a metal ferrule. Sable is very expensive and squirrel hair may be used instead. Used for decorative detail.

sable writer [pai] A brush like a *sable pencil* but longer, with a flat edge or a point, used for sign writing.

saddle (1) or **saddle piece** A piece of *flexible metal* about 46 × 46 cm (18 × 18 in.) dressed to shape and fixed under *slates* or *tiles* at vulnerable points such as the intersection of a *dormer* ridge and the main roof.

(2) A large roll on a flat roof which divides it into bays.

(3) [carp.] A *bolster*.

(4) A fixing lug which passes round a pipe or *conduit* and is screwed down each side of it.

(5) To make a connection to an existing pipe run or sewer, a saddle may be bolted over the pipe after breaking into or drilling it. The saddle makes a watertight or gastight joint to the branch drain or pipe.

saddle-back board [carp.] A narrow floor board chamfered on both edges, fixed under a door and projecting above the boards on each side of it. If the door is hung to close on this board, it will clear the carpets when open.

saddle-back coping A *coping* with a sharp apex and flat flanks which slope down from it each side.

saddle bar A horizontal metal bar which stiffens *leaded lights*.

saddle bead A *glazing bead* for fixing the glass to each side of a curved *glazing bar*.

saddle joint or **water joint** A *joint* between stones in a *cornice* in which the stones meet in a saddle shape to throw the water away from the joint.

saddle piece A *flexible-metal* saddle.

saddle roof A *pitched roof* with gables.

saddle scaffold A *scaffold* built over a *ridge* and sometimes supported by *standards* at the side of the building. It is used for repairing chimneys.

saddle stone The *apex stone* of a gable.

safe (1) A *lead safe*.

(2) A strong box.

safety arch A *relieving arch*.

safety glass (1) Glass containing thin wire reinforcement, such as *Georgian glass*.

(2) or **armourplate** or **toughened glass** Glass toughened by heat treatment. When struck, it granulates instead of splintering.

(3) Sheets of glass glued to transparent plastic material used in cars to prevent splinters flying.

safety lintel A *lintel*, which carries load, to protect another more decorative lintel.

sagging [pai.] *Curtaining*.

sailing course or **oversailing course** A *string course*.

Saint Andrew's cross bond *English cross bond*.

Saint Petersburg standard or **Petrograd standard** [tim.] 165 ft^3 (4·67 m^3) of timber.

sal ammoniac (NH$_4$Cl, ammonium chloride) A *flux* in soldering and in making *iron cement*, found as a mineral near volcanoes.

salient Description of an external projecting corner, the opposite of a *re-entrant corner*.

sally [carp.] A re-entrant angle cut into a timber (birdsmouth).

salt glaze A glaze formed on *glazed ware* drain pipes and similar ware by shovelling salt into the hot kiln.

sample [tim.] Sheets of *sliced* or *half-round veneer* taken from the top, middle, and bottom of the *flitch* as an indication of the *figure*. These veneers are kept in the manufacturer's sampling trunk with a record of the log number and the area of veneer in the flitch. *See* **swatch**, and **C**.

sand box A device which enables *centering* to be quickly and easily struck. The posts supporting the centering are set on small timbers bearing on dry sand in boxes. When the post is to be removed, one side of the box is slipped out and the sand removed and the post can then drop out. *Compare* **folding wedges**.

sand-dry surface [pai.] A surface on which sand will not stick. *See* **drying**.

sanded bitumen felt A *bitumen felt* for roofing which is saturated with bitumen. It weighs about 2·5 kg/m^2 ($\frac{1}{2}$ lb/ft^2).

sanded fluxed-pitch felt *Bitumen felt* saturated with fluxed coal-tar pitch, coated with it, and sanded both sides.

sander [joi.] A *sanding machine*.

sand-faced brick A *facing brick* coated with sand to give it an attractive rough surface. *See* **rustic brick**.

sanding [joi.] Smoothing wood surfaces with *glasspaper* by machine or by hand. In painting, the same process of smoothing paint surfaces is called rubbing or *flatting down*.

sanding machine or **sander** [joi.] An electrically driven or mechanically powered machine used for cleaning up *joinery* or for smoothing walls or floors with *glasspaper* or other *abrasives*. Sanders may either be portable sanders or *bench sanders*.

sanding machinist [tim.] A *tradesman* who fits an abrasive belt or disc on to a sanding machine and works the machine either in the shop or on the site, smoothing a timber floor.

sanding sealer [pai.] A specially hard first *coat* which seals or fills but does not hide the *grain* of wood. The surface can be sanded after the sanding sealer is put on. Compare **filler, sealer, surfacer.**

sand-lime bricks *Bricks* made by compressing a mixture of damp sand with 7 % of slaked lime and maturing them in a steam oven for seven hours. They are as heavy as clay bricks and can be made with a crushing strength up to 41 N/mm^2 (6000 psi). They are pale in colour (nearly white) and so reflect more light than clay bricks. *See* **calcium silicate.**

sandpaper [joi.] *Abrasive* paper used by hand or on machines. *See* **glasspaper.**

sand rubbing Applying a surface layer of sand or broken rock to a roof covered with *roofing felt.*

sandwich beam A *flitched beam.*

sandwich construction Composite construction of *light alloys, plastics,* plasterboard, etc., generally of a hard outer sheet glued to an inner core of foam plastics or paper honeycomb. For its weight it is extremely strong, particularly if purposely arched or warped. *Expanded polystyrene* and *polyurethane* are often used in cores.

sanitary shoe [joi.] US term for a *congé.*

sap or **quarry sap** The moisture in a freshly quarried stone. If, as is generally believed, stone is harder when the sap has dried out, this is probably because the sap contains the *matrix (C)* of the stone dissolved in it.

saponification [pai.] The action of alkalis (such as the lime in cement) on oils. The oil is converted to soap, which is dissolved by water. Any paint containing oil can therefore be destroyed by saponification. *See* **alkali resistance.**

sap stain [tim.] *See* **blue stain.**

sapwood or **alburnum** [tim.] The outer wood in a tree, lighter in colour than the *heartwood.* It usually decays more easily than heartwood but it also absorbs preservative more thirstily and is not always weaker than the heartwood.

sarking (1) [carp.] (Scotland) Roof boarding up to 19 mm (¾ in.) thick.

(2) Elsewhere, *sarking felt.*

sarking felt or **underslating felt** Bituminous flax felt laid under *slates* or *tiles* sometimes without roof boarding under it but usually on roof boarding. *See* **reinforced bitumen felt.**

sash [joi.] (1) A sliding *light* of a double-hung *sash window* (BS 565).

(2) (Scotland) Any *light* of a casement or sash window.

sash and case window [joi.] (Scotland) A *sash window.*

sash and frame [joi.] A *cased frame* and *sash window.*

sash balance [joi.] A spring which operates a *sash window* and thus eliminates sash weights, pulleys, and cords.

sash bar [joi.] A *glazing bar*, in USA often called a *muntin*.

sash chain [joi.] A chain used instead of *sash cord* in the best (or the heaviest) windows. It must be used with a cogged *axle pulley*.

sash chisel [joi.] A *pocket chisel*.

sash cord or **sash line** [joi.] A cord nailed to the side of a *sash window*, passing over the pulley and held tight by the sash weight. The window is easy to slide up and down, but chains last longer. *See* **sash ribbon**.

sash cramps [joi.] *Cramps* between 0·6 and 1·5 m (2 and 5 ft) long used for clamping *sashes* during gluing.

sash door or **half-glass door** [joi.] A door of which the upper half is glazed.

sash fastener or **sash lock** or **(USA) sash fast** [joi.] A fastening on the *meeting rail* of one sash which swings across to the meeting rail of the other sash and engages with a spur on it.

sash fillister [joi.] A *fillister*.

sash lift [joi.] A *pull* or hook on a *sash window* for opening it.

sash lock [joi.] A *sash fastener*.

sash pocket [joi.] *See* pocket.

sash pulley [joi.] Pulleys at each side of a *sash window*, set in *pulley stiles* and carrying the *sash cords*.

sash ribbon [joi.] A steel tape which, attached to a spring instead of to a sash weight, holds the window up.

sash run [joi.] (USA) A *pulley stile*.

sash stop The term recommended by BS 565:1963 for the side and head members of the *guard bead*. The sill member should be called the draught stop.

sash tool [pai.] A round *brush*, smaller than a *ground brush*, made in various sizes, bound with metal or string, and used for painting sashes, frames, and other small areas.

sash window or **balanced sash** or **(USA) vertical sash** [joi.] A window in which the *opening lights* slide up and down in a *cased frame*, balanced by *sash cords* passing over a sash pulley. This window came to England from Holland in the seventeenth century. *Compare* **sliding sash**.

saturated roofing felt A *roofing felt* which is saturated with bitumen or fluxed coal-tar pitch but has no surface dressing of either.

saturation coefficient (1) (Britain) The ratio between the volume of water absorbed by a brick in 5 hours boiling and the volume of its pore space.

(2) (USA) The ratio of the water absorbed by a brick soaked in cold water for 24 hours to that absorbed when the brick is boiled for 5 hours. It is a measure of the frost resistance of the brick.

saucer dome A glass or plastics dome moulded in one piece like an inverted saucer and used as a roof *light*.

saw A steel blade for cutting wood (or stone or metal). The main power-driven saws for wood are *circular*, *band*, or *jig saws*. *See* **handsaws, sawyer.** For cutting metal the *hacksaw* is used. For hard stone, frame-saws like large hacksaws or circular saws with diamonds set in the cutting edges are used.

saw bench [tim.] A steel table through which a *circular saw* passes. The *rise-and-fall* table is being superseded by the stationary saw bench, in which the saw is raised or lowered.

saw doctor [mech.] A skilled mechanic who looks after the mechanical saws in a sawmill and sometimes also the *handsaws*.

sawdust cement A concrete made with cement and sawdust. It is sometimes used for making fixing blocks but has very high *moisture movement (C)* that of timber added to that of cement. In addition, the material of the wood may react with the cement and weaken it.

saw horse [carp.] A four-legged stool made on the job by a carpenter for hand sawing.

sawn veneer [tim.] In USA *quarter-sawn* oak is still cut on large diameter *segment saws*, which make a cut 1 mm thick, with a *veneer* of about the same thickness or slightly thicker. Even with this narrow kerf the loss of wood in sawing can amount to half the log volume, but the method is still used because it makes sheets of strong veneer which polish well. In Britain where timber is more costly and scarcer, sawn veneer has not been made since about 1935. Veneer was certainly sawn in Battersea, London, as early as 1805, and probably earlier. *See* **plywood.**

saw pit [tim.] A pit in which a bottom sawyer used to stand holding the lower end of a rip saw for converting timber. Now superseded by power-operated saws.

saw set [joi.] An adjustable tool which gives the correct *set* to the teeth of the saw.

saw-tooth [joi.] Each small cutting blade in a *saw* from *point* to root. *See* **angle of saw-tooth, back, face, fleam, gullet, pitch, set, space.**

sawyer [tim.] A craftsman who sets and operates mechanical saws. They may be of many types and the sawyer can thus be described according to his saw, for instance as a band-mill sawyer, frame sawyer, circular sawyer, or chain-cross-cut sawyer.

sax A slater's *zax*.

Saxon shakes Hand-split western red cedar *shingles* 60 cm (2 ft) long, of random widths varying in butt thickness from 19 to 32 mm ($\frac{3}{4}$ to $1\frac{1}{4}$ in.). They may be halved in thickness by sawing, to provide a flat under surface.

scabbing hammer or **scabbling hammer** A *hammer* with one pick point used for rough dressing, for example, of *nidged ashlar*.

scabbling Dressing building stone roughly with a pick or hammer. *See also C.*

scaffold or **scaffolding** A temporary steel, *light alloy*, or timber erection for supporting men and materials during building, and other work. *See* **bracket scaffold, bricklayer's scaffold, cradle, mason's scaffold, saddle scaffold, tubular scaffolding.**

scaffold boards *Softwood* boards which form the working platform of a *scaffold*. They measure 23×4 cm ($9 \times 1\frac{1}{2}$ in.) and may be up to 3·7 m (12 ft) long. They are also used for forming *barrow runs* and are bound with hoop iron round the ends to prevent them splitting.

scaffolder A *tubular scaffolder* or *timber scaffolder*.

scaffold pole A *larch* pole, 6 to 15 cm ($2\frac{1}{2}$ to 6 in.) dia. at the middle of its length, used as a *standard* by *timber scaffolders*.

scagliola [pla.] Plasterwork coloured to imitate marble. It is made of plaster like *Keene's* with added *pigments* and chips of coloured material. It is then polished with a stone and coated with linseed oil. It was used in the seventeenth and eighteenth centuries. *Compare* **marezzo marble.**

scale [plu.] The deposit of magnesium or calcium salts formed in a *furred* pipe by *hard water*.

scallops (1) Short *withies* placed to hold *thatch* at the *verges* of a roof and held down by *spars*.

(2) Decoration like a scallop shell.

scant [tim.] The term preferred by BS 565 for a shortage in one dimension.

scantle or **gauge stick** or **size stick**. A strip of wood 4×2 cm ($1\frac{1}{2} \times \frac{3}{4}$ in.) with two nails projecting from it to mark the position of the hole measured from the *tail* of the slate.

scantle slating Cornish slating with small *random slates*.

scantling [tim.] (1) In softwood, a piece of square-sawn timber 48 mm ($1\frac{7}{8}$ in.) to under 102 mm (4 in.) thick, and 51 mm (2 in.) to under 114 mm ($4\frac{1}{2}$ in.) wide (BS 565).

(2) In *hardwood*, timber converted to an agreed specification, such as waggon oak scantlings. Otherwise any square-edged pieces of odd dimensions not conforming to other standard terms.

(3) A loose term for the dimensions of the cross section of a timber.

scarcement (Scotland) The narrowing of a brick wall as it rises from the *footings*.

scarf (1) or **scarf joint** or **scarfed joint** [carp.] A joint made by fixing together two bevelled ends of timber so that their centre lines are in line. Usually the joint is fished and bolted, sometimes also glued.

(2) [tim.] A special sort of scarf joint is made in *veneers* by cutting the ends of the two sheets at a slope of 1 in 12 to 1 in 20. They are then glued together and are almost as strong as if there were no splice.

scent test or **smell test** A *drain test*, made by pouring strong scent such as peppermint oil into a drain and plugging it.

schedule of dilapidations A list of repairs to be done by a tenant when a lease expires. It is drawn up by an *architect* or surveyor acting for the landlord.

schedule of prices [q.s.] A document that takes the place of a *bill of quantities* in a certain type of *fixed-price contract* in which no *quantities* appear in the tender documents. Each firm invited to *tender* writes down its rates to a list of *items* (the schedule of prices) which are described only. For small-scale work such as general repairs this may be satisfactory, but some idea of the quantity involved must be given to the *contractor* or his prices will be too high.

scissors truss [carp.] A simple *truss* formed of four main members of which two are *rafters* from wall plate to ridge. The other two extend from the rafter at *wall-plate* level to the middle of the opposite rafter. These two members intersect at the middle of the span, giving a scissors-like appearance to the truss. A scissors truss may also be of steel. It has the advantage of giving good ceiling height in the centre of the span.

score To scratch a surface so as to obtain a better mechanical bond for glue, plaster, etc.

scotch A bricklayer's *scutch*.

Scotch bond *American bond*.

Scotch bracketing [pla.] *Laths* cut to length and fixed at a slope from wall to ceiling as a base for a hollow plaster *cornice*, a form of *cradling*.

Scotch glue [joi.] *See* **animal glue.**

scouring [pla.] To give plaster a smooth hard surface by working it with a *cross-grained float* in a circular motion. Sometimes water is sprinkled on with a brush at the same time. The finishing process, with a trowel, is called 'ironing in'. Not all plasters are suitable, but *Keene's* and Sirapite can be scoured.

scraped finish [pla.] A finish to lime-cement stuccos, used in Central Europe. The finishing coat *stucco* is scraped with a steel straight edge just as it begins to harden, 1·5 to 3 mm ($\frac{1}{16}$ to $\frac{1}{8}$ in.) of plaster being removed. The aggregate is well exposed and the finish can be patterned if the scraping tool is serrated.

scraper or (USA) **scraper plane** A *cabinet scraper*.

scratch awl [joi.] An *awl*.

scratch coat or **pricking-up coat** [pla.] The first coat of plaster on lathing. *Compare* **rendering.**

scratcher [pla.] A *drag* or *nail float*, or *fingers*.

scratching [pla.] Laying the first coat of plaster and roughening its surface for the next coat.

scratch tool [pla.] A *small tool* for completing plaster enrichments.

scratch work [pla.] *Graffito*.

screed (1) A band of plaster or concrete carefully laid to the correct surface as a guide for the *rule* when plastering or for the *screed rail* when concreting. The bands are level with the final surface.

(2) A wood or steel *straight edge* fixed temporarily to a wall, floor, or ceiling as a guide to the rule or screed rail instead of (1) above.

(3) A layer of *mortar* 2·5 to 6 cm (1 to 2½ in.) thick, laid to finish a floor surface in forming a *jointless floor*.

screed rail A heavy *rule* used for forming a *concrete* or *mortar* surface to the desired shape or level.

screen A rectangular frame about 2×1 m (6×3 ft) with a mesh built into it for separating pebbles from sand. *See also C.*

screw [carp.] A steel or rustproof metal fastening with a spiral thread and pointed end (wood screw) or a helical thread and blunt end for threading with a nut or screwing into metal. Wood screws are described by their length and diameter. The diameter is described in numbers called the screw gauge, which vary from 0 (1·5 mm or 0·06 in.) up to 32 (12·7 mm or ½ in.). Generally it is impossible in hardware stores to find screws thicker than No. 14 (6·3 mm or ¼ in.) and there are no standard sizes of any type of screw thicker than No. 20 (8·4 mm or ⅓ in.). The British screw gauge is little different from the US screw gauge. See also **bolt** (*C*), **coach screw**, **dowel screw**, **grub screw**, **nylon**, **recessed head screw**.

screw anchor A metal *Rawlplug* sleeve for a *bottomless* hole.

screw auger Any *auger* for carpentry or *soil mechanics* (*C*).

screw clamp [joi.] A *handscrew*.

screw-down valve [plu.] A cock which, like the domestic water tap, works by screwing down a jumper, and so has a higher resistance to flow than a full-way valve.

screwed pipe [plu.] Gas or water pipe, made previously of *wrought iron* (*C*) now of *dead-mild steel* (*C*), usually in lengths of about 6 m (20 ft). These pipes are screwed outside and thus joined to fittings which are screwed inside, both having a *taper thread*. This is the cheapest domestic pipe but it is heavy. It looks less neat than *light-gauge copper* or *stainless steel tube* with *capillary* joints, but because of its weight and strength it is often used for semi-structural purposes, e.g. railings.

screw eye [joi.] A metal loop at the head of a screw, used with a *cabin hook* to form a door or window fastening.

screw gauge [carp.] *See* screw.

screw nail [carp.] A *drive screw*.

screw plug or **disk plug** A rubber ring which expands to block the end of a length of drain when the two steel plates each side of it are screwed together in a *drain test*.

screw thread or **thread** [mech.] A helical ridge on a metal, plastics, or wooden rod, formed by cutting, rolling or casting. The shape of the thread may be of various standard types, **such** as *Whitworth* for building fixings, *buttress* (*C*) for force transmission, fine threads for mechanical work, but it is usually related to the bar diameter; *see* **screw, tap, thread rolling.**

scribe (1) To cut a line on a surface with a pointed tool to the outline of a *templet.*

(2) [joi.] To shape one member to the surface which it touches, thus to fit a board snugly to a surface which is not straight.

scribe awl [joi.] An *awl.*

scribed joint [joi.] A *coped joint* between *mouldings.*

scriber [joi.] A pointed tool or a tool like a pair of dividers for marking out scribing lines.

scribing plate [plu.] A metal plate used with a pair of dividers to draw an ellipse on a pipe for a branch connection.

scrim [pla.] Coarse canvas, or cotton or metal mesh, used for bridging the joints between board, sheet, or slab *coverings* before they are plastered, and as reinforcement for *fibrous plaster. Compare* **joint tape.**

scrimming [pla.] Setting *scrim* over joints in *gypsum plasterboard,* masonry, or carpentry, as a base for plastering, painting, or papering.

scrub board [joi.] (USA) A *skirting board.*

scrub plane [carp.] A *plane* with a rounded *cutting iron* for removing a considerable depth of shaving at each stroke.

scumble glaze or **scumble stain** [pai.] Transparent paints used in *scumbling* for enriching or softening or otherwise modifying the *colour* of the coat below the finish coat. *See* **ground coat.**

scumbling [pai.] The achievement of broken *colour* effects by removing or texturing a wet coat to expose part of the coat beneath it.

scutch or **scutcher** or **scotch** A *bricklayer*'s tool for cutting bricks, also used by the *walling mason.* It resembles a small *hammer* except that there is no striking face but a *cross peen* on both ends of the head.

scutcheon [joi.] An *escutcheon.*

seal [plu.] (1) The water contained in the *trap* of a drain or gulley. It prevents foul air from passing out of the drain.

(2) In a drain trap, the vertical measurement from the level of the water down to the crown of the U-shaped part of the pipe. In 10 cm (4 in.) drains the seal is about 5 cm (2 in.). *Compare* **deep-seal trap.**

sealant *See* **sealing compound,** *also C.*

sealed flue or **sealed boiler** A *balanced flue* of a gas-fired boiler.

sealed system [plu.] Sealed heating circuits, sometimes using a closed *diaphragm tank* instead of an open expansion pipe and tank, have been introduced to Britain from USA and Europe. Since air cannot enter the system, corrosion is less likely than with the *expansion tank.* Sealing also allows the water temperature to be higher and this makes for economy. *See* **Langman cylinder.**

sealer or **sealing coat** [pai.] (1) Liquids, some of which are transparent, used like *size* as priming coats to close the pores of wood, plaster, and other building materials, for instance to prevent the wood of a *glazing bar* absorbing oil from *mastic*. Clean pine floorboards which have been coated with a modern sealer can be beautiful and are easily kept so.

(2) A liquid laid over bitumen, or creosote, etc., to prevent it bleeding through other paints. The sealer must be insoluble in the paint laid over it, and must not dissolve the material under it. (Also called **buffer coat**.)

sealing compound (1) Material used to fill and seal the surface of a *contraction joint* (*C*) to prevent water and grit entering. It can be applied like *mastic* with a *pressure gun*. See **sealant** (*C*).

(2) or **lap cement** A fluid bitumen applied cold to *roofing felt* for sealing the laps to each other. See **bonding compound**.

seam or **welt** [plu.] A joint between two sheets of *flexible-metal roofing* made by turning the edge of each vertically upwards, bringing the two upturned parts together, doubling them over, and dressing them flat. See **cross welt, standing seam**.

seamer or **seaming pliers** [plu.] A pair of *pliers* with jaws specially shaped for forming *seams* between *flexible-metal sheets*.

seam roll. A *hollow roll*.

seasoning (1) [tim.] The drying of timber to a specified *moisture content*. It may be *natural* or *kiln seasoning*.

(2) The drying and hardening of stone. See **quarry sap**.

seasoning check [tim.] A *check* like most checks, formed in seasoning.

seat cut [carp.] A *foot cut* in a *rafter*.

seaweed See **eel grass**.

secondary circulation [plu.] See **indirect cylinder**.

second fixer [carp.] A *carpenter* or *joiner*, fixing *joinery*.

second fixings [joi.] *Joinery* which is fixed after the plastering, e.g. *skirting boards*, linings, picture rails, cupboards, and in fact most joinery except doors. Plumbing and electrical wiring may also be included in the second fixings.

second-growth timber [tim.] A term used to distinguish wood which has grown later than the first growth of a virgin forest and usually has less *clear timber*.

seconds Second-quality material, particularly clayware such as *bricks* and drain pipes.

secret dovetail [joi.] A *mitre dovetail*.

secret fixing [joi.] Fixing of *joinery* which cannot be seen at the surface of the joinery. One method is *secret screwing*, another *secret nailing*.

secret gutter or (USA) **closed valley** A nearly hidden *valley gutter*, in which the lead is hidden by the slates. It has the disadvantage of being easily blocked by leaves. *Compare* **open valley**.

secret joggle See **joggle**.

secret nailing or **blind nailing** [joi.] Nailing which is not seen on the surface. Tongued and grooved, or *rebated* boards can be slant-nailed through the tongue or rebate. Another method is to shave off a thin chip of wood with a sharp knife, drive the *nail* in where the chip was taken out, and *glue* the chip back over it.

secret screwing [joi.] A joining method used for making a wide board when only two narrow ones are available. *See* **secret screw joint.**

secret screw joint [joi.] A strong glued joint between the planed edges of boards. *Screws* are inserted in one edge and left with their heads projecting about 10 mm ($\frac{3}{8}$ in.). Slots are drilled opposite them in the board to which they are to be fitted. The two boards are fitted together first, being pressed towards each other, then one is given an endways blow to drive the screws to the end of the slots. The same board is then given a blow on the other end, to release the screws from the slots. The joint is unmade, each screw screwed in a quarter turn, and the surfaces are glued and re-assembled. Stout screws must be used, any screw less than 6 mm ($\frac{1}{4}$ in.) dia. being liable to bend.

secret tack A strip of lead soldered on the back of a lead sheet. It is passed through a slot in the roof boarding and screwed to the inside of it, and is used for holding large vertical sheets of lead.

secret wedging [joi.] Fixing a *stub tenon* by inserting wedges into sawcuts in its end before it is inserted into a *blind mortise*. As the *tenon* is driven in, the wedges are forced in by the pressure of the blind end of the *mortise* on them. If the mortise is *dovetail*-shaped the assembly is called foxtail wedging.

section *See* C.

sectional insulation [plu.] *See* **moulded insulation.**

section mould A *template* of *plywood* or similar material cut to the cross-section required for a member.

sedge (Cladium mariscus) A tall grass from marshy places which forms the ridge of thatch. It is placed wet across the roof and fixed by *liggers*. *See* **straw.**

SE-duct or **appliance ventilation duct** [plu.] In multi-storey buildings, a vertical duct open at both ends and used both as flue and as air intake for all the *balanced-flue appliances* in its sector of the building. Only some appliances are suitable for connection to a SE-duct and a check should be made with the gas authority before the appliance is chosen. (Developed by the SE Gas Board in the 1950s.)

seediness [pai.] A rough, sandy-looking paint film is called seedy. This fault may be caused by lack of *compatibility*.

segment saw [tim.] A special *circular saw* of up to 3·6 m (12 ft) dia. which makes a very thin kerf and is still used in USA for cutting *sawn veneer*.

seize To tie together with ropes.

selenitic lime or **selenitic cement** Lime with 5% to 10% of *plaster of Paris* added. *See* **gauged stuff.**

self-build housing society A *cooperative housing society* which has become popular in Britain since 1945. Its members subscribe the usual proportion of cost (which may be as little as £100) but do not send out for *tender* from builders because they build their houses themselves in their spare time. Lots are drawn for the family to occupy the first house and all the members help to build until the last house is built and the last family housed.

self-centering formwork *See* **telescopic centering.**

self-centering lathing *Expanded metal* (*C*) stiffened by sheet steel ribs which enable it to be used as formwork for floors or as a *lathing* for a plaster wall without other stiffening.

self-faced stone Stones like flagstones which split cleanly and need no dressing.

self-finished roofing felt *Roofing felt* which is saturated with bitumen, coated on both sides with it, and sometimes also dressed with fine talc on both sides.

self-priming cylinder [plu.] An *indirect cylinder* which does not have to be fitted on the heating coil either with an *expansion pipe*, or with an *expansion* tank.

self-supporting scaffold Any scaffold like the *mason*'s and unlike the *bricklayer's scaffold.*

self-supporting wall (USA) A *non-bearing wall.*

semi-detached house or (USA) **double house** One of a pair of houses (usually) built simultaneously, with a common wall, called a party wall, between them. *Compare* **terrace house.**

semi-engineering brick A *brick* with a crushing strength higher than *common bricks* but rather below the second grade of *engineering bricks*, 48 N/mm² (7000 psi).

semi-hydraulic lime A *lime* intermediate between *high-calcium lime* and *eminently hydraulic* limes, for example, greystone lime. When run to putty it is nearly as workable as non-hydraulic lime but its *hydraulic* properties are much reduced by *fattening up*. If hydrated to *dry hydrate* at the factory it does not lose its hydraulic properties.

semi-skilled man A workman such as a *rigger* (*C*), *rivet catcher* (*C*), or *builder's labourer*, who has learnt his work by helping others, and is rather less skilled than a *skilled man* or a *tradesman*. He has risen from *labourer* in his *trade* merely by working at it and may become a skilled man.

separate application [tim.] A method of using *urea-formaldehyde glue* in which the *accelerator* is spread on one side of the joint and the *resin* spread over the other side. The two parts are then clamped together. The method is simple and economical since *mixed glue* is not wasted.

septic tank A purifier for sewage where no *sewer* (*C*) is available. It is a tank through which sewage flows slowly enough for it to be purified by decomposition. It is divided into two or more chambers separated by scum boards with an opening below. The receiving compartment capacity is made equal to the house-water supply of at least seven, and at most twenty-four hours. The best method of removal of effluent from a septic tank is by automatic siphons which periodically discharge a quantity of the water into pipes buried in a field, in which the water passes to plant roots.

sequence of trades The order in which (originally) the various trades carried out their work in a new building. This sequence is followed in enumerating the items in a bill or *specification*, since it is thus easy to find any particular trade. The sequence below is fairly full, most sequences are much shorter.

1. preliminaries
2. demolition and shoring
3. earthwork
4. piling
5. concrete work
6. precast concrete and hollow floors
7. brickwork and partitions
8. drainage and sewage
9. asphalt work
10. pavings
11. masonry
12. roofing
13. timber and hardware
14. structural steelwork
15. metalwork
16. plastering, wall tiling, terrazzo
17. sheet metal
18. rainwater goods
19. cold-water supply and sanitary plumbing
20. hot-water supply
21. gas and water mains
22. heating
23. ventilating
24. electrical
25. glazing
26. painting and decorating
27. provisional sums; items by specialists

service cable The power cable joining the supply authority's main to the *house service cutout*.

service core The *mechanical core* of a building.

service ell or **service tee** [plu.] An ell-shaped or tee-shaped screwed *fitting* with an external thread outside it at one end (as usual) and its other end enlarged to form a screwed socket.

service layer *See* **pipe layer**.

service pipe A gas or water pipe between the main and the premises receiving the supply. *See* **communication pipe, supply pipe,** *also* BSCP 331.

service riser A gas pipe which rises to supply an upper floor.

services Supply or distribution pipes for cold or hot water, steam, or gas; also power cables, bell and telephone cables, lift machinery, transformers, drains, ventilation trunks, and so on. (USA utilities.)

set (1) [joi.] A *nail punch*.

(2) The slight overhang given to the point of a saw-tooth to make the *kerf* just wider than the saw so that the saw can run easily in it. *See* **setting.**

(3) The first hardening (initial set) of concrete, mortar, or plaster.

(4) [pla.] To apply a *finishing coat*.

(5) [pai.] A paint or varnish *film* which has ceased to *flow* is said to have set.

setback A withdrawal of the *building line* in the upper floors of a tall building to a distance about one storey height back for each floor above a certain level. This is required by city *by-laws* to ensure that enough light reaches the street.

set screw (1) [joi.] A *grub screw*.

(2) [mech.] A *screw* designed to fit, not into a nut, but into a hole tapped into a block, plate, or machine part, like a *grub screw*.

setter out or **marker out** [joi.] A skilled *joiner* or *carpenter* of many years' experience who marks out work for other men to shape. *See* **setting out.**

setting (1) The first hardening of *concrete, mortar, plaster, glue,* or *paint* after which they should not be disturbed.

(2) The laying of stones, *bricks, lintels,* etc., in a wall.

(3) The brickwork which supports and encloses a furnace or a boiler.

(4) [joi.] The third operation after *topping* and *shaping*, preceding the *sharpening* of a saw. It is the bending over of the teeth alternately to one side and then to the other of the saw so that the saw-cut is (for *handsaws*) about 1·5 times the thickness of the saw. The commonest mistake in sharpening saws is to give them too much set. *See* illustration overleaf.

setting coat [pla.] The *finishing coat* of plaster.

setting-in stick [plu.] A hardwood tool for bending sheet lead.

setting out Putting pegs in the ground to mark out an excavation, marking floors to locate walls or preparing dimensioned rods for *carpentry* or *joinery*. Large scale setting out is done by a civil engineer, small scale work by a *setter out* or *foreman* of a *trade*.

setting stuff [pla.] A *finishing coat* of *hydrated lime* or *lime putty* gauged with plaster and mixed with washed sand, for plastering walls and ceilings (now used less than formerly).

setting-up [pai.] The gelling of a *paint* during storage. If the paint cannot be thinned by stirring, the fault is called *feeding*. *See* overleaf.

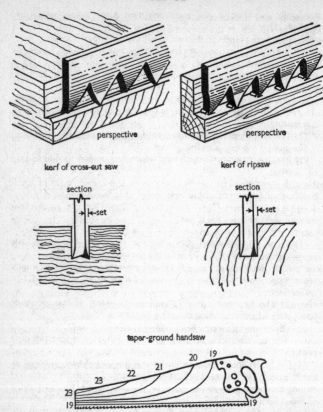

perspective   perspective

kerf of cross-cut saw   kerf of ripsaw

section   section

taper-ground handsaw

Setting *handsaws*. The saw is taper-ground, and the contours on it show the tapering thickness of the blade in *Standard wire gauge*.

set up [plu.] (1) To bend up the edge of lead sheet.

(2) To caulk a joint with lead, driving the lead in with a blunt chisel. *See also* **upsetting** (*C*).

sewer brick (USA) Low *absorption*, abrasion-resistant bricks, used for the same purposes as British *engineering bricks*.

SfB (Samarbetskommitten för Byggnadsfrågor) An international classification of building information subjects, that originated in Sweden but is now sponsored by CIB (International Council for Building

Research and Documentation). The English language version, known as *CI/SfB*, is the responsibility of the *RIBA*.

sgraffito [pla.] *Graffito*.

shake [tim.] Separation of wood fibres along the *grain* of the wood in the log due to stresses while standing, felling or seasoning. *See* **check, heartshake, ring shake.**

shakes [tim.] Hand split *shingles*. *See* **processed shakes.**

shaped work [joi.] Curved *joinery*.

shaping [tim.] The second operation in putting a cutting edge on a saw. It is the filing of the teeth to a uniform shape and size after they have been topped. The subsequent operations are *setting* and *sharpening*, *topping* being the first.

sharp coat [pai.] A *coat* of *white lead* in oil, thinned liberally with turpentine or *white spirit*. When put on fresh plaster, it is called sharp colour.

sharpening saws [tim.] The last operation, preceded by the *topping*, *shaping*, and *setting*, in putting a cutting edge on a *handsaw*. Like shaping and topping, it is done with a taper saw file. The saw is fixed in the vice with not more than 6 mm ($\frac{1}{4}$ in.) of *saw-tooth* projecting from it. The file is held horizontally, and starting at one end each alternate tooth face is given one or two strokes. The other faces are then sharpened after reversing the saw in the vice. The saw should then be sharp and can be dressed on the side with a smooth stone to remove burred metal. A rip saw should be sharpened with the file perpendicular to the saw, a cross-cut saw with the file at about 70° to the saw.

sharp paint A quick-drying paint with a flat finish, usually strongly pigmented, drying by *evaporation* and used for *priming* or *sealing*.

sharp sand Sand grains which are angular, not rounded. The term has also been used for sand which is gritty and contains no clay, and is therefore clean and suitable for making *concrete*.

shave hook [plu.] A *plumber*'s tool used for shaving lead pipes before *soldering* or by painters for scraping off burnt paint.

sheariness [pai.] The oily opalescent surface of a painted *film* which should have a high *gloss* and has failed because of greasy surface, foul air, or lack of *compatibility* in the *paint*.

shear plate [carp.] A timber *connector* of diameter 6·7 cm ($2\frac{5}{8}$ in.) held by a 19 mm ($\frac{3}{4}$ in.) dia. bolt. *See* **hole saw.**

sheathing or (USA) **boxing** [carp.] *Close-boarding* nailed to a timber frame or to *rafters* as a base for wall or roof cladding. *See also* C.

sheathing felt Impregnated flax felt made specially as an *underlay* for asphalt roofing.

sheathing paper *Building paper*, often called sheathing paper when it is laid on sheathing boards.

shed roof A *lean-to roof*.

sheen [pai.] A *gloss* seen at glancing angles on an otherwise flat *finish*.

sheet A description of copper, aluminium or zinc which is thicker than 0·15 mm (0·006 in.), thinner than 9·5 mm ($\frac{3}{8}$ in.), and more than 46 cm (18 in.) wide. *Foil* is thinner, *strip* is narrower. Flat lead sheet is rolled, not cast, and is usually described as milled sheet or strip. For steel, sheet is material thinner than *plate* (*C*), (3 mm or 0·12 in.).

sheeter or **iron roofer** A *skilled man* who fixes corrugated steel, zinc, or *asbestos-cement* sheeting on roof and walls with *hook bolts*, *screws*, or clips. He may cut sheeting to shape and is described according to his material as a zinc roofer or asbestos-cement sheeter.

sheet glass Ordinary window glass, not of such uniform thickness as *float* or *plate glass*. For most window purposes the thinnest sheet is 3 mm but 2 mm thickness can be used for very small panes. Glass 3 mm thick weighs 7·5 kg/m² or 34 oz/ft² and so is equivalent to the old 32 oz glass.

sheeting *Sheathing*.

sheeting clip A cranked clip fixed like a *hook bolt* to the lower end of an *asbestos-cement* sheet.

sheet-metal work The working of steel or flexible sheet, by *plumber* or *sheet-metal worker*.

sheet-metal worker [mech.] A *skilled man* who shapes sheet steel with the *hammer* or with machines. He joints sheets by riveting, seaming, welding, *soldering*, or *brazing* and may mark them out or cut them by guillotine. He may be specialized as an air-duct maker, chimney-cowl maker, or ventilator maker. Generally, in Britain, workers in metal thicker than 5 mm ($\frac{3}{16}$ in.) are called sheet-iron workers, those who work very thin metal less than 1·5 mm ($\frac{1}{16}$ in.) being called tinsmiths, regardless of the metal they work.

shelf life The time for which a *glue* or paint can be stored before it becomes unusable. *Compare* **pot life**.

shelf nog A piece of wood built into a wall and projecting out from it to support a shelf.

shellac or **lac** [pai.] An incrustation formed by an insect on the trees of India and neighbouring regions. It is a natural *resin*, the orange variety being commoner, white shellac being bleached. It is soluble in alcohols and is much used in spirit *varnishes*, *French polish*, *knotting*, *abrasives*, etc.

shellac and white-lead mortar A *mortar* used for brickwork which is to be carved.

shell gimlet [carp.] An old type of *gimlet* with a parallel hollowed shank.

shelling [pla.] *Crazing*.

shell shake [tim.] A *ring shake* which shows on the surface of sawn timber (shelly timber).

sherardizing [mech.] Coating small iron or steel articles with zinc by heating them with zinc dust in a revolving drum at about 350° C. It gives a penetrated coating which is more durable than *galvanizing*

(*C*) or which at least does not peel off. There is also said to be less increase in size or distortion of the article than with galvanizing.

shim [tim.] An *insert* in *veneer*.

shingle (1) Rounded aggregate of variable size and shape, without sand.

(2) A thin rectangular piece of timber about $40 \times 13 \times 0.6$ cm ($16 \times 5 \times \frac{1}{4}$ in.) used like a *tile* for covering walls or roofs. It usually reduces in thickness from *tail* (butt) to *head*. For the different sizes of shingles, *see* **royals, Saxon shakes, takspan, perfection, five x, dimension**. In ancient Rome shingling was one of the earliest roofing methods, as it was on English houses of the Middle Ages (oak shingles). In USA metal shingles are also used, of bronze, aluminium, copper, galvanized steel, zinc, painted tinplate, etc. *See* **tip, weather**.

shingling hatchet or **claw hatchet** A hatchet like the plasterer's *lath hammer*, with a notch in the blade for drawing *nails*.

shiplap boards or **shiplap siding** [tim.] *Weather-boarding* of rectangular cross section with a *rebate* cut on each edge, fitting into corresponding rebates on the neighbouring boards.

ship spikes or **boat spikes** [carp.] Steel spikes for fixing large timber, forged from square bar, having a wedge point.

ship worm [tim.] *Marine borers* (*C*), the teredos, of which teredo navalis is the best known. The largest borers found on the British coast, they attack timber through small holes which increase to 6 mm ($\frac{1}{4}$ in.) dia away from the entrance. *See* **preservatives** (*C*).

shoddy work (1) Inferior work of any sort.

(2) Squared granite less than 30 cm (12 in.) thick.

shoe (1) A short length at the foot of a *downpipe* bent to direct the flow away from the wall.

(2) An iron or steel socket enclosing the end of a *rafter* or other loadbearing timber.

(3) In *patent glazing*, a fitting which holds the lower end of a glazing bar on to a roof member such as a *purlin* and prevents the glass sliding.

(4) A *base shoe*.

shoe mold A *base shoe* (USA).

shook [tim.] A small piece of sawn timber or *sliced veneer*, or a set of such pieces for making a box.

shoot (1) *See* **drain chute**.

(2) [joi.] To true the edge of a board with a *jointing plane*.

shooting board [joi.] A board framed to steady another board while its edge is being shot (planed).

shooting plane [joi.] A *jointing plane*.

shop and office fitter [joi.] A *joiner* specialized in making shop fronts, and who therefore is accustomed to the use of *hardwoods*, metals, and metal-faced *plywood*.

shop drawing A manufacturer's drawing which will be worked to in his workshops and is usually first submitted for approval to a consulting *architect* or engineer.

shopwork Work done in the manufacturer's shop, not on the site.

shore Generally a sloping prop (*raking shore*) but sometimes a vertical (*dead shore*) or horizontal one (*flying shore*).

shoring Giving temporary support with *shores* to a building being repaired, altered, or *underpinned* (*C*).

short (1) or **short circuit** [elec.] An electrical fault caused by two power leads being electrically joined and causing a very high current to pass. It should cause the *fuses* to melt or other *cutouts* to come into action.
(2) [plu.] A short piece of pipe.

short grain [tim.] *See* **brash**.

short oil [pai.] A low ratio of oil to *resin* in a *varnish*. *Compare* **long oil**.

short-oil alkyd [pai.] An *alkyd resin* containing not more than 40% of oil as a modifying agent (BS 2015). *Compare* **long oil alkyd**.

short-oil varnish [pai.] An *oleo-resinous varnish*, other than an *alkyd* containing not more than $1\frac{1}{4}$ parts by weight of oil to 1 part by weight of *resin* in the finished *varnish* (BS 2015). *Compare* **long oil**.

short-working plaster Plaster which is difficult to mix, and from which the sand separates when the mix is squeezed on the board. It is caused by deterioration of plaster which has been stored in a damp place.

shot hole [tim.] *Worm hole* between 1·5 and 3 mm ($\frac{1}{16}$ and $\frac{1}{8}$ in.) dia. *See* **powder-post**.

shot sawn A description of the smooth surface given to building stone sawn by chilled shot, which is fed into the groove during cutting.

shoulder [carp.] The surface at the root of a *tenon* which abuts on the wood beside the *mortise*.

shouldered architrave An *architrave*, round a door, which widens at the top.

shouldering (1) or **half-torching** A thin bed of *haired mortar* inserted under the head of each *slate* in exposed places to keep the *tail* down. *See* **torching**.
(2) The small diagonal pieces cut from the bottom left-hand and top right-hand corners of *single-lap tiles*.
(3) The splay cutting of the top corners of slates from Westmorland and some other quarries.

shoulder nipple [plu.] (USA) A *nipple* (pipe *fitting*) with a space of about 19 mm ($\frac{3}{4}$ in.) between threads at the middle.

shoved joints (USA) Bricklaying by *buttering* the end of the *brick* and pushing it against the last brick laid. The *cross joints* are usually badly filled. This is a common bricklaying method in Britain also. *Compare* **slushed joints**; *see* **western method**.

shread head A *jerkin-head roof*.

shrinkage [tim.] The maximum fractional shrinkage of timber during drying is about 0·001 along its length, about 0·1 parallel to the *growth rings*, and about 0·05 at right angles to the rings. The radial shrinkage is therefore half the tangential shrinkage. *See also C and flat-sawn log* (illus.).

shutter A wooden (now sometimes steel) cover that fastens over a window at night to protect the occupants. It is hung either inside or outside the window.

shutter bar A pivoted bar for fixing shutters in their position over a window.

shuttered socket [elec.] A *socket outlet* with two holes for the insertion of power prongs and one for an earth prong. The earth prong is longer than the others and must therefore be inserted first. The power holes are ordinarily blocked by small shutters within them. These shutters are opened up when the earth prong is inserted and the whole plug can then be pushed in. They are therefore not dangerous to children. Standard shuttered sockets are rated at 13 amp (3KW at 240 volts) and often used for *ring mains*.

shutter hinge [joi.] An *H-hinge*.

shutting post The fixed post against which a gate shuts.

shutting stile or **lock stile** or **slamming stile** [joi.] That *stile* of a door which carries the lock. In a folding door it is called the *meeting stile*. *See* **hanging stile**.

side [tim.] A broad surface of *square-sawn timber*, usually called a *face*.

side cut or **side board** or **siding** [tim.] A piece of wood sawn to exclude the *heart centre*.

side flights *Double-return stairs*.

side gutter A small gutter down a roof slope at the intersection of a *chimney* or *dormer* or other vertical surface with the main roof.

side hook [joi.] A *bench hook*.

side jointing [joi.] (USA) The *topping* of *saw-teeth* at the side of the saw with a *hone*.

side lap (1) The amount by which *single-lap tiles* cover each other at the side.

(2) The amount by which the vertical joint in one *course* is covered by the *slate, tile*, or *shingle* in the course above.

sidelight or **winglight** [joi.] A *margin light* or a *flanking window*.

side posts [carp.] *Princess posts*.

side rabbet plane [joi.] A small metal *plane* for planing the wall of a *rabbet* or groove. Its *cutting iron* is therefore in the side of the plane, not in the *sole*. Normally a right- and a left-hand plane are needed, but some are built with both a right- and a left-hand cutting iron.

siding [tim.] (1) *See* **side cut**.

(2) *See* **weather-boarding**.

(3) (USA) Any wall *cladding* except *masonry* or *brick*, for instance *asbestos-cement*, metal, or other sheets.

sienna [pai.] *Pigment* obtained from hydrated mineral iron oxide, and described as raw sienna when it is the clean crushed mineral, or burnt sienna when it is calcined. Sienna changes from yellow brown to rich orange brown when calcined. It is not very opaque, but its transparency can be made use of in oil *stainers*. *See* **ochre, umber**.

sieve A round tray about 38 cm (15 in.) dia., with one or more meshes per linear cm, used for separating different sizes of sand and gravel. *See* **screen**.

sight size The width and height of an opening in a window which admits light. *See* **daylight width**.

silicate cotton *Slag wool*.

silicate paints Water paints containing sodium silicate (waterglass). They are non-inflammable but alkaline and therefore must not be used under paints which have no *alkali resistance*.

silicates Very common and chemically complex rock-forming minerals containing the radical SiO_3. Sodium silicate, Na_2SiO_3, is used for waterproofing bricks, stone, or concrete, for hardening concrete floors, and for improving the *fire-resistance* of timber. *See* **Joosten process** (*C*), **clay** (*C*), **calcium silicate**.

silicon bronze Probably the most durable but one of the most costly metals used for making nails, screws, and similar fixings for situations like boats or roofs where steel rusts quickly. It contains 1 per cent manganese, 3 per cent silicon, and 96 per cent copper, but is considerably harder than copper.

silicone [pai.] Silicone *resins* are *polymers* containing a substantial amount of silicon. They are chemically inert, have a high heat resistance, and are used for painting walls above ground to repel water. They are not entirely waterproof, so cannot be used as *tanking* below ground.

silking [pai.] *See* **lining**.

sill or **cill** or **sole plate** or (USA) **abutment piece** [carp.] The lowest horizontal member of a *framed partition*, of *frame construction*, or of a frame for a window or door. A wooden door or window sill is usually fixed over a stone, concrete, or brick sill. *See* **lug sill**.

sill anchor or **plate anchor** A bolt anchored in the concrete or masonry foundation to a timber house. It passes through a hole drilled in the *sill* to hold the frame down against the wind.

sill bead [joi.] A *deep bead*.

sillboard [joi.] (Scotland) A *window board*.

sill cock (USA) A *hose cock*.

sill-drip moulding A small *moulding* like a *sub-sill*.

silver brazing or **silver soldering** [plu.] An economical, neat way of joining *light-gauge copper tubes* to each other, introduced generally to the British plumbing trade about 1960. With the set of tools specially designed for the job, very neat joints can be formed which are barely visible. But even with improvised tools, silver brazed joints are neater, cheaper, and offer less resistance to flow than most other types of joint in copper tube. The solder is an alloy of copper, silver, and phosphorus, its melting point is appreciably lower than most *brazing* alloys, and it needs no flux because the phosphorus in it is a flux.

silver fir [tim.] (Abies alba) Many firs go by this name, most of them from central and southern Europe. It is the largest European tree, growing up to 60 m (200 ft) high. Good quality silver fir is used for general building work, including floorboards and *joinery*. It is generally slightly weaker and lighter than *redwood*. *See* **whitewood**.

silver grain [tim.] The fine, pale-grey, shining flecks of *wood ray* seen in *quarter-sawn* oak or beech.

silver-lock bond Another name for *rat-trap bond*.

single bridging [carp.] *Herring-bone strutting* at the midspan of floor *joists*.

single Flemish bond Brickwork laid in *Flemish bond* which is seen on one face only, as opposed to *double Flemish bond*. There is generally *English bond* within the body of a wall faced in single Flemish bond.

single floor [carp.] A timber floor in which the *joists* span from wall to wall. *Compare* **double floor**.

single-hung window A *sash window* of which only one sash, usually the bottom one, is movable and balanced by *sash cord* and weights.

single-lap tiles Curved *roofing tiles* of which several standardized types exist in Britain (*pantiles*, *interlocking tiles*, etc.). They are so called because they overlap only the tiles in the course immediately below, unlike *slates* and *plain tiles*.

single-lock welt [plu.] A *cross welt*.

single-pitch roof A *lean-to roof*.

single-point heater [plu.] An *instantaneous (sink) water heater*.

single Roman tile *See* **Roman tile**.

single roof [carp.] A roof carried by *common rafters* without roof *truss*, *purlins*, or *principals*. It may be tied by a 15×5 cm (6×2 in.) timber tie at eaves level (*close-couple roof*) or higher (*collar-beam roof*) or it may be a lean-to or flat roof. The common rafters are generally of 10×5 cm (4×2 in.) timber at 38 cm (15 in.) centres. In the best work they are *close-boarded*. *Compare* **double roof**.

single-stack system [plu.] One vertical pipe with waste water and soil water flowing down it, and no anti-siphon pipe, a system first tentatively allowed in the 1950s, a development of the *one-pipe system*.

sinkage (1) (USA) A depression in a floor, wall, or ceiling, for example, a *mat sinking*.

(2) [pai.] The defect of blotchiness.

sink bib or **bib nozzle** (USA) The tap at a kitchen sink.

sinking [joi.] A recess cut below the surface of wood, for example, where *butt hinges* are fixed on a door. *See* **mat sinking**.

sinking in or **sinking** [pai.] Loss of *gloss* of a finish coat due to the absorption of its *medium* by the *undercoat*.

site Land which is used, or will be used, for construction; in USA often called a lot.

Sitka spruce [tim.] (Picea sitchensis) A *softwood* from the island of Sitka on the north west coast of Canada. It is the finest *spruce* and is obtainable in large sizes, pinkish white to pale brown with a silky sheen which is absent from the commoner *Canadian spruce*. It is used for *blockboard* cores, for *joinery*, and for *ladder* stiles.

size [pai.] A liquid *sealer*, usually transparent, with which wood or plaster is coated so that *varnish* or *paint* or *glue* applied over it will not be too much absorbed. Size is also used as a *binder* in *distempers*. Of the two sorts most commonly used, glue size is glue diluted with water and put on to plaster, varnish size is made from oil, resin, and *thinner* and put on below paint. *See* **gold size**.

sized slates *Slates* of uniform dimensions, not *random slates*.

size stick A slater's *scantle*.

sizing or **pre-sizing** [tim.] Spreading dilute *glue* (called *size*) on to a surface before gluing it. Since size penetrates more than glue, it reduces the absorption of glue by the wood during gluing. Woods of different densities should be sized before gluing, but the sizing should not be hard before the glue is applied.

skeleton construction A steel or reinforced-concrete frame with floors of reinforced concrete. No wall rises higher than one storey without its load being transferred to a *column* through a floor slab or beam, except occasionally, when reinforced concrete loadbearing walls are used to carry the floor loads instead of columns.

skeleton core [joi.] The hidden internal frame of a *hollow-core door*.

skeleton steps *Treads*, often of metal, with no risers.

skew (1) Oblique.

(2) A *kneeler*.

skewback The upper surface of a *springer* (or the springer itself). It is so called because the surface slopes to carry the first of the uniform *arch-stones*. *See* **clean back**.

skewback saw [carp.] An expensive *handsaw* with a *back* slightly curved instead of straight, as it is in the cheap but very slightly heavier saws. It is usually taper ground. *See* **setting** (illus.).

skew corbel A *gable springer*.

skew flashing A *flashing* between a *gable* coping and the roof below it.

skew nailing or **toe nailing** or **tusk nailing** [carp.] Driving *nails* in, not perpendicularly, but obliquely to the surfaces to be joined. Where possible, alternate nails are driven in opposite directions.

skew table A *kneeler* in a *gable* coping.

skid road A road for quarrying, made of partly embedded round logs set 1 to 2 m (3 to 6 ft) apart across the road. Sledges or *stone boats* are the only users of the road. It is greased in summer but in frosty weather needs no grease, being naturally slippery.

skids Short lengths of wood used for packing walling stones to the correct height when laying them.

skiffling or **knobbing** Dressing stones roughly in the quarry.

skilled man Either a *tradesman* or a leading hand in a modern *trade* which has expanded too quickly for the apprenticeship system to become general (general riggers, bar benders, rivet testers, *steel erectors* (*C*)). Many such men, who are not called tradesmen, may nevertheless have more skill and adaptability than most tradesmen. Most of them work to drawings. Since there is no apprenticeship, many skilled men have risen from *labourer*.

skimming coat or **skim coat** [pla.] (1) A normal *finishing coat* about 3 mm ($\frac{1}{8}$ in.) thick.

(2) A very thin, therefore inferior, finishing coat.

skin [tim.] In *flexible bag moulding* a thin piece of *moulded ply-wood* or a thin *hardwood* covering over the mould to protect it. *See* **caul.**

skinning [pai.] The formation of a skin on the surface of a *paint* or *varnish* in its tin, caused by *oxidation* of the *drying oil* in the skin. It occurs when the tin is left to stand or when a high proportion of *driers* or a viscous *medium* or a very rapidly-drying oil is used. *See* **anti-skinning agent, inhibitor.**

skips (1) or **planing skips** [tim.] Areas of timber which the surfacing machine missed (skipped).

(2) or **holidays** [pai.] Areas left unpainted by mistake.

skirting or **skirting board** or (Scotland) **base plate** or (USA) **mopboard** or **washboard** or **scrub board** or **base** Originally a wooden board set on edge round the foot of a wall to protect it from kicks. It may now be built of *wall tiles*, metal trim, *terrazzo*, asphalt, or other material which is more durable than wood, and can also be a low upstand at a roof *abutment*.

skirting block An *architrave block*.

sky factor The *daylight factor*.

skylight A *roof light*.

skylon A metal pylon, tapered to a point at top and bottom, hung on steel-wire ropes in a vertical position at the Festival of Britain 1951. *Compare* **trylon.**

skyscraper A multi-storey steel-framed building, typical of New York where the bedrock is only 15 m (50 ft) below ground and makes an excellent foundation for a 50-storey building.

slab [tim.] A piece of timber with one sawn face and one *waney*, curved face.

slabbing [tim.] Squaring a log; cutting *slabs* off it.

slab floor (1) A *reinforced concrete* (*C*) floor.

(2) A floor covered with slabs of marble, slate, limestone, granite, York stone, *cast stone*, *terrazzo*, etc.

slack side [tim.] The *loose side* of *sliced veneer*.

slag bricks Bricks made from *blast-furnace slag* crushed and mixed with *lime*. They are thus a sort of *sand-lime* or *concrete* brick.

slag strip [carp.] (USA) A strip of wood nailed round the edges of a felt roof to keep the gravel covering in place and to give a finished appearance to the edge of the roof. *See also* **no-fines concrete**.

slag wool or **silicate cotton** Filaments of metallurgical slag used for their high *fire resistance*, thermal insulation, and acoustical *absorption*. *See* **mineral wool**.

slaked lime (1) *Hydrated lime* ($Ca(OH)_2$), *quicklime* which has been submerged in water to make *lime putty*.

(2) *Dry hydrated* lime made from quicklime without submersion by adding to it only the calculated amount of water. It is sold in sacks as powder.

slaking The hydration of *quicklime*, thus: $CaO + H_2O = Ca(OH)_2$.

slaking box [pla.] A wood tank above ground in which lime is slaked. A gate at one end allows the *slaked lime* and water to flow out to the *screen* over the *maturing pit*.

slamming stile [joi.] A *shutting stile*.

slamming strip [joi.] An edge *banding* to the *shutting stile* of a *flush door*.

slap dash *Rough cast*.

slash grain [tim.] The *grain* of *flat-sawn timber* (slash-sawn timber).

slat (1) [tim.] A thin strip of wood in a *louvre*, blind, etc.

(2) A *stone slate*.

slate A fine-grained soil (clay, silt, shale, or volcanic ash) which has been compressed by earth movements like those which formed the mountains of North Wales. This compression has given the slate a cleavage which is across the bedding. The bedding is lost but the new cleavage turns it into slate. Slates are quarried in square-sawn blocks about 76 mm (3 in.) thick, which are then split to about 6 mm (¼ in.) thickness. Slates are usually sold by the *tally* of 1000. Ton slates are either *randoms* or very large ones, and are sold not by count but by weight. There are three thicknesses of slate, the thickest being 13 mm (½ in.). Normal lengths are 20 to 61 cm and widths 14·5 to 40·5 cm (8 to 24 in. long by 6 to 16 in. wide). Apart from thickness they are graded in three types: uniform length and thickness; uniform length and random widths; random lengths and widths. BS 680:1971 quotes 27 sizes from 'princesses', the largest, 61×36 cm (24×14 in.) to 'units', the smallest, 25×15 cm (10×6 in.). About 74 units are needed to cover 1 m² but only 11 princesses because they are so much larger. *See* **peggies, slate nails**, *also* BSCP 142, Slating and Tiling.

slate-and-a-half slate A slate half as wide again but of the same depth as the others in the roof. It is used at *valleys*, *hips*, and *verges*. *See* **laced valley, swept valley**.

slate axe A slater's axe or *zax*.

slate batten A *slating and tiling batten*.

slate boarding [carp.] Close *boarding* over a roof under slates or tiles.

slate cramp A piece of slate cut to a shape which in plan is waisted like an hourglass, having dovetailed ends. It is about 5 cm (2 in.) long by 2·5 cm (1 in.) wide at the ends, and 13 mm ($\frac{1}{2}$ in.) wide at the waist. It is from 2·5 to 15 cm (1 to 6 in.) deep, and fits into vertical *mortises* in neighbouring stones to lock them together.

slate hanging or **weather slating** Slating over a wall.

slate nails *See* **roofing nails.**

slate powder [pai.] Very fine, impalpable powder of slate used as an *extender* in paints. It is fairly opaque but is too dark to be used in any pale paint.

slater-and-tiler A *tradesman* who lays *slates*, *tiles*, and lead *soakers*, punches holes in slates by hand or machine, cuts and trims them, and lays *sarking felt*. He may also nail *close-boarding* on to *rafters* which have been fixed by the *carpenter*. He usually nails his own *slating and tiling battens* (and *counter battens* if necessary). Where torching is still done he does it. He also lays *ridge tiles* in *mortar*.

slate ridge or **slate roll** A ridge made of a circular rod of slate with a V-cut beneath. The roll is bedded on a heavy slate each side called a wing.

slater's iron A light *dressing iron* with a single spike, used by the slater while he is on the roof.

slating-and-tiling batten Square-sawn timbers between $1 \times 2 \cdot 5$ cm ($\frac{1}{2} \times 1$ in.) and 3×8 cm ($1\frac{1}{4} \times 3$ in.), nailed horizontally over *common rafters* or *counter battens* as bearers on which *slates* or *tiles* are hung.

slatter A *mason* who shapes stone slates (slats).

sleeper (1) [carp.] A *valley board* carried on the *rafters* of the main roof. It carries the feet of the *jack-rafters* and replaces a *valley rafter*. (2) [carp.] A *sleeper plate*. *See also* (C).

sleeper clip US term for a *floor clip*.

sleeper plate or **sleeper** [carp.] A *wall plate* on a *sleeper wall* or a similar wood plate on a concrete floor on to which floor boards or *joists* are fixed.

sleeper wall A *honeycomb wall* (2).

sleepiness [pai.] A reduction of *gloss* as a film dries, a defect sometimes akin to *seediness*.

sleeve *See* **expansion sleeve.**

sleeve piece or **thimble** or **ferrule** or **liner** [plu.] A short, thin-walled brass or copper tube used in *soldering* a lead or copper pipe to one of different metal. *See* **taft joint, union.**

sliced or **flat-cut** or **knife-cut veneer** [tim.] *Veneer* made by a machine with a knife up to 5 m (16 ft) long which cuts straight slices as a hand *plane* does. The *flitch* is first heated in a *cooking vat* so that when cut it is wet and steaming. The slices can be as thin as 0·2 mm but are usually about 1 mm. This method produces no sawdust but may make a slightly weaker veneer than *sawn veneer*. It is the usual method of cutting figured veneer and is more precise than *rotary cutting*. Sliced veneer has a *loose side* and a *tight side*.

slicing [tim.] *See* **sliced veneer**.

sliding-door lock [joi.] A *lock* designed for sliding doors. It has a hook-shaped bolt which engages with the back of the *striking plate* and prevents the door opening.

sliding sash A *sash* which opens by moving horizontally. *Compare* **sash window**.

slip (1) A *fixing fillet*.

(2) [joi.] A *parting slip*.

(3) [pla.] Fluid grout made with *gypsum plaster* or *cement*.

(4) [pai.] A paint which is so little *gummy* that it seems to be lubricated is said to have slip.

slip feather [joi.] A tongue for a joint such as a *feather joint*.

slip-joint conduit or **slip conduit** or **light-gauge conduit** [elec.] Metal *conduit* for house wiring which is slipped into sockets without screwing or other mechanical grip. There is therefore no perfect electrical connection through the conduit for *earthing*.

slip mortise [joi.] An *open mortise* or a *chase mortise*.

slipper [pla.] A metal shoe on a *horsed mould* which slides along the *running rule*.

slipper guide [pla.] A *running rule*.

slip sill A *sill* which fits or slips between the *jambs* of an opening and is not built into the walls, like a *lug sill*.

slip stone [joi.] An *oilstone slip*.

slip tongue or **false tongue** [joi.] A tongue in a *feather joint*.

slope of grain [tim.] The angle between the axis of a timber and the general direction of the *grain*. Timber for structural purposes should not have a slope more than 1 in 8 for beams (or struts less than 10 cm (4 in.) thick). For thicker struts, the slope should be below 1 in 11. *See* **gross features**.

slop sink or **housemaid's sink** A low sink, large enough to take a bucket under the tap, often installed in hospitals.

slot mortise [joi.] An *open mortise*.

slot screwing [joi.] Screwing of boards (such as a drawing board on to its stiffening battens) through slots which allow some movement. Any *shrinkage* cracking can be closed up by slacking off the screws in the slots, pushing the boards together, and tightening the screws again. *Compare* **secret screwing**, which can never be undone.

slow-burning construction (USA) (1) Heavy-timber construction which is more *fire-resisting* than bare steel frames.

(2) Construction with materials treated with fire-resistant surfacing or impregnation.

slurry A fluid cement-water mix, sometimes containing sand.

slurrying Protection of the finished surface of a stone facing by covering it with a weak mix of lime and stone dust, which is washed off when the building is handed over to the client by the builder.

slushed joints or **slushed-up joints** (1) Brickwork in which the vertical *joints* are filled with *mortar* poured from a watering can (grouting).

(2) In USA this term can mean joints filled by throwing in mortar with the edge of the trowel, a more usual method than grouting, but one which fills the joints less well.

small-bore system or **small-pipe system** [plu.] Central heating with radiators (or other heat exchangers) fed by pairs of pipes, originally 12·7 mm (0·5 in.) nominal bore, through which the water is driven by a pump. Because of the changes of standard dimensions of copper tube at the time of metrication, the nearest tube to this is now 15 mm outside diameter with a wall thickness of 0·7 mm, consequently a bore of 15 minus 1·4 = 13·6 mm. Small mass-produced pumps of high head have facilitated the development of *minibore* systems.

small lime or **lime ashes** [pla.] The *quicklime* which remains when *best hand-picked lime* has been removed from a run of kiln lime. It consists of small lumps of *lime* with ash or clinker and is therefore less pure than lump lime. It is used in many districts for making *black mortar*.

small tool or **spoon** [pla.] A small curved steel tool for moulding, mitring, or scratching an indentation in a *moulding*. It is used for finishing off mouldings by hand and is therefore made in several different shapes.

smell test The *scent test* for drains.

smith's hammer [mech.] A *hammer* weighing from 0·7 to 2 kg (1½ to 4 lb), shaped like a sledge, with a *cross peen*.

smoke outlet or **smoke extract** An opening, or a fire-resisting shaft or duct provided in a building to act as an outlet for smoke and hot gases produced by an outbreak of fire (BS 4422). *See* **fire venting.**

smoke rocket A *rocket tester*.

smooth ashlar A block of squared smooth-faced stone.

smoothing plane [joi.] A relatively short *bench plane* about 20 cm (8 in.) long used after the *trying plane*. *See* **carpenter's tools.**

smudge (1) [plu.] or **soil** A mixture of *lamp black* and *size* put over a lead surface to prevent *solder* sticking to it.

(2) Waste paint used for coating *formwork* (C) or the insides of iron *gutters*, but not for painting.

snake [plu.] A tool like a long *bending spring*, used for unblocking drains.

snakestone or **Water of Ayr stone** or **Scotch stone** A smooth stone used for polishing *Keene's cement* to a glass-like finish when making *scagliola* or *marezzo marble*.

snap header or **blind header** A half-brick, not a *header*.

sneck In *snecked rubble* a small squared stone, not less than 8 cm (3 in.) deep, which has its top surface level with the top of a *riser*.

snecked rubble (Scotland) A *rubble wall* built of squared stones of irregular size. Its peculiarity lies in the use of *snecks*.

snips *See* **tin snips.**

snow boards or **snow slats** (1) Horizontal wood slats nailed with gaps between them on short bearers over a *box gutter* to allow melting snow to drain away, known in Scotland as **snow cradling.**

(2) or **snow guard** or **roof guard** A wire fence or a board on edge about 20 cm (8 in.) high fixed at least 10 cm (4 in.) along the roof slope above the eaves gutter to prevent heavy masses of snow slipping off the roof and breaking the eaves gutter or its supports. BSCP 142:1971 states that the clearance between the roof surface and the bottom of the guard should be at least 5 cm (2 in.). From this it is evident that they cannot be relied on to hold up loose slates slipping off a steep roof.

snow cradling *See* **snow boards** (1).

soaker A small piece of *flexible metal* cut to shape by the *plumber* and laid by the *slater* to interlock with *slates* or *tiles*. It makes a watertight joint at a *hip* or *valley* or at an *abutment* between a roof and a wall. It is bent at right-angles for an abutment or at an obtuse angle for a *valley*.

soap A *queen closer*.

socket (1) [plu.] An enlarged end of a pipe (often of cast iron) into which a similar pipe (the *spigot*) is fixed by pouring in molten lead (or cement mortar for a stoneware pipe). In USA called a bell.

(2) [plu.] A *coupling*.

(3) [joi.] The cavity into which the pivot of a pivoted sash fits.

(4) [joi.] The *mortise* of a *dovetail* joint. *See* **capel** (*C*).

socket chisel [carp.] A massive *chisel*, with a socket formed at the base of the *tang* into which the wood handle fits. It is used for mortising and is strong enough to be struck with a *mallet*.

socket former [plu.] A tool that is inserted into an end of copper tube to widen it to the bore needed for a socket of a *capillary joint*. It is driven into the tube with a hammer and thus opens out the end. The tool is expensive but one can be improvised from a short piece of stainless steel tube of the same diameter as the copper tube to be expanded. Joints made in this way are less conspicuous than any other type and the method enables short ends of tube to be used up, but probably takes more time than other methods. If the copper hardens or begins to split, the widening must be stopped, the split piece cut off, and the copper softened (annealed) by heating it to a dull red heat.

socket inlet or **plug** [elec.] An electrical part, fixed generally to the *flex* of an appliance for connecting it to the supply through a *socket outlet*, or sometimes through an *adaptor*.

socket outlet (1) [elec.] or (USA) **receptacle** An electrical fixture on a wall, containing two or three holes into which the metal prongs of an *adaptor* or *socket inlet* are inserted. *See* **shuttered socket, ring main.**

(2) [plu.] That part of a *plug-in connector* for gas which is fixed to the wall.

soda-lime process A *water softening* process for dealing with large quantities of water. A large reaction tank is required, containing the water to be softened, to which sodium carbonate (Na_2CO_3) is added to remove permanent hardness, and lime (CaO) to remove temporary hardness. The water must be filtered or allowed to settle, so that the precipitated salts can be removed.

soffit or **soffite** (1) the under-surface of a *cornice, stair,* beam, arch, *vault,* or rib or the uppermost part of the inside of a drain, sewer, or culvert (also called the crown). Generally, it is any under-surface except a *ceiling.*

(2) [joi.] The *lining* at the *head* of an opening.

soffit board or **planceer piece** [carp.] A horizontal board nailed to the underside of rafters, forming the *soffit* under an *overhanging eave.*

soft-burnt A description of clay *bricks, tiles,* etc., which have been fired at a low temperature and therefore have high *absorption* and low compressive strengths.

softener [pai.] A flat multi-knot *brush* of hog's hair used for softening markings in *oil paint* or *glaze coats.*

soft sand Sand with fairly small grains, rounded and uniform in size, smooth to the hand when squeezed, quite unlike *sharp sand.* It is used for making mortar for rendering, pointing, bricklaying, etc., but not for making concrete.

soft solder [plu.] *Fine solder, plumber's solder,* etc., made mainly of lead and tin, both of which melt below red heat, unlike *hard solder.*

softwood [tim.] The conventional, but universally used, description of timbers of the botanical group gymnosperms, most of which in commerce are conifers. They are sometimes harder than *hardwoods.* The softwood yew is nearly as hard as *oak* and harder than many other hardwoods. *See* **balsa.**

soil [plu.] (1) *Smudge.* (2) Sewage as opposed to dirty water.

soil branch A branch pipe leading to the *soil pipe.*

soil cement (USA) *Cob* construction. *See* **soil stabilization** (*C*).

soil drain or **foul drain** [plu.] A drain for carrying sewage or trade effluent to the *sewer* (*C*). It is of at least 10 cm (4 in.) dia. A soil drain is tested before use by the *smoke rocket* or by the *hydraulic test* (*C*).

soil pipe or **soil stack** A vertical pipe which takes sewage down from the parts of a building above ground into the *soil drain.* Its upper end passes above the roof and is left open to ventilate the drain. Cast iron is now the usual material for these pipes in Britain except at very awkward bends, where copper is used. *Plastics* pipe also is being widely introduced.

soil ventilation pipe [plu.] *See* **ventilation pipe.**

solar reflecting surface or **spar finish** A special finish to a flat roof covered with asphalt or other dark material, consisting of materials such as white rock (spar) chippings which considerably reduce the heating of the roof in sunlight.

solder [plu.] (1) Any alloys used for joining metals except those which consist mainly of the pure parent metal. These are called filler rods or electrodes and are used in *welding* (*C*). *Plumber's solders* and *fine solders* are lead–tin alloys. *Brazing* or *hard solders* are copper–zinc alloys. *See* **silver brazing**.

(2) To unite metal parts with solder. Soldering is much used for joining copper, brass, or lead pipe; brazing for uniting iron, steel, copper, etc.

soldered dot [plu.] *See* **dots**.

soldering iron or **copper bit** or **soldering bolt** [plu.] A small copper block fixed on one end of a thin steel rod to which a wooden (or other thermally insulating) handle is attached. The copper bit may be heated electrically or by other means, but the handle is always kept cool by this construction. *See* **plumber's tools** (illus.).

soldier (1) [joi.] A short upright *ground* for fixing a *skirting board*.

(2) An upright brick, a brick on end. *See also C*.

soldier arch A *flat arch* of uncut bricks on end.

soldier course A course of bricks on end, laid usually as a *coping*.

sole [joi.] The smooth under-surface of a *plane*, out of which the *cutting iron* projects.

sole plate (1) or **sole piece** or **mudsill** or **ground sill** or (USA) **abutment piece** [carp.] The timber laid on the ground under the feet of *raking shores* and at right-angles to them.

(2) A *sill*.

solid [joi.] Timber without *planted* mouldings.

solid bossing [plu.] The *bossing* of lead to shape instead of *soldering* it.

solid bridging or **solid strutting** or (USA) **block bridging** [carp.] *Strutting* of floor *joists* by rectangular short lengths of *joist* cut to fit tightly between, and at right-angles to the joists at midspan. Solid bridging needs more precise work than *herring-bone strutting* but is stiffer and forms a valuable *fire stop*, for which reason it may also be used in hollow wooden *partitions*. *See* **bridging**.

solid door [joi.] (1) A *flush door* with a solid core resembling *blockboard*, not a *skeleton core*. *Compare* **hollow-core door**.

(2) A *fire-resisting* door built of three thicknesses of tongued and grooved boarding, the inner one horizontal, the outer two vertical. Sometimes this door is plated with sheet metal.

solid floor (1) A *concrete* floor slab, built without *hollow blocks* but with solid concrete throughout its thickness. The finish may be wood boards, mortar *screed*, or any other finish.

(2) [carp.] *See* **plank-on-edge floor**.

solid frame [joi.] An expensive door frame in which each *stile* with its *stop* is a single rebated piece of wood and the stop is therefore not *planted*.

solid masonry unit (USA) Any *brick* whether *perforated* or cellular, provided that its cross-sectional area in every plane parallel to the bed, is 75% or more of the bed area. A hollow masonry unit has less solid than 75% of the bed area. *See* **hollow blocks**.

solid moulding or **stuck moulding** [joi.] A *moulding* cut on the timber, not planted.

solid-moulding cutter [tim.] A single piece of steel used on the *spindle moulder* which combines the functions of *cutter block* and knives. These cutters are perfectly balanced and can therefore run at high speeds and do not need frequent sharpening; they are accurate in cutting, but costly to buy. The knives cannot be changed, but this is an advantage since they are in perfect balance.

solid-newel stairs *Spiral stairs* of stone in which the inner end of each step is shaped to form a nearly continuous cylinder with the inner ends of the other steps. Partly because they take up very little space they were the commonest means of access to upper floors until about 1550 when the framed wood stair was introduced. They were also popular because they were easily defended by a single swordsman. If the builder of the house was right handed, he built his stair with the *newel* on the left looking down. *See* **turret step**.

solid partition A *partition* which has no cavity. It may be of *bricks*, or *blocks*, or be plastered both sides on the same *lathing*.

solid plasterwork [pla.] Plaster with a solid *core* which is formed in place, not in the shop, like *fibrous plaster*.

solid punch A hard steel bar shaped like a *nail punch*, but with an end which is even blunter. It is used for driving a bolt out of a hole, etc.

solid roll A joint between sheets of *flexible-metal roofing* which are folded together over a *wood roll*. *Compare* **hollow roll**.

solid stop [joi.] A *stop* rebated in a *solid frame*.

solid strutting [carp.] *Solid bridging*.

solid wood floor [carp.] A *plank-on-edge* floor.

soluble driers or **liquid driers** [pai.] *Driers* which are soluble at ordinary temperatures in *drying oils* or hydrocarbon *solvents*. They are usually resinates, linoleates, or naphthenates of lead, manganese, or cobalt.

solum The ground area within the outside walls of a building.

solvents (1) Liquids like *acetone* which will dissolve solids.

(2) [pai.] *Volatile* liquids used to dissolve or disperse the *binders*. They evaporate during *drying* and are therefore absent from the *film*, but they control the character of the *finish* and the ease of application. *See* **thinners**.

solvent-welded joint [plu.] A permanent joint between plastics tubes, made after smearing both spigot and socket with a solvent, by merely inserting the spigot into the socket.

soot door or (USA) **ashpit door** or **cleanout door** A door at the foot of a *chimney* through which soot is removed.

soot pocket An extension of a *chimney* about 30 cm (12 in.) below the smoke inlet. This extension is usually fitted with a *soot door*.

Sorel's cement *See* **magnesite flooring.**

sound absorption *See* **absorption.**

sound boarding [carp.] Horizontal boards fitted closely between *joists* and resting on them in the thickness of the floor. They carry *pugging*, which increases the sound insulation of the floor.

sound knot [tim.] A tight, solid, undecayed *knot* which is at least as hard as the wood round it. *Compare* **unsound knot.**

soundness of cement The freedom from cracks or expansion of a cement sample when set.

soundproofing A term which has very little meaning, and is therefore falling out of use in the same way as '*fireproofing*'. *See below, also* **absorption, discontinuous construction, pugging,** *also* BSCP 3, chapter 3, Sound insulation and noise reduction.

sound-reduction factor or **acoustical reduction factor** A value which can be expressed for a given wall or *partition* in *decibels*. This value gives a measure of the reduction in intensity of the sound of any given frequency which passes through the wall. It is found that the sound-reduction factor can be increased by 5 decibels if the weight of the wall is doubled. This relationship is the opposite of that for thermal *conductivity*.

south-light roof A *north-light roof* of the southern hemisphere, with the steeper slope glazed and facing south.

soya glue [tim.] *Glue* made from soya bean meal after its oil has been extracted. It resembles *casein glue* except that it is a *vegetable glue*.

space [carp.] Of a saw, the length from one point of a *saw-tooth* to the next.

spaced slating *Open slating*.

space heating The heating of the area within a building by *direct* or *indirect heating*.

spacing The distance between centre lines of *columns*, beams, roof *trusses*, *purlins*, and so on. *See* **pitch.**

spall (1) To break off the rough edges of stone with slanting blows of a *chisel*.

(2) or **gallet** A flake of stone used for filling the spaces in *rubble walls* or in pitching *revetments* (*C*).

spandrel (1) A roughly triangular space between the *extrados* of an arch, one abutment, and the road carried by the arch.

(2) or **apron wall** (USA) The rectangular infilling in a multi-storey building between a window *sill* and the window *head* below.

(3) The triangular infilling under the *outer string* of a stair.

spandrel step A solid stone step of triangular shape, carved to make, with other similar steps, a flush *soffit* to the *stair*.

Spanish mahogany [tim.] *Cuban mahogany.*

Spanish tile (1) or (USA) **mission tiles** *Roofing tiles* shaped like a half-cylinder slightly wider at one end than the other. The *over-tile* and *under-tile* are of roughly the same shape, the under-tile being larger. These tiles form a beautiful roof which is flexible both in *side lap* and end lap. However, the amount of timber needed is considerable, as 8 × 5 cm (3 × 2 in.) timbers must be fixed up the slope of the roof (over felt on *close-boarding*), to secure the under-tiles. *See* **Italian tiling.**

(2) (USA) A *single-lap tile* which slightly resembles the tile called in Britain a single *Roman tile.*

span piece [carp.] A *collar beam.*

span roof [carp.] An ordinary *pitched roof.*

spar (1) or **brotch** or **buckle** In *thatching*, a split piece of hazel or willow 60 cm (24 in.) long. Eight spars are obtained from a 5 cm (2 in.) dia. branch. They are pointed at both ends, doubled, and driven into the thatch over *runners*, *liggers*, or *scallops.*

(2) A *common rafter.*

sparge pipe [plu.] A perforated pipe for flushing a urinal stall.

spar piece or **span piece** A *collar beam.*

sparrow peck (1) A texture given to plaster by pitting it with a stiff broom.

(2) A texture given to stone by *picking.*

spar varnish [pai.] A boat *varnish* with outstanding resistance to water and sunlight. It usually consists of *tung* or *linseed oil* with durable *resins.*

spatterdash A wet, rich mix of *cement* and sand (1 : 1½ to 1 : 3) which is thrown hard on to a smooth *brick* or *concrete* surface and allowed to harden, and thus provides a *key* for, or reduces the suction of, the first coat of plaster. If these hardened rough splashes of *mortar* do not give enough adhesion, *concrete-bonding* plaster may have to be used.

special A bend or tee or similar *fitting* of a pipe, or a bullnose brick, squint quoin, splay brick, etc., which needs to be specially ordered, usually by number. If it is a piece kept in stock it is called a 'standard special', otherwise it has to be specially made.

specification A detailed description prepared by a consulting engineer or *architect* to tell the *contractor* everything about the workmanship which cannot be shown on the drawings. It is written in the *sequence of trades. See* **National Building Specification.**

spigot [plu.] An end of a pipe inserted into a *socket* in the next pipe to form a *spigot-and-socket joint.*

spigot-and-socket joint or (USA) **bell-and-spigot joint** [plu.] A joint in glazed ware or other pipes which is made tight by caulking one end of pipe of normal diameter (the *spigot*) into an enlarged end, the *socket*. For cast iron the caulking is with lead wool or molten lead, while for *glazed ware* it is done with cement *mortar*, plugged with old rope at the lower end.

spike knot [tim.] A *splay knot*.

spindle (1) [mech.] A small axle.

(2) [joi.] A turned piece of wood such as a circular *baluster*.

(3) [tim.] A *spindle moulder*.

spindle moulder or **spindle** [tim.] A *moulding machine* which has a vertical shaft rotating at about 7500 rpm carrying a *cutter block* with two or more specially shaped knives or a *solid-moulding cutter*. Since it is a very versatile machine it usually needs a specialist to operate it. *See* **woodworking machinist**.

spiral grain [tim.] Fibres growing spirally round the trunk of a tree, thus making the wood difficult to work. *See* **interlocked grain**.

spiral stair A circular *stair* in which the *treads* are all *winders*. (The shape is helical, not spiral.) The central parts of the winders are laid on top of each other and form a continuous cylindrical *solid newel*. These stairs can be made of stone, as was customary in the past, or of cast iron. Recently, elegant precast concrete spiral steps have been made. *See* **turret step**.

spiriting off [pai.] The final operation in French polishing, in which the last traces of oil are removed by a rag damped with *methylated spirit*, drawn quickly and often over the surface. *See* **bodying in**.

spirit level *See* **level**; *also* **mason's tools** (illus.).

spirit stain [tim.] A dye dissolved in alcohol (methylated spirit), usually with *shellac* or other dissolved *resin* as a *binder*. It is used for darkening a wood surface, but emphasizes the *grain* of the wood less than a *water stain*.

spirit varnish [pai.] *Varnish* made by dissolving a gum or *resin* in alcohol or other spirit. It dries by *evaporation* of the spirit and resembles *knotting* in composition.

spit The depth of one spade blade, about 23 cm (9 in.).

spitter *See* **lead spitter**.

splashboard (1) A board placed on edge against a wall beside a *scaffold* to keep the wall clean.

(2) A *weather moulding* planted at the foot of an outside door.

splash lap That part of the *overcloak* of a *flexible-metal* drip or roll which extends on to the flat surface of the next sheet.

splat A cover strip over the joints of *wallboards*.

splay A slope across the full width of a surface, often at 45°, a large *chamfer*.

splay brick or **slope** or **cant brick** A special brick which is bevelled at about 45° at one end (splay header) or at 45° along one edge (splay stretcher).

splayed coping A *feather-edged coping*.

splayed grounds [joi.] *Grounds* with bevelled or rebated edge to provide a *key* for the plaster where the ground also acts as a *screed*.

splayed heading joint [carp.] A joint between the ends of floorboards which are not cut vertically but at 45°, so that one overlaps the other.

splayed skirting [joi.] A *skirting* with its top edge bevelled, not moulded.

splay knot or **spike knot** [tim.] A *knot* cut nearly parallel to its length. *Compare* **round knot.**

splice [carp.] or **fished joint** A joint between *halved* timbers which are covered on each side by steel or wood plates (*fishplates* (*C*)), bolted together through the timbers.

spline (1) [carp.] A timber strip nailed (with another, in pairs) down a rectangular timber *sheet pile* (*C*) to form a groove into which the next pile fits.

(2) [joi.] A wood strip nailed or screwed to a window frame for *glazing.*

(3) A *feather.*

split [tim.] A crack in wood or *veneer* which passes through it. *Compare* **check.**

split course A *course* of bricks cut lengthwise to reduce their depth to less than that of an ordinary course.

split head [pla.] A steel tripod carrying a standard notched at the top to hold a *scaffold board* on edge. This makes a fairly strong platform for plasterers, but of course at least four split heads are needed to carry two boards on edge.

split pipe A pipe cut lengthwise, a channel.

split-ring connector [tim.] A timber *connector* inserted in a pre-cut ring-shaped groove in both timbers, made with a *hole saw.*

split shakes [tim.] Wood *shingles* which have been split, not sawn.

split system (USA) Space heating, by blowing warm air into a room and using *radiators* simultaneously.

splocket A *cocking piece* (sprocket).

spokeshave [joi.] A two-handed light *plane* for shaping surfaces which are curved like the spokes of a wheel.

sponge-backed rubber floor Solid rubber sheet about 3 mm ($\frac{1}{8}$ in.) thick, backed with sponge rubber 3 to 9 mm ($\frac{1}{8}$ to $\frac{3}{8}$ in.) thick, obtainable in rolls 1·8 m (6 ft) wide, 23 m (75 ft) long. It can be laid on most smooth surfaces which are perfectly dry. A *screed* of about 3 mm of latex cement is a good foundation over wood, metal, or concrete floors. Adhesives based on natural rubber must be used for fixing the sponge to the floor and to the solid rubber surface layer. *See* **cement-rubber latex.**

spoon bit [carp.] A *dowel bit.*

spot board or **gauge board** or **spot** [pla.] A plasterer's board about 1 m (3 ft) square on which he works up the plaster before he puts it on. It rests on a stand about 68 cm (27 in.) high.

spot gluing [tim.] Gluing wood in small separate areas.

spotting [pai.] The defect of small areas of a painted surface which have a different *colour* or *gloss* from the rest.

spotting in or **spot finishing** [pai.] *Rubbing* down and refinishing small defective patches in a coating (BS 2015).

sprayed asbestos or **limpet asbestos** Thermally insulating *asbestos* blown on to a surface by a spray gun, in thicknesses from 0·6 to 5 cm ($\frac{1}{4}$ to 2 in.). It has the low thermal *conductivity* of 0·046 W/m deg. C. (0·32 Btu in./ft²h deg. F.), is jointless, continuous, adhesive, rotproof, vermin-proof, and wholly *incombustible*. One cm of it is equivalent in *fire protection* to 2 cm of concrete, and its weight is only about $\frac{1}{14}$ that of concrete. *See* BSCP 299, Sprayed Asbestos Insulation.

spray gun or **air brush** or **spraying pistol** or **paint spray**, etc. [mech.] A compressed-air operated tool for ejecting a fine mist of paint or powder or *metal coating* or cement mortar or plaster. *See* **atomization, gunite** (*C*), **spray unit.**

spray painter A craftsman who sprays paint through a *spray gun*, which may be worked by hand pump, electric motor, or by an extension from a vacuum cleaner. If the spray is hand pumped, the painter's labourer must pump continuously.

spray painting or **spraying** [pai.] Putting paint on with a *spray unit*. This is the only good way of applying *lacquer* and is also suitable for *water paint* or *oil paint* on large areas. It is not a good way of putting on *priming*, because brushing makes the priming penetrate the surface better. Spraying of *lead paints* is not allowed in Britain owing to the danger of breathing the spray.

spray unit [pai.] An air compressor together with a small pressure tank for air, another for *paint*, and a *spray gun* connected by hose to the tanks. The paint container is not under pressure in every type. *See* **spray painting.**

spreading rate [pai.] The surface area covered by unit volume of mixed paint (or unit weight of dry distemper or paste paint). This term is now preferred to covering power, which is ambiguous.

sprig [joi.] A small wire *nail* with no head, such as a *glazing sprig.*

sprig bit [joi.] A *bradawl.*

spring (1) or **edge bend** or (USA) **crook** [tim.] A variety of *warp* which consists in the curving of a board in a plane parallel to its face.

(2) [plu.] *See* **bending spring.**

springer or **springing** The first stone laid in an arch. It is bedded on the *springing line* and its upper surface is called the *skewback.*

spring-head roofing nail A galvanized drive screw for fixing *sheet.*

springing line The horizontal line in an arch joining the intersections between the surfaces of the walls and the surface of the arch *intrados* at each side. *See* **springings.**

springings The intersections at each side of an arch between its lower surface (*intrados*) and the faces of the walls or piers which carry it.

spring snib [joi.] A sash fastener which is spring controlled.

springwood or **earlywood** [tim.] That part of the *annual ring* which is formed in spring and is usually paler, less dense, and weaker than *summerwood*.

sprinkler system A *fire-extinguishing* system consisting of pipes installed in the ceiling throughout a building. Branches from the pipe project through the ceiling 3·3 m (10 ft) apart. The lower end of the branch is sealed with a plug of metal which melts at 68° C. (155° F.), so that water is released from the sprinkler when the air temperature reaches 68° C. The principle of the *drencher* system is similar. The modern tendency is to replace the metal plug by a plastics plug containing liquid which expands and bursts it at a temperature which can be predicted within 5° C. Sprinklers may be 'wet' or 'dry' or 'alternate-wet-and-dry'. The dry sprinkler is fitted with upward-turned sprays, its pipes being laid to a fall of about 1 in 200 and filled with air at a pressure of about 275 kN/m² (40 psi). This system is used in Russia, Canada, and other countries with hard winters, since the water is only admitted to the pipes when the air is released from them. The wet sprinkler is permanently filled with water, and operates more quickly when there is a fire, but is not proof against frost. The alternate-wet-and-dry sprinkler can be water-filled in warm weather and air-filled in frosty weather. It is used in temperate countries with occasional frost. The best-quality sprinkler installation requires two independent water supplies, of which at least one is a public main or large reservoir, and the highest spray must have a pressure of 15 m (50 ft) of water or 173 kN/m² (25 psi). *See* **emulsifier.**

sprocket A *cocking piece*.

sprocketed eaves *Eaves* given an outward tilt by *sprockets*.

spruce [tim.] Many species of picea, softwoods which are exported from North America and are very light, weighing only 432 kg/m³ (27 lb/ft³) at 12% *moisture content*. *See* **Canadian spruce, Sitka spruce.**

sprung [tim.] A description of timber which has warped by *spring*.

spud [joi.] A *dowel* in the foot of a door post to fix it to the floor.

spur [elec.] A *socket outlet* connection from a *ring main*. As its name implies it is a single cable branching off the ring main. If the ring main has its full allowance of about ten outlets, some authorities will allow only one spur.

square (1) [carp.] A measure of area for floors and roofs, 100 sq. ft (9·3 m²), used before metrication in England south of Birmingham.

   (2) or **try square** An L-shaped metal or metal-and-wood tool for setting out right-angles. *See* **mason's tools.**

   (3) A timber of square cross section.

   (4) A *pane* of glass of any shape, cut to size for glazing.

square chisel [joi.] *See* **mortising machine.**

squared log [tim.] (1) A *baulk*.

   (2) A *half-timber*, that is, a baulk sawn down the middle, at least 13 × 25 cm (5 × 10 in.) in cross section.

squared rubble *Rubble walling* of stones of varying size which are squared and not snecked. It is usually *coursed* at every third or fourth stone and may be built almost as carefully and with as thin joints as *ashlar.*

square-edged timber Timber without *wane. See* **square sawn.**

square joint [carp.] A *butt joint.*

square roof [carp.] A roof in which the rafters rise at 45° and therefore meet at a right-angle at the *ridge.*

square-sawn timber Timber sawn to a rectangular cross section, with or without *wane* (BS 565). *See* **square-edged.**

square shoot [carp.] A wooden *downpipe.*

square staff [pla.] A rectangular wood *fillet* fixed at a salient corner of a room as an *angle bead.*

square-turned baluster or **newel** [joi.] A *baluster* or *newel* post with *mouldings* cut on four faces by any method except *turning.*

squaring or **squaring up** or **working up** [q.s.] Calculating areas, a process which follows *taking off* from drawings or measuring up work.

squeeze [pla.] To take a squeeze of a *moulding* is to press wet plaster over it and thus to take a cast of it from which a *template* can be cut.

squint quoin A projecting corner of a building which is not a right angle.

squirrel-tail pipe jointer [plu.] A *pipe-jointing clip.*

stability The resistance of a structure to sliding, overturning, or collapsing. *See C.*

stable door or (USA) **Dutch door** [joi.] A door cut through horizontally at about half its height and having each half separately hung.

stack (1) A *chimney stack.*

(2) A rainwater *downpipe.*

(3) A *soil pipe.*

staff *See* **angle staff.**

staff bead (1) [pla.] A moulded external angle normally run in Keene's plaster (BS 4049).

(2) [joi.] A *guard bead.*

Staffordshire blues An extremely hard and dense, deep-blue *brick* which can be obtained *wirecut* or *pressed.* It has the remarkably high crushing strength of 110 N/mm² (16 000 psi) and is therefore one of the best *engineering bricks* in Britain.

stagger To arrange rivets, bolts, or other *building elements* alternately so that, for example, rivets do not come opposite other rivets in the next row. *See* **staggering.**

staggered courses *Courses* of *shingles* laid with their butts not in one horizontal line.

staggered-stud partition [carp.] A *partition* formed of two rows of vertical timbers (*studs*) which are separated from each other, each being surfaced by its own *covering.* The rows of studs each fit into the gap in the opposite row and may be separated by an insulating *blanket*. This is a *discontinuous construction* used for reducing sound transmission between rooms.

staggering An arrangement of joints which are spread out to distribute their weakness or of supports which are spread to give uniform strength. For instance, in *flexible-metal roofing*, *cross welts* are staggered so that they intersect only one *standing seam* at a time.

staging A *mason's scaffold*.

stain (1) [tim.] *See* **blue stain.**

    (2) [pai.] A solution or suspension of dye or other colouring matter in a *vehicle* designed to *colour* a surface by penetrating rather than hiding it. True stains are *water*, oil, or *spirit stains*, according to the vehicle. *Varnish* stains are not true stains, since they do not penetrate the surface but merely leave a coloured coating on it.

stained-glass windows Windows of glass which is coloured during the making. The name is misleading since the colouring is not stain, but is fired into the glass.

stainers or **tinters** [pai.] Coloured *pigments*, ground in a paint *vehicle*, which can be added in small amounts to ready-mixed *paints* to modify their colour. Stainers have intense *staining power* but are not always very opaque.

staining power [pai.] The amount of colour given to a *white pigment* by a certain amount of coloured pigment. It corresponds to the *reducing power* of a white pigment.

stainless steel An *alloy steel* (*C*) with varying amounts of alloying metals, but often with chromium and nickel, commonly 18 per cent Cr and 8 per cent Ni. It resists corrosion well, and so, when polished, is used for shop fronts, door furniture, etc., but it is not invariably stain-free. Water-supply or heating tube made of it is compatible with fittings made for copper tube, and as copper tube increases in price, stainless steel tube becomes more popular, though it is stiffer than copper and more difficult to bend. To obtain a good finish, the polished, more expensive grade of tube must be used.

stair [joi.] (1) A series of steps with or without landings, including necessary *handrails* and *balustrades* and giving access from floor to floor (BS 565). This BS does not mention *staircase*, which is more often used than stair in England in this sense. The minimum recommended dimensions of house stairs are 90 cm (3 ft) overall width, 24 cm ($9\frac{1}{4}$ in.) *treads*, and 21·5 cm ($8\frac{1}{2}$ in.) *going*. In a straight flight, twice the rise plus the *going* should not add up to more than 70 cm ($27\frac{1}{2}$ in.) nor to less than 55 cm ($21\frac{5}{8}$ in.).

    (2) One step consisting of tread and riser.

staircase Originally the space within which a stair was built, this word has now come to mean *stair*, which is dropping out of conversational use, in favour of staircase or *stairs*.

stair clip or **stair rod** A clip or rod which holds a stair carpet in place.

stairhead The top of a *stair*.

stair horse [carp.] A *carriage*.

stairs A series of steps between floors, a more commonly used term than *stair*. Internal stairs are now often of reinforced concrete or timber; *escape stairs* are of steel, cast iron, or reinforced concrete. Stairs may be *bracketed*, *dog-legged*, *double-return*, *geometrical*, *spiral*, or *straight flights*, among others. *See also* **flier, nosing, tread, winder**.

stairway A *staircase*, or a stair *well*.

stair well *See* **well**.

stake [carp.] A timber pointed at one end for driving into the ground.

stallboard [joi.] A strong *sill* and its framing beneath a shop window over the *stall riser*.

stallboard light [joi.] A *pavement light* near a stallboard.

stall riser [joi.] The vertical surface of polished granite, armour *plate glass*, *tile*, wood, marble, or similar material from the pavement up to the *stallboard*.

standard (1) or **scaffold pole** An upright of a *scaffold* whether of wood or metal.

(2) [tim.] *See* **Petrograd standard**.

(3) *See* **British Standard**.

standardization Agreement between producer and consumer under the authority (in Britain) of the *British Standards Institution* on certain tests, dimensions, tolerances, and qualities of a certain product for certain purposes. When agreement is reached it is published as a *British Standard*.

standard knot [tim.] (USA) A knot of $1\frac{1}{2}$ in ($3\cdot8$ cm) dia. or less. *Compare* **pin knot**.

standard method of measurement [q.s.] The metric method of measuring builders' work, approved by the Royal Institution of Chartered Surveyors and the National Federation of Building Trade Employers, and published by them in a book of that title (metric, 1968). For engineering quantities, civil engineers usually prefer the method of the Institution of Civil Engineers. *See* **measurement**.

standard special A piece made to standard dimensions and quality, such as a *bullnosed* coping *brick* or a *bend* of pipe, which is always stocked but not sent with an order for bricks or drains unless ordered. *See* **special**.

standard specification *See* **British Standard**.

Standard wire gauge (SWG) or **imperial standard wire gauge** An old established way (likely to be superseded by metrication) of specifying the thickness of steel saws, sheet, wire, tube, cut nails and some non-ferrous metals such as copper. The Birmingham gauge (BG), used for sheet steel, generally differs from the same number of the SWG by less than 20 per cent. There are many other British and American 'gauges' for sheet metal and wire, including the Birmingham wire gauge which is not the same as the Birmingham gauge.

SWG diameters in millimetres and inches

SWG	(mm)	(in.)	SWG	(mm)	(in.)	SWG	(mm)	(in.)
7/0	12·700	0·500	13	2·337	0·092	32	0·274	0·0108
6/0	11·786	0·464	14	2·032	0·080	33	0·254	0·0100
5/0	10·973	0·432	15	1·829	0·072	34	0·234	0·0092
4/0	10·160	0·400	16	1·626	0·064	35	0·213	0·0084
3/0	9·449	0·372	17	1·422	0·056	36	0·193	0·0076
2/0	8·839	0·348	18	1·219	0·048	37	0·173	0·0068
1/0	8·230	0·324	19	1·016	0·040	38	0·152	0·0060
1	7·620	0·300	20	0·914	0·036	39	0·132	0·0052
2	7·010	0·276	21	0·813	0·032	40	0·122	0·0048
3	6·401	0·252	22	0·711	0·028	41	0·112	0·0044
4	5·893	0·232	23	0·610	0·024	42	0·102	0·0040
5	5·385	0·212	24	0·559	0·022	43	0·091	0·0036
6	4·877	0·192	25	0·508	0·020	44	0·081	0·0032
7	4·470	0·176	26	0·457	0·018	45	0·071	0·0028
8	4·064	0·160	27	0·417	0·0164	46	0·061	0·0024
9	3·658	0·144	28	0·376	0·0148	47	0·051	0·0020
10	3·251	0·128	29	0·345	0·0136	48	0·041	0·0016
11	2·946	0·116	30	0·315	0·0124	49	0·031	0·0012
12	2·642	0·104	31	0·295	0·0116	50	0·025	0·0010

(With grateful acknowledgement to *Specification*)

standing ladder A *ladder* with rectangular *stiles*, as opposed to a *builder's ladder*. An *extending ladder* is built of two or more standing ladders.

standing leaf [joi.] A leaf of a *folding door* which is bolted in a closed position, as opposed to the *opening leaf*.

standing seam A *seam* in *flexible-metal roofing*, usually running from *ridge* to *eaves*. See **cross welt**.

standing timber Growing trees. Their volume in timber is estimated by a *cruiser*.

stand oil [pai.] *Drying oil*, such as *linseed* or *tung oil*, which has been *polymerized* by heat treatment. It is so called because it was originally made by standing it in the sun for a time.

stand sheet A *dead light*.

staple (1) A metal U-shaped loop for padlocking a door or gate. *See* **hasp and staple**.

(2) A U-shaped nail with two points.

star drill (USA) A star-shaped *plugging chisel*.

star shake [tim.] According to BS 565 several *heartshakes*. Others regard it as a heartshake which does not reach the pith.

starved [pai.] *Hungry*.

stat Abbreviation for *thermostat*.

station roof or **umbrella roof** A roof carried on a single row of stanchions. It is therefore *cantilevered* (*C*) to one or both sides.

statutory undertaker An organization, in the United Kingdom, with a duty laid down by law to provide a service to the general public or to a section of it such as shipping interests. Some statutory undertakers are British Rail, electricity boards, gas boards, passenger transport authorities such as London Transport, water authorities and dock, harbour or inland navigation boards. Until 1973 all were immune from noise legislation and could be as noisy as was convenient to them but in that year it was proposed to the Department of the Environment that they should lose this immunity.

stave [carp.] A *rung* of a *ladder*.

stay bar (1) A horizontal bar which strengthens a *mullion* or a leaded light.

(2) A bar which holds together the two opposite walls of a building and prevents them falling apart.

(3) A *casement stay*.

stay log [tim.] A large timber fitted to a *veneer* lathe. It is used as a base on to which *flitches* are screwed when *half-round veneer* is cut from them.

steam-brush cleaner A *skilled man* who cleans the outside stone or brick faces of buildings with a steam jet and wire brush.

steam-stripping appliance [pai.] A portable boiler connected to a flat metal distributor which blows steam on to a wall and makes it easy to strip *wallpaper* or *distemper*.

steel See *C*.

steel casement A *casement* made of steel. It may be fixed to the wall directly or to a wood or steel *sub-frame*.

steel core In *patent glazing* a specially shaped rolled steel member enclosed within a *lead sheath*.

steel lathing [pla.] Expanded-metal or steel-wire mesh used as *metal lathing*.

steel square or **roofing square** or **framing square** [carp.] An L-shaped graduated steel plate with legs at right angles to each other, like a carpenter's *square*. Its legs are about 5 cm (2 in.) wide and 45 or 60 cm (18 or 24 in.) long. The graduations on it include data for working out the lengths and angles of *rafters* from their known span and rise, and for setting out the lines along which they should be cut, to fit *wall plate* or *ridge*.

steeplejack A *tradesman* who repairs and builds steeples and other tall brickwork or *masonry*, including *lightning conductors*, *weathercocks*, clock faces, and *chimneys* whether of brick or steel. He usually erects his own staging and ladders and does any incidental carpentry, painting, gilding, roofing, steel erection, cutting, welding, or *grout* (*C*) injection.

steeplejack's mate A helper to the steeplejack. He works mainly on the erection of *ladders*, staging and tackle.

step One unit of a *stair*, consisting of a *riser* and a *tread*. It may be a *flier* or *winder*.

step flashing A *stepped flashing*.

step joint [carp.] A joint between *rafter* and *tie-beam* in which the tie-beam is notched with a *birdsmouth* to receive the end of the rafter. To reduce the depth of the cut a *double step* may be used.

step ladder A wooden *ladder* built with rectangular *stiles* and *treads* (not *rungs*) which are designed to be horizontal in use. *See* **steps**.

stepped flashing [plu.] A *flexible-metal* or roofing-felt cover *flashing* let into the joints of brickwork to make a watertight joint between a wall (often a chimney) and the sloping part of a roof. The flashing steps down occasionally so that the vertical part of the flashing is kept to about 8 cm (3 in.) height and the metal is not wasted. *See* **raking flashing, upstand**.

stepped skirting An asphalt *skirting* at an inclined intersection with a roof.

steps or pair of steps A *step ladder* with a framed stay hinged to the top so as to make it self-supporting.

step turner [plu.] A hardwood tool for shaping *stepped flashings*. It has a sawcut 6 mm ($\frac{1}{4}$ in.) wide in one edge. The flashing is inserted into this and it is then turned through 90° so that the flashing is correctly shaped for insertion into the bed joint of the brickwork.

stick-and-rag work [pla.] *Fibrous plaster*.

sticker [tim.] A small separator laid with others of uniform thickness between sheets of freshly cut *plywood* in the direction in which warping is most likely. They encourage the circulation of air and the drying of the wood.

sticker machine [joi.] A machine for cutting *mouldings* out of the *solid*.

sticking [joi.] The shaping of a *moulding* with a *plane* or a *sticker machine*.

sticking board [joi.] A framed board in which small pieces are held steady while they are being moulded with a *plane* (or stuck).

stiffened expanded metal *See* **self-centering lathing**.

S-tile (USA) A strongly curved *pantile*.

stile [carp.] An upright end-framing member *mortised* to enclose a *ladder* rung or a *tenon* of a *rail*. *Compare* **muntin**.

stile end A junction between the end bar of a *stretch* of *patent glazing* and the roof, including the flashing at this junction.

Stillson [plu.] A *pipe wrench*.

stipple [pai.] (1) To dab a coat with a *stippler* and thus to remove *brush* or other marks immediately after a coat is put on.

(2) To break up the colour of a coat with spots of a different colour, or to break up its texture with a bristle or rubber *stippler*.

stippler [pai.] (1) A *brush* with many tufts of soft bristles set in a flat stock, with the bristle tufts all ending in the same plane. It is used to even up the coat of paint, to remove brush marks, and to leave the wet surface with a uniform, slightly granulated finish.

(2) A rubber tool for breaking up the texture of a coat.

stirrup strap or **stirrup** or **hanger** or (USA) **bridle iron** [carp.] A steel strap built into brickwork or fixed to a post or beam and holding a horizontal member up to it. One example is the steel strap which holds the tie beam up to the king post in a *cottered joint*. *See* **wall hanger.**

stock (1) [tim.] *Converted timber*, also called stuff.

(2) [carp.] The wooden part of a *plane*, the body or handle of a tool.

(3) [pla.] A wood backing to the zinc profiles made for running a *moulding* with a *horsed mould*. It is firmly housed into a *horse* and braced with short wooden stays which are used as handles.

(4) [plu.] A tool which holds a *die* for cutting an external thread.

stock brick The *brick* which is most commonly available in any district is the stock brick of the district. The former London stock brick (a yellow Kentish brick which blackens with age in towns) has become so well known that it is now generally referred to in the south of England as a stock brick, or stock, but is no longer the commonest London brick. *See* **Fletton.**

stock brush [pla.] A *brush* used for wetting a wall before plastering, to prevent it absorbing too much water from the plaster.

stock lumber [tim.] (USA) Wood sawn to market sizes.

stone (1) A walling material; either cut natural rock such as sandstone, limestone, and granite, or *cast stone*.

(2) [joi.] A carborundum or other natural or artificial *hone* for putting a cutting edge on to a *chisel*, *plane iron*, etc.

stone boat (USA) A wooden or sheet steel tray, preferably mounted on sledge runners, used for hauling stone short distances over a *skid road* or similar track.

stone dresser or **stone cutter** or **scabbler** or **squarer** or **block chopper** A man who roughly dresses stone blocks at the quarry.

stone lime Any *lime* except lime made from chalk.

stone-machine hand or **stone sawyer** or **sawman** or **stone polisher** or **stone turner** A *mason* who sets and operates stoneworking machines.

stone saw A long reciprocating blade, with no teeth, fed with water and an *abrasive* such as sand or carborundum, for cutting stone.

stone slate A thin stone used for roofing. Stone slates are very heavy and therefore an expensive roofing material but some types used in England and elsewhere are beautiful (Horsham, Collyweston, Cotswold).

stone surfacer A heavy *stone tool* for dressing large areas of stone.

stone tongs *Nippers.*

stone tool A percussive tool held in the hand for dressing or carving stone.

stoneware Hard *ceramic* material from which drains, channels, and similar drainage fittings are made. Stoneware is nearly always *salt glazed*. *See* **earthenware.**

stoning [tim.] With reference to a *circular saw*, running the saw and pressing an abrasive stone in a direction at right angles to the axis of the saw. It corresponds to the *topping* of a handsaw.

stooling or **stool** (1) The upper surface of the end of a concrete or stone *lug sill*. It is horizontal, to form a bed for the masonry over it.

(2) (USA) A *window board*.

stoothing [carp.] *Common grounds* for joinery, plaster *lathing*, etc.

stop (1) A *bench stop*.

(2) A decorative conclusion to a stuck moulding.

(3) A rectangular *fillet stuck* or *planted* on a door frame, against which the door closes.

stop bead [joi.] US term for a *guard bead*.

stopcock [plu.] A *cock* in a gas- or water-supply pipe, used for closing or opening, but not for regulating the supply.

stop moulding or **stopped moulding** [joi.] A *stuck moulding* which ends at a *stop*, and does not continue to the end of the member.

stopped chamfer or **stop chamfer** [joi.] A *chamfer* which dies away, gradually merging into a sharp *arris*.

stopped mortise [joi.] A *blind mortise*.

stopper or **stopping** [pai.] (1) *Filler*.

(2) *Hard stopping*.

stopping knife A *glazier*'s knife for smoothing putty, also used by painters for putting *hard stopping* into holes or cracks. It is like a *chisel knife* with one rounded edge and one splayed edge, meeting at a point. *See* **house painter's tools** (illus.).

storage tank [plu.] *See* **cistern**.

storage water heater [plu.] A self-contained gas-fired appliance that heats water under thermostatic control and stores it until needed.

storey or (USA) **story** The part of a building between one floor and the next above it. Thus the seventh storey is the part of the building from seventh floor level to eighth floor level. *See* **first floor**.

storey rod or (USA) **story pole** A *batten* cut to the exact height of a *storey*, often having dimensions marked on it such as the levels of certain brick courses or of window *sill*, window *head*, and *stair* treads. *See* **going rod**.

storm cellar or **cyclone cellar** A cellar in which the people of the house take shelter against the violent cyclones and tornados of central USA.

storm clip A saddle-shaped metal clip, fixed outside a *glazing bar*, to hold the glass down in *patent glazing*.

storm door [joi.] (USA) An additional inner door, used in winter, to insulate a house from hard weather.

stormproof window [joi.] A wood *casement window* with additional protection against rain, for example, hood mouldings and throatings and lips in the joints.

storm sheet An *asbestos-cement* roofing sheet which is curved down at one edge to protect an eave against rain.

storm window (1) or **double window** [joi.] An additional outside *window* with an air space between it and the inner sash. It reduces the noise passing through the windows and keeps the room warm in cold weather. *See* **double glazing.**

(2) An *internal dormer*.

story (USA) *Storey*.

stoving or (USA) **baking** [pai.] Drying by heat, generally above 65° C. (150° F.). In convection-oven stoving the heat reaches the painted surface largely by convection. In radiant-heat stoving or infra-red drying the heat reaches the painted surface mainly by radiation. *Compare* **forced drying.**

straddle pole A sloping *scaffold* pole laid along a roof in a straddle scaffold from a *standard* to meet the other straddle pole at the ridge.

straddle scaffold A *saddle scaffold*.

straight arch A *flat arch*.

straight courses *Shingles* laid with their *butts* in line like tiles or slates.

straight edge A long piece of planed, seasoned softwood with parallel and straight edges, used by most of the building *trades* for *setting out*.

straight flight [joi.] A *stair* consisting of *fliers* only, with no *winders*.

straight grain *Grain* which is parallel with the length of the timber.

straight joint (1) [carp.] A *butt joint*.

(2) A brick joint which is above another *joint* and is thus a mistake in the *bond* (if the brickwork is structural).

straight-joint tiles *Single-lap tiles* made so that their edges in successive courses run in one line from *eaves* to *ridge*.

straight-line edger [tim.] A modern, mechanically-fed *circular saw* which straightens the edges of veneer and smooths them so well that they can be glued without further treatment.

straight-peen hammer [mech.] Any *hammer* which has, opposite the striking face of the hammerhead, a blunt wedge, parallel to the shaft of the hammer. *Compare* **ball peen, cross peen.**

straight tongue [joi.] One edge of a board made thinner by rebates each side of it so that the tongue so made can fit into a groove to match it. *See* **cross tongue, matchboard.**

straining beam [carp.] Any horizontal *strut* (C), particularly that between the heads of the *queen posts* in a *queen-post truss*.

straining piece or **strutting piece** [carp.] A horizontal timber dogged or bolted to the middle of the central, horizontal *flying shore* as an abutment from which the shorter, sloping *struts* (C) at each end of it obtain their thrust.

straining sill [carp.] A timber lying on and dogged to the *tie-beam* of a *queen-post truss* between the *queen posts* or between the queen and *princess posts* to keep them in place.

S-trap [plu.] A trap in which the second vertical leg of the U-shaped seal is followed by a third vertical leg, the outlet, so that it resembles a letter s on its side: ∽ . *See* **P-trap.**

strap anchor [carp.] A steel plate joining two floor *joists* which butt at a support. It is fixed on those *joists* which are anchored to the walls with *wall anchors*. In this way the walls are given good lateral support by the floor joists.

strap bolt [carp.] A fastening for heavy timbers, consisting of a metal strap with holes drilled through it at one end and a threaded rod (bolt) at the other end.

strap hinge A *band-and-hook* hinge or *cross-garnet* hinge, or one formed of two metal *straps* (*C*) of equal length.

strapped elbow [plu.] A *drop elbow*.

strapping *Common grounds* on a wall, used as a base for *lath* and plaster.

straw Straw is sometimes used for *thatch*, wheat or rye straw being the best. *Sedge* may be called **sedge straw.**

strawboard *Compressed straw slab.*

stress-graded timber [tim.] Converted timber which has been divided into various classes according to the *rate of growth* and the amount of *knots*, *shakes*, *slope of grain*, and *wane*. In practice 80% of good *softwood* can be stressed to 6·9 N/mm² (1000 psi) in bending. Stress grading was introduced because strength tests have always been done on *clear timber*, now a very scarce material. Therefore to enable wood structures to be economically designed, it became necessary for engineers to have some rational means of assessing the strength reduction due to any particular defect. *See* **gross features.**

stretch A surface of single-tier *patent glazing*. Its area is measured by the length overall the end bars times the depth, and expressed in m². The depth is the length of the *glazing bar*.

stretcher A brick or stone laid with its length parallel to the length of the wall. *Compare* **header.**

stretcher face The long face of a brick seen after it is laid.

stretching bond or **stretcher bond** Brickwork half a brick thick (about 10 cm (4 in.)) in which each brick is laid as a *stretcher*. This bond is also seen on the face of *cavity walls* 28 cm (11 in.) or 38 cm (15 in.) thick.

stretching course A course of *stretchers*.

strike plate [joi.] A *striking plate*.

striking off lines [pla.] Lines drawn on wall or ceiling for setting *fibrous plaster*.

striking plate or **keeper** or **strike plate** or **strike** [joi.] A plate with a rectangular hole in it, screwed to the *mortise* in a door post. When the door closes, the *latch* of a *mortise lock* slides against it and holds the door shut by passing into the rectangular hole. *See* **box staple.**

striking wedges *Folding wedges.*

string or (USA) **stringer** [joi.] A sloping board at each end of the treads housed or cut to carry the treads and risers of a stair. A string is either a *wall string* or an *outer string*, and either a *close string* or a *cut string*. *See* string piece.

string course A decorative, usually projecting, thin horizontal course of brick or stone, often continuing the line of the window *sills* or *dripstones*. In modern construction its place is sometimes taken by the floor slab projecting and throwing water away from the walls.

stringer [joi.] (USA) A *string*.

stringing mortar (USA) Spreading enough mortar on the *bed joint* to lay several bricks.

string piece [carp.] The horizontal *tie-beam* of a *Belfast truss*.

strip [tim.] (1) Softwood: a piece of square-sawn timber under 48 mm (1⅞ in.) thick and under 102 mm (4 in.) wide.

(2) Hardwood: a piece of square-sawn timber usually 51 mm (2 in.) and under thick and 51 mm (2 in.) to 140 mm (5½ in.) wide, (BS 565).

(3) (USA) *Lumber* less than 2 in. (5 cm) thick and less than 8 in. (20 cm) wide.

[mech.] (4) Copper, zinc or aluminium which is thicker than foil (0·15 mm or 0·006 in.), thinner than 9·5 mm (⅜ in.) and narrower than 45 cm (18 in.).

(5) A description of lead *sheet*.

strip board [tim.] A term sometimes used for *blockboard* in which the core blocks are not glued to each other.

strip flooring *Parquet strip*.

strip heating [tim.] Heating of wood joints by a bare metal strip heated by a heavy current at low voltage, when gluing with *synthetic resins*. *Compare* radio-frequency heating.

stripper A *skilled man* in the precast concrete or building industries who dismantles moulds or *formwork* (*C*) for concrete. He may also clean them, oil them, and insert reinforcement.

stripping (1) Clearing a site of turf, brushwood, etc.

(2) [pai.] Removing old *paint*, *distemper*, etc., with a blow lamp, *stripping knife*, or other means. *See also* C.

stripping knife or **broad knife** [pai.] A knife with a stiff steel blade, widening from the base to a square edge from 2·5 to 10 cm (1 to 4 in.) wide (BSCP 231). It is used for removing *wallpaper* and loose *distemper*. *See* chisel knife, house painter's tools (illus.).

strip slates or (USA) **asphalt shingles** Lengths of mineral-surfaced *bitumen felt* laid in horizontal strips along a roof. They are cut to look like slates.

strip soaker A thin strip of waterproof material laid under each course of *shingles* at a *swept valley*. *See* soaker.

stripe figure [tim.] *Ribbon grain*.

strongback or **lifting beam** A beam designed to be amply strong enough to be lifted by crane while it is connected to a pile, beam, column, etc., which it can safely lift. Strongbacks are needed because concrete piles, for example, without one would receive the most severe stressing of their lives while being slung into position by crane. Careful location of the lifting points on the pile raised by the strongback very greatly reduces these stresses. Usually each strongback is designed to lift a particular unit and its expense is quite justified by the assurance it gives that this will suffer no damage during lifting.

struck joint (1) A *weather struck joint*.

(2) A joint pressed in at the lower edge, sloping in the reverse direction from a weather struck joint. This joint is only suitable for interior work, in which rain does not need to be thrown off.

structural A description of a part of a building which carries load in addition to its own weight, as opposed to *partitions, joinery, plaster*, etc. which only carry their own weight. In USA the term is used in a wide sense; *see* **structural clay tile, structural glass.** *See also C*.

structural clay tile (USA) American term for what might generally be called in Britain non-structural burnt-clay *hollow blocks*. Many different sorts exist: side construction tile with cavities horizontal, end construction tile with cavities vertical, partition tile, facing tile, floor tile (hollow floor blocks), *furring* tile for lining the inner face of outside walls, header tile providing cavities in the backing blocks for *headers* from the *facing brickwork*, and so on.

structural glass (USA) Rectangular panels or tiles of glass, used for facing walls.

structural lumber [tim.] (USA) Sawn timber 5×10 cm (2×4 in.) or larger, usually *stress graded* and under calculated stress.

structural steelwork *Rolled-steel joists* (*C*) or built-up members fabricated as building frames, by riveting, welding, or bolting, or all three.

structure (1) The loadbearing part of a building.

(2) Anything built by man, from a *hydraulic-fill dam* (*C*) built of earth or a *pyramid* of stone to a hydro-electric power station. A structure is not necessarily roofed, a building is.

strutting (1) Using *struts* (*C*) as temporary supports, for example, *dead shores*.

(2) [carp.] *Solid bridging* or *herring-bone strutting*.

strutting piece [carp.] (1) A *straining piece*.

(2) A piece of joist cut for *solid bridging*.

stub A *nib* of a *tile*.

stub tenon [carp.] A *tenon* which is inserted into a *blind mortise*. If it is wedged this must be done by *secret wedging*.

stuc Plasterwork made to look like stone.

stucco [pla.] A term originally used for all plasterwork, but now confined to smooth plastering to the outside of a wall, which may be of many sorts. Stucco made with *lime* and sand was the only sort used in Britain before the nineteenth century, when *cement* began to be used. (In France *gypsum plaster* was and still is used.) *Rough cast* and *pebble dashed* surfaces are not stucco. Recommended mixes for stucco are as follows: undercoat 3 volumes of *lime putty* or white *hydrated lime*, 1 of cement, 10 of sand; *finishing coat* the same as the undercoat with 12 instead of 10 volumes of sand. These mixes are less likely to crack or craze and are therefore more durable and weatherproof than cement–sand mixes. *See* **cement rendering, scraped finish.**

stuck moulding [joi.] A *moulding* cut out of the solid by a plane or *sticker machine*, the opposite of a *planted* moulding.

stud (1) or **studding** [carp.] Intermediate vertical members in a framed partition (about 8×5 cm (3×2 in.)) or in *frame construction* (about 10×5 cm (4×2 in.)), usually placed about 45 cm (18 in.) from studs on each side. The end members are called posts and are heavier. On a *partition* or the inner face of a building frame, *lathing* or *wallboards* are usually nailed to the studs. Timber close-boarding may be nailed to the outside of studs in a building frame to stiffen it.

(2) [mech.] A threaded rod, a *screw* (*C*) with no head; *see* **stud gun, stud welding.**

stud gun An explosive-operated gun which shoots hard steel studs of 6, 9·5 or 14 mm ($\frac{1}{4}$, $\frac{3}{8}$ or $\frac{9}{16}$ in.) dia. (male- or female-threaded) into concrete, brickwork, or mild steel. A strong hold is obtained in steel by 12 to 19 mm ($\frac{1}{2}$ to $\frac{3}{4}$ in.) penetration. The end of the stud which is driven in is pointed. A safety device prevents the gun firing except when it is pressed against a wall, and no licence is needed for it in Britain. *Compare* **stud welding.**

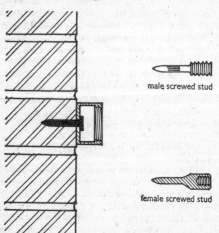

male screwed stud

female screwed stud

Rivet driven by an explosive-operated stud gun to hold an electrical junction box to brickwork.

344

stud partition [carp.] A *partition* built of *studs*, a *framed partition*.

stud shooting The driving of threaded studs with a *stud gun*.

stud welding Fixing a metal stud on to a steel frame by *resistance welding* (*C*) after the frame is built. Studs were used as a fixing for *joinery* in the UNO building in New York. The welding is rapid, the stud being held in place with a sort of gun during welding. *Compare* **stud gun**.

stuff (1) [pla.] Plaster; *see* **coarse stuff, finishing coat**.

(2) or **stock** [joi.] *see* **square-sawn timber**.

stump foundation or **brush block foundation** The usual foundation for the wooden *bungalows* built in Australia where wood is cheap. It consists of pits dug at 1 to 2 m (4 to 6 ft) spacing both ways, each pit containing a 10 × 10 cm (4 × 4 in.) 'stump' or round timber upright, resting on a 23 × 15 × 4 cm (9 × 6 × 1½ in.) sole plate on the ground. The wood must be proof against decay, termites, etc., since the pits are back filled.

stump veneer [tim.] *Veneer* from the butt of a tree. That from *walnut* butts is particularly prized for its fine *figure*. *See* **burr**.

sub An advance on wages due, before pay day.

sub-base or **base** (USA) A *skirting board*.

sub-basement The second *storey* below the ground, the storey below the *basement*.

sub-casing (USA) A *blind casing*.

sub-circuit [elec.] A *branch circuit*.

sub-contract A part of a *contract*, often specialist work such as asphalting, which is done by a separate firm from the *main contractor*. The main contractor is responsible for the work, pays the sub-contractor, and is paid by the *consultant* or *client* for it.

sub-contractor A specialist employed by a *main contractor* to perform a *sub-contract*. Sub-contractors must usually be approved by the *architect* before the contractor engages them.

sub-floor A wooden or concrete floor which carries load but is not seen, being covered by a finish of wood blocks or other material. In USA it is called a blind floor, or if wooden a *rough floor*.

sub-frame (1) A frame attached to the main building frame as a fixing for the *cladding*.

(2) A frame built into a wall as a fixing for a door or window, often of *pressed steel*. *See* **door buck**.

subletting A *main contractor* can sublet most of his *contract* to *sub-contractors*, but this sub-division of responsibility involves possible shoddy work. Subletting should therefore be allowed only with the architect's permission in writing.

subsidence or **settlement** Downward movement of the ground surface *See also* C.

sub-sill or **sill drip moulding** [joi.] An additional *sill* fitted to the outside of the sill of a window after manufacture. Its purpose is to increase the distance from the wall at which rain is thrown off.

sub-station [elec.] A room or building containing electrical equipment such as switches, usually with transformers to reduce high-voltage incoming power to a *voltage* at which the consumer can conveniently use it. It may be provided by the electricity authority or by the consumer.

substrate [pai.] A *ground*.

subway (1) A passage below ground for people to walk through, sometimes (when under a building) containing cables and other building services, and mainly provided for their maintenance.

(2) (USA) An underground railway.

suction (1) [pla.] The adhesion of bricks to wet *mortar* or of wet plaster to a wall. *See* **absorption rate**.

(2) [pai.] The absorption of liquid from paint by a porous surface. It can be prevented by applying a *sealer*.

Suffolk latch A *thumb latch*.

sulphate of lime [pla.] *Gypsum*, anhydrite, or calcium *hemihydrate*.

sulphur cement Equal parts of pitch and sulphur heated together and poured into holes in concrete in which metal *balusters* are set.

summerwood or **latewood** [tim.] The denser, darker wood formed in summer. *Compare* **springwood**.

sump pump A pump of small capacity for occasionally emptying a *sump* (*C*) in a part of a building which is below the level of the drains.

sunk draft A margin of a building stone set below the rest of the face.

sunken joint [tim.] A defect in veneering, small depressions in the surface of *veneer* above joints in *coreboard* or *cross bands*, caused by gluing faults or variation in thickness of the cross bands.

sunk face An *ashlar* face which is cut below the margins of the stone.

sunk gutter A *secret gutter*, sunk below the roof surface.

super [q.s.] Abbreviation for superficial, and thus for 'area'.

superficial measure [tim.] *Face measure*.

super hardboard The most dense and water-resistant *hardboard*. It is used as a lining to *formwork* (*C*), as a floor finish, for boat building, and so on.

supply pipe That part of the *service pipe* which belongs to the consumer. It is between the building and its boundary or the stop valve, whichever is nearer to the main. *Compare* **communication pipe**.

surbase [joi.] (1) A *dado capping*.

(2) (USA) A *moulding* which crowns a *skirting board*.

surface coefficient In steady-state conditions, the amount of heat transmitted to or from a surface in contact with air or any other heating (or cooling) fluid from all causes (conduction, convection and radiation) divided by the difference between the surface temperature and the effective ambient temperature. Its reciprocal is the surface resistance. From this value, the difference between *U-value* and *conductance* can be calculated.

surface dry [pai.] A stage in *drying* when a paint is dry on the surface but wet beneath.

surfaced timber or **dressed timber** or **wrot timber** [joi.] Timber planed on one or more surfaces.

surface measure [tim.] *Face measure*.

surface of operation A plane surface on a stone prepared as a datum from which to work the rest of the stone.

surface planer or **surfacer** [tim.] A steel bed plate with its upper surface in two halves at different levels, but parallel and carefully ground plane. Between the two halves the adzing *cutter block* rotates, so that its blades project above one half and are level with the other half. The levels of the halves can be adjusted. *See* **thicknessing machine**.

surfacer (1) [tim.] A *surface planer*.

(2) [pai.] A thin, pigmented *filler*, or *sealer*, or both, for smoothing slightly uneven surfaces before painting. It is usually sanded smooth after drying. *See* **sanding sealer, guide coat**.

(3) A *dunter* machine.

surface resistance The reciprocal of the *surface coefficient*.

surface retardant [pla.] A liquid put on to *formwork* (*C*) to make it easier to strip, and to give the concrete good bonding properties for later plastering. The surface concrete drops away or can be brushed off, leaving a rough surface on which plaster will stick.

surface spread of flame *See* **flame spread**.

surface water drain Any pipe for rainwater in the ground or elsewhere. *See* **agricultural drain** (*C*).

surface waterproofer A liquid, sometimes based on *silicone* resins, which makes the surface of *masonry*, *concrete*, etc., slightly water-repellent but will not prevent water under pressure from passing through it. It is not visible on the masonry. Paints are also used, as well as cement slurry. *See* **waterproofing**.

surfacing materials *Gypsum plasters* and plasterboards, *asbestos-cement* sheets and wallboards, *fibre boards*, *woodwool slabs*, *chipboards*, *laminated plastics*, *corkboard*, *plywood*, *blockboard* and timber are all used for surfacing the walls of rooms. Examples of floor surfacings are *linoleum*, *parquet*, *rubber sheet*, *magnesite* composition, clay or concrete *tile*, *granolithic* (*C*) or ordinary sand-cement mortar screeds. Many others exist which are not listed in this dictionary.

surround Material placed around something to protect or decorate it, e.g. the concrete round a drain, the bricks round a fireplace.

surveyor A vague term which may mean a member of several very different professions which fall into two groups, the first of *mining* or *land surveyors* (*C*), the second of *quantity* or *building surveyors*.

suspended ceiling All ceilings are suspended, but this term refers usually to a *false ceiling*. *See* BSCP 290.

suspended floor Any floor which is supported at its ends, not in the middle.

suspended scaffold A *cradle* or a *projecting scaffold*. See BSCP 97, Metal Scaffolding.

suspended shuttering *Formwork* (*C*) for a floor which is carried on the supports of the floor and not propped from below.

suspension [pai.] (1) Small particles of solid, distributed through a liquid. Most paints are suspensions of this sort.

(2) An emulsion in which one liquid carries the other liquid suspended within it in small separate drops. *See* **emulsion paint.**

Sussex garden-wall bond *Flemish garden-wall bond.*

swage [mech.] A smith's tool for shaping hot or cold metal, particularly rivet heads. Swages are made in pairs, one male and one female, of which one acts as the hammer and the other, the swaging block (usually hollow), as the anvil. A special sort of swage is used for setting *circular saws.*

swage-setting [tim.] A method of *setting* circular saws for ripping (not for cross cutting). Such saws are often very thin, e.g. 1 mm (0·040 in.). No sideways set is given, but the point of each tooth is spread by a hammer in such a way that it looks fishtailed. The point is symmetrical and each side projects an equal amount beyond the blade. An average set is 0·25 mm (0·01 in.) each side, so that the swage-set saw makes a cut of about 1·5 mm (0·06 in.) thickness. Swage-set saws take a faster feed than other saws. More even than other saw setting, swage-setting is highly skilled.

swan-neck (1) An S-bend, particularly a junction between a *downpipe* and an *eaves gutter* under overhanging eaves.

(2) A combination of *ramp* and knee in a handrail.

swan-neck chisel [carp.] A strong *socket chisel* curved so as to extract the chippings of a *mortise. Compare* **corner chisel.**

swatch [tim.] A pile of samples of *veneers*, each sheet being taken from the centre of its *flitch*. The term is also used for a pile of samples of *linoleum* hinged into book shape.

sway A 2 cm (¾ in.) dia. hazel or willow sapling 2·7 m (9 ft) long used for holding *thatch* down. It is laid horizontally across the *rafters* under the thatch and fixed to them by iron hooks or tarred cord.

sweat [plu.] To unite metal parts by holding them together while molten solder flows between them. *See* **capillary jointing.**

sweating (1) [plu.] *See* **sweat.**

(2) [pai.] The separation of the liquids in a paint with the result that one of them appears at the surface of the film.

(3) [pai.] A *gloss* which may develop in a dry paint or varnish film after *sanding.*

(4) *See* **condensation.**

sweat joint [plu.] (USA) A *capillary joint.*

sweat out [pla.] Plaster which, after it has set, appears damp and mushy.

It may be caused by cold weather, by imperviousness of the backing brick, or by dirty sand which needs an excessive amount of mixing water.

sweep tee [plu.] A *tee* for copper or screwed pipe, in which the branch is not, as in the normal tee, precisely perpendicular to the run, but curves gently away from it.

swept valley A *valley* formed of *shingles*, *slates*, or *tiles* cut or made to a taper so as to eliminate the need for a *flexible-metal* valley. A *tile-and-a-half* tile is used and cut to shape so that its *tail* is narrower than its *head. Compare* **laced valley**; *see* **strip soaker.**

SWG [mech.] *Standard wire gauge.*

swing door [joi.] A door which can open in both directions and therefore has no *stop* in the frame. It can be hinged on a *floor spring* or on a *helical hinge. See* **revolving door.**

swinging post The *hanging post* of a gate.

swirl [tim.] The irregular *grain* round knots or *crotches.*

switch and fuse [elec.] A switch usually built into a cast-iron fitting containing *fuses.* The fuses are not built into the switch.

switchboard [elec.] Any group of switches which is hand-operated and with or without instruments.

switch fuse [elec.] A switch containing a *fuse.*

synthetic paint A very vague term which sometimes means paints containing *synthetic resin* in the *medium.*

synthetic resin *Urea-* and *melamine-* and *phenol-formaldehyde* glues and casting *resins* (and other synthetic resins) have been in commercial use since the 1930s. They are immune from attack by moulds or bacteria and are all highly water resistant. Although more expensive they are replacing the older *glues* for exterior work in *plywood* for aircraft, houses, or bridges. *See* **accelerator, alkyd, epoxide, film glue, glassfibre, plastics, separate application.**

synthetic-resin-bonded paper sheet *See* **laminated plastics.**

synthetic-resin cement [tim.] (1) A *synthetic resin* used as a *glue.*

(2) The synthetic resin without its *accelerator.*

(3) *See* **Perspex.**

synthetic stone (Scotland) *Cast stone.*

system building *Industrialized building methods.*

T

tack (1) A sharp, short nail for fixing *linoleum*.

  (2) *See* **lead tack.**

  (3) [pai.] Stickiness of a paint film which is *drying*.

tack rag [pai.] Cheese cloth or other cotton fabric damped with slow-drying varnish to remove dust from a surface after *rubbing* down and before putting on the next coat. Tack rags should be kept in an airtight tin so that they do not harden.

tacky [pai.] Sticky, a stage in *drying*.

taft joint or **finger-wiped joint** [plu.] A joint between lead pipe and brass *liner* made by *wiping* in *plumber's solder* instead of pouring in fine solder as in the *blown joint*. It therefore uses slightly more solder of a less expensive sort.

tag A strip of copper folded in several thicknesses and used as a wedge for holding copper sheet into a masonry joint. *See* **lead wedge.**

tail (1) The lower edge of a *slate*. *Compare* **head.**

  (2) The built-in end of a stone step. *See* **tailing in.**

tail bay [carp.] An end span of a timber floor or roof.

tail bolts Bolts for fixing the ends of *asbestos–cement* roof sheets.

tailing in or **tailing down** Fixing the end of a member which is cantilevered from a wall by laying stones or bricks or any heavy weight on it.

tailing iron A steel section built into a wall to hold down the end of a member which is *cantilevered* (*C*) out below it.

tail joint [carp.] A *joist* which rests on a *tail trimmer*.

tailpiece [carp.] (USA) A *trimmed joist*.

tail trimmer [carp.] A *trimmer joist* close to, and parallel to, a wall on which it is undesirable to bed the floor *joists*.

taker-off [q.s.] An experienced *quantity surveyor* who specializes in reading drawings, taking dimensions from them, and writing them on *dimensions paper*. *See* **taking off.**

taking off [q.s.] Taking off quantities is the first step in working out *quantities* from a drawing when compiling a *bill of quantities*. It involves writing down the dimensions of each *item* systematically on sheets of *dimensions paper* so that the next operation (squaring up or working up) can be done without misunderstanding by someone else. *See* **dotting on, measurement, timesing.**

takspan Swedish pine roof *shingles* 43–51 cm (17–20 in.) long, 10–25 cm (4–10 in.) wide, 3 mm ($\frac{1}{8}$ in.) thick, made like *sliced veneer*.

tall boy A hood about 1·5 m (5 ft) high, made of steel sheet 0·56 mm ($\frac{1}{44}$ in.) thick or more, galvanized after it is made, fixed over a chimney to prevent downdraughts.

tally slates *Slates* sold, as usual, by number, not by weight. *Compare* **ton slates.**

talus wall A wall to hold back an earth slope, therefore built at a *batter*.

tamarack [tim.] (USA) *Larch*.

tambour (1) A circular wall carrying a dome.

(2) or **vestibule** A ceiled, circular, wooden lobby which encloses *revolving doors* and prevents draughts through them.

(3) Any drum shape such as a stone in a circular column.

tampin or **turning pin** [plu.] *See* **boxwood tampin**.

tang The pointed part of a steel tool such as a file, knife-blade, or *chisel* which is driven into the wooden handle. (In the *socket chisel* the handle is also driven into the socket.)

tangential shrinkage [tim.] The *shrinkage* of timber parallel to the *growth rings*. *See* **flat-sawn** (illus.).

tank [plu.] *See* **cistern**.

tanking A waterproof skin, usually of 19 mm ($\frac{3}{4}$ in.) asphalt laid beneath a *basement* floor and up the basement walls. The wall skin is protected outside by a half-brick (or thicker) wall from stones which might pierce it during *back filling*. On the inner face it is in contact with the basement *retaining wall* (*C*). The floor skin is also laid between two slabs, the upper being the *loading coat*. *See* **damp course**.

tap [plu.] A screwed plug, accurately threaded, made of hard steel and used for cutting internal threads. *Compare* **die**.

tape [tim.] *See* **joint tape**.

tapeless splicer or **tapeless jointer** [tim.] A machine for gluing sheets of *veneer* together at their edges without *joint tape*.

tapered parapet gutter A *box gutter*, usually of *flexible metal*, behind a *parapet*. It becomes narrower towards the lower end because of the roof slope.

tapered-roll pantile A *pantile* in which the roll width increases slightly from head to tail of the tile.

taper ground [joi.] *See* **setting of saws** (illus.).

taper pipe or **diminishing pipe** [plu.] An *increaser* or *reducer*.

taper thread [plu.] A standard *screw thread* used on pipes and their *fittings* to ensure a watertight, gastight, or steamtight joint. Taper threads are now used on all usual pipes and fittings except *connectors*, on which the thread is too long to be tapered. The amount of the taper is 1 in 16, that is, if continued to a point, the taper would form a cone 1 unit dia., 16 units long. *Compare* **parallel thread**.

taping strip A strip of *roofing felt* laid over the joints between precast slabs in a roof before it is covered with roofing felt and *bonding* or *sealing compound*.

tar-gravel roofing (USA) A low-cost roof covering of *roofing felt* mopped with hot tar or pitch and covered with gravel or sand.

tarpaulin A large, waterproof, canvas sheet used during building or repairs for protecting materials from frost or rain.

teak [tim.] (Tectona grandis) A *hardwood* from Burma, India, Siam and (of less good quality) Java, and Indo-China. It is *fire-resisting* and acid-resistant, and for these reasons used for making sinks and laboratory benches. It is used for outdoor *carpentry* and good *joinery*. In *workability* it is about equal to *oak* or higher.

tear [pai.] *See* **run**.

technical assistant [q.s.] A *quantity surveyor*'s assistant. He may be described as a worker-up or as a *taker-off* according to the work he is doing.

tee [plu.] A short pipe with three openings of which one is a *branch* at right angles to the other two and half way between them. The *run* is the length which carries the two openings in line. *See* **sweep tee**, **fittings** (illus.).

tee hinge [joi.] A *cross-garnet* hinge.

Teflon *See* **polytetrafluorethylene**.

tegula The under-tile of *Italian tiling*.

telescopic centering or **collapsible pans** or **self-centering formwork** Floor *formwork* (*C*) made of *pressed-steel* sections which fit into each other telescopically. The end sections are laid on the wall or beam which carries the floor. Generally no posts are placed under the formwork.

temper To toughen *steels* (*C*) and non-ferrous metal by heat. *See* **dead-soft temper**, *and C*.

tempera [pai.] A mural painting method practised in the Middle Ages, using a *medium* of gum, egg, and water on *gesso*. *Size* or other *media* are now used, and the result is called *distemper*.

template or **templet** or **profile** (1) A full size pattern of wood or metal used for testing or forming the shapes of building stone, plaster, concrete, etc.

(2) A *pad* to carry a concentrated load.

templating [tim.] Cutting *veneers* to shape for *flexible-bag moulding* work in which very difficult compound curves are used.

templet A *template*.

tender or (USA) **bid** [q.s.] An offer from a *contractor* to do certain work for a price which he names, usually in the *price bill*.

tendering [q.s.] The procedure of sending out drawings and a *bill of quantities* to *contractors* for them to state their prices for all the *items* of one contract. Tendering may be open, in which case the contract is advertised and every contractor who wishes may tender, or it may be limited. 'Limited' tendering is adopted for unusual work by inviting only a few contractors of known ability for the work.

tenement An apartment generally of low quality occupied by poor people, sometimes sharing W Cs.

tenon [joi.] An end of a *rail* or similar member, reduced in area at its end, to enter a *mortise* in another member, often a *stile*. The width of a tenon should be about four times its thickness. The word, like mortise, is from the French, but tenons were also used by the ancient

Egyptians in their wooden beds. (*See* plywood.) *See* abutting, haunched, lapped, stub, tusk tenons.

tenon saw or **mitre saw** [joi.] A saw which is about 30 cm (12 in.) long, has about 5 points per cm (14 per in.), and is stiffened like the *dovetail* saw with a fold of steel or brass along its back.

teredo [tim.] *See* ship worm.

terminal (1) A *finial* or similar decorative ending.

(2) The end of a lightning conductor, gas flue, etc.

(3) [elec.] The end of a power line, or the connection for leading power into, or out of, a piece of electrical plant.

termite An insect which behaves like a white ant in that it lives in colonies. It does very much more damage than an ant and eats most sorts of wood (*see* **termite shield**). It is found in tropical countries and in some temperate parts of USA.

termite shield (USA) A metal sheet resembling a *damp course* with overhanging lip to throw off water. It prevents termites climbing into the house and eating the woodwork. It must be placed below any wood in the building and must also be fixed round any pipes entering the house.

terne plate Steel plate coated with an alloy containing up to 10% tin the remainder being lead. It is used for roofing. *See* tin roofing.

terrace (1) A raised, level, earth platform with at least one upright or battered side.

(2) A flat roof.

terrace house One of a row of houses touching each other. *Compare* semi-detached.

terra cotta A yellow to brownish-red burnt clay, like *brick* or *tile* but of more uniform, finer texture, used for making moulded *cornices*, vases, statuettes, and *building blocks*. In Britain the limits of size of terra cotta are 150 litres (5 ft^3) total volume and a maximum face area of 0·19 m^2 (2 ft^2). Terra cotta may be unglazed but is more usually covered with a clear glaze or an opaque *colour* and should then be called *faience*. About twenty colours are obtainable. Most terra cotta, even unglazed, is very durable. It was used about 1500, revived in Victorian times and now has again relapsed into the dictionaries. *See* ceramic veneer.

terrazzo or **Venetian mosaic** [pla.] Coloured stones, laid in cement mortar, in a layer about 2 cm (¾ in.) thick over level concrete or a screed. When the mortar has set, the stones are abraded to smooth them and improve their colour. The finish can be pleasingly bright in colour, easily cleaned, and in plain geometrical patterns. *See* Roman mosaic.

terrazzo layer [pla.] A *tradesman* who lays *terrazzo*, fixes expansion joints, polishes the terrazzo with a stone or by machine, and on small jobs prepares the stones. He sometimes fixes precast terrazzo.

tessellated pavement *Roman mosaic*.

tessera (plural, **tesserae**) A small cube or square of marble, glass, stone, or pottery for forming *tessellated pavement*.

texture [tim.] The distribution and relative size of the cells of wood. It may be *coarse*, *even*, *uneven*, or *fine*.

texture brick A *rustic brick*.

textured finish (1) [pai.] A rough *finish* formed by sand or stone chippings mixed with a paint or by working the *plastic paint* into patterns.

(2) [pla.] A rough finish to outside plaster, obtained by throwing on the plaster or by other means. A textured finish resists the wet better than a smooth finish.

thatch A roof covering of *reed*, *straw* (or heather), laced with *withies*. A *reed* roof can last sixty or seventy years but the life of a good *straw* roof is not more than twenty years. About every seven years, thatch must be cleaned and patched to lengthen its life. It has a high insulating value but its main disadvantage (apart from its short life, the unpleasantness of thatching work, and the difficulty of finding thatchers) is the fire risk. This can now be lessened by soaking the reed or straw in a fire-resisting solution instead of in water before laying. *See* **liggers, sways, yelm, spar.**

thatcher A *tradesman* who prepares and bundles *reed*, *straw*, or heather, soaks it in fireproofing solution, and fixes it while damp with tarred string, wire, straw rope, or osier strips over *sways*. He combs and trims the surface, cuts the eaves square with shears, and can thatch strawstacks or haystacks. The craft is dying out because the work of handling wet thatch is unpleasant for the hands.

thatcher's labourer or **thatcher's server** A helper to a thatcher who usually soaks the *reed* or *straw* and hands it up to him.

therm For many years the heat unit on which the sale of piped gas has been based, equivalent to 100 000 Btu, expected to be replaced in 1975 by a new unit of sale, one hundred megajoules (100 MJ = about 95 000 Btu). The French 'thermie' is 1000 kilogram calories, about 4000 Btu or 4187 kilojoule (kJ).

thermal conductance or **conductivity** *See* **conductance** or **conductivity**.

thermal movement Movement due to expansion or contraction caused by temperature change, as opposed to *moisture movement* (*C*).

thermal transmittance or **air-to-air heat-transmission coefficient**. The *U-value*.

thermal wheel *See* **heat-recovery wheel.**

thermoplastic Description of a *synthetic resin* or other material which softens on heating and hardens again on cooling. *Plywoods* made with thermoplastic glue must cool before the pressure is released from them. *See* **thermo-setting resin.**

thermoplastic putty *Glazier's putty* which is made plastic by adding tallow to it. This enables it to move with the glass which it touches.

thermoplastic tiles *Flooring tiles* made from asphalt, *asbestos* fibre, thermoplastic *resins*, and similar materials. They are about 3 mm ($\frac{1}{8}$ in.) thick, are glued down, and should be fixed only to a *sub-floor* which is as rigid as *concrete*.

thermo-setting resin A *resin* such as *film glue* which hardens on heating and does not soften when reheated. *Plywoods* made with thermo-setting glue do not need to be cooled before they are released from pressure. *See* **caul, cold-setting, thermoplastic.**

thermo-siphon *Gravity circulation.*

thermostat A device, usually electrical, for maintaining temperature between limits. It often incorporates a *bi-metal strip* (*C*) which bends on heating (or cooling) and breaks (or makes) a circuit or turns off (or on) a gas supply. Wherever *central heating* is automatically controlled a thermostat is part of the controls. *See* **froststat.**

thickening [pai.]. *See* **fattening.**

thicknessing machine or **thicknesser** or **panel planer** [tim.] A *planing machine* which reduces, to the desired thickness, wood which may have already had the face made true on the *surface planer*. The two faces can be made parallel and planed to the correct thickness by the setting of two feed rollers at a certain level above the table.

thimble [plu.] A *sleeve piece.*

T-hinge [joi.] A *cross-garnet* hinge.

thinner [pai.] or **solvent** Any *volatile* liquid which lowers the viscosity of a *paint* or *varnish* and thus makes it flow easily. It should have complete *compatibility* with the *medium*. Thinners occasionally contain non-volatile liquid. The best known thinners are *white spirit* and *turpentine*, toluol being often used in *lacquers*.

thinning ratio [pai.] The proportion of *thinner* recommended for a particular use.

third fixings or **second fixings** [joi.] The final *joinery* fixings such as door hanging and *door furniture*, which follow the first painting.

thousand slates or **1000 actual** Now 1000 slates, though previous to 1940 a thousand slates were 1200.

thread [mech.] A *screw thread.*

thread rolling [mech.] Forming a *screw thread* on copper or similarly ductile metal tube or rod by pressure, without cutting metal away.

three-coat work (1) or **render, float, and set** (on walls) or **lath, plaster, float, and set** (on lathing) [pla.] The best quality plastering, in which the first coat (*rendering coat* on walls or pricking-up coat on laths) fills the rough places, and the second forms a smooth surface for the third. The total thickness is about 2 cm ($\frac{3}{4}$ in.), of which the *finishing coat* is 3 mm ($\frac{1}{8}$ in.) or less.

(2) Paint or mastic asphalt applied in three layers. *See* **priming coat.**

three-foot rule A three-feet-long *fourfold rule.*

three-ply (1) [tim.] The commonest sort of *plywood*. It is built of a core *veneer* with one veneer each side, generally in *balanced construction*.

(2) *Built-up roofing* or a *damp-proof* membrane formed of three layers of felt lapping over and bonded to each other.

three-prong plug [elec.] A plug with three prongs for insertion into a *socket-outlet*. Two plugs provide power, the third is an earth connection for safety reasons. *See* **shuttered socket.**

three-quarter bat A *brick* cut straight across to reduce its length by one quarter. *Compare* **king closer.**

three-quarter header A *header* of length equal to three quarters of the wall thickness.

three-way strap [carp.] A steel tee-plate with its three arms shaped so as to anchor together at a node three members of a wooden *truss*. It is fixed to them by *coach screws* or through-bolts. This is an old fixing which is now being used again, with modifications.

threshold or **sill** or (USA) **saddle** [joi.] A horizontal timber at the foot of an outside door.

throat (1) The undercut part of a *drip*.

(2) The top of a *flue gathering*.

(3) [carp.] The opening from which shavings come out of a *plane*.

through and through [tim.] or **flat-sawing** Sawing a log by parallel lengthwise cuts, usually called *flat-sawing*.

through bonder A *bond stone*.

through lintel A *lintel* which is of the full thickness of the wall.

through stone A *bond stone* which is seen on both faces of the wall.

through tenon [carp.] A *tenon* which passes beyond the mortised member.

thumbat A *wall hook* for fixing sheet lead.

thumb latch or **Canadian** or **Norfolk** or **Suffolk** or **Garden City latch** or (USA) **lift latch** [joi.] A steel fall bar, under which a lifting lever, worked by the thumb, passes through a slot in the door. When the lifting lever is raised it lifts the fall bar and unfastens the door.

thumb screw (1) [mech.] A *wing nut*.

(2) [joi.] A metal screw which passes through the *meeting rail* of one *sash* and screws into the meeting rail of the outer sash to form a burglar-proof fastening.

tie (Scotland) A *tingle* for fixing *flexible-metal roofing* sheets.

tie-beam [carp.] The horizontal, lowest timber of a roof *truss*, equal in length to the full *span* (C) of the roof. It ties together the feet of the *rafters* and is held to them in large trusses by a *heel strap*.

tie iron A *wall tie*.

tier (USA) A leaf of brickwork half a brick thick. *See* **withe.**

tight [mech.] A description of a dimension which is *bare*.

tight cesspool (USA) A cesspool which needs to be pumped out when full, as opposed to a *leaching cesspool*.

tight knot [tim.] A *knot* held firmly in the wood around it. *Compare* **loose knot.**

tight sheathing [carp.] Diagonal *matchboards* nailed to *studs* or *rafters*.

tight side [tim.] The side of a sheet of *sliced veneer* which was not in contact with the knife when it was cut off. *See* **loose side**.

tight size or **full size** or **rebate size** The size of the rebated opening for glass, about 3 mm ($\frac{1}{8}$ in.) more than the *glazing size*.

tile (1) A thin, burnt-clay, concrete, aluminium, plastics, or asbestos-cement plate for roofing (either a *plain tile* or a *single-lap tile*) or for flooring or wall covering. There is much more variety in the materials for wall or *flooring tiles* than for *roofing tiles*. *See* **wall tiles**.

(2) *Agricultural drains* (*C*).

(3) (USA) *See* **structural clay tile**.

tile-and-a-half tile A *plain tile* of width one and a half times that of the tiles with which it courses, used at *swept valleys*, *verges*, and *laced valleys*. It is normally 25 cm ($9\frac{3}{4}$ in.) wide.

tile batten A *slating-and-tiling batten*.

tile creasing *See* **creasing**.

tiled valley A *valley* covered with purpose-made *valley tiles*.

tile fillet An arrangement of *tiles* cut and set in mortar at an angle to a wall over a roof, to eliminate the *flashing* there. *See* **tile listing**.

tile floors *See* **flooring tiles**.

tile hanging or **weather tiling** or **vertical tiling** Fixing *plain tiles* on a wall to keep out the rain.

tile listing The use of *tiles* to make a tile fillet.

tile pins *Oak* pegs used instead of *nails* for fixing *tiles*.

tile slabber [pla.] A *floor-and-wall tiler*, who builds up hearths, curbs, and fireplaces in the workshop by setting *tiles* in a mould to a design, and filling the mould with *concrete* or cement *mortar*, reinforced with steel.

tiling batten A *slating-and-tiling batten*.

tilting fillet or **tilting piece** or **eaves board** or **doubling piece** or **arris fillet** or (USA) **cant strip** [carp.] (1) A board of triangular cross-section nailed to the *rafters* or roof boarding under the *double eaves course* to tilt it slightly less steeply than the rest of the roof and to ensure that the *tails* of the lowest tiles bed tightly on each other. *Compare* **cocking piece**.

(2) A similar but sloping strip at an open *valley gutter* or along an *abutment* or *verge*.

(3) An *eaves fascia* rising above the feet of the rafters, used for the purpose of (1).

timber (1) Wood for building, generally large in section (*see* **round timber**) and converted to *baulks*, *battens*, or *boards*, which are known as *lumber* in North America. *See* BSCP 112, Structural Use of Timber.

(2) In North America a timber is lumber of 4×6 in. (10×15 cm) section or larger. *See also* **timbers**.

timber brick A *fixing brick*.

timber connector [carp.] *See* **connectors**.

timber framing [carp.] A load-carrying frame of timber, used in the early medieval cruck house and in the *frame construction* of the present day. Nowadays house walls in Britain are rarely of timber, since brick walls are cheaper, but the roof is usually of wood. *See* Belfast truss, hammer-beam, king post, queen post, scissors truss.

timber in the round *Round timber*.

timbers [tim.] Out of the hundreds of timbers now in commercial use, the following are the best known in the United Kingdom:

alder	elm	oaks
ash	Douglas fir	redwood
balsa	silver fir	Canadian spruce
beech	greenheart	Sitka spruce
birch	hickory	teak
boxwood	larch (tamarack)	walnut
western red cedar	mahoganies	whitewood
chestnut		

timber scaffolder A man who builds *scaffolds* by lashing wooden *standards*, *putlogs*, and *ledgers* with fibre rope or steel rope. *Tubular scaffolding* is replacing wood scaffolding in industrial countries. *See* whip.

time limit or **completion date** The date specified in the *contract* for completion of the work. It may be postponed because of bad weather or other difficulties.

time-and-a-half The 50% additional payment usual in Britain for *overtime* worked later than two hours and less than four hours after normal finishing time. After this, *double time* is paid. Time-and-a-half is also paid for working after finishing time on Saturdays until 4 pm. *See below*.

time-and-a-quarter The 25% additional payment for the first two hours of *overtime* on the first five days of the week in the British building industry. *See* time-and-a-half.

timesing column [q.s.] A column on a sheet of *dimensions paper* for the purpose of showing how many times the same quantity must be taken.

tin The main use of tin is as a very thin but protective covering to sheet steel, sometimes also to copper. Steel which has thus been tinned is tinplate, colloquially called tin. *See* tin roofing.

tingle (1) or **cleat** or **tab** A strip of lead, copper, or zinc about 4 cm (1½ in.) wide, used for tying down the edges of *flexible metal*, for fixing *panes* of glass in *patent glazing*, for stiffening a *hollow roll* or *seam*, or fixing *slates* which replace broken slates.

(2) A support at the middle of a long *line* used by *bricklayers*.

tinker's dam A clay or *solder* dam which prevents solder overflowing.

tinning Coating steel, copper, or other metal with a film of tin or tin alloy to reduce corrosion.

tinman's solder [plu.] *See* **fine solder**.

tinplate Sheet steel about 0·12 mm (0·005 in.) thick coated with a thin film of tin both sides by dipping it in molten tin. It is used in *tin roofing*.

tin roofing Roofing with *tinplate* or *terne plate*. It must be kept painted to prevent rust and even then is not used in wet climates.

tin saw A *bricklayer*'s saw with which he cuts bricks.

tinsmith [mech.] *See* **sheet-metal worker**.

tin snips [mech.] Strong scissors with blades from 15 cm (6 in.) in length upwards, used for cutting thin metal sheet. *See* **plumber's tools**.

tint [pai.] A *colour* made by mixing much *white pigment* with a little coloured pigment.

tinters [pai.] *Stainers*.

tinting [pai.] The final adjustment of the *colour* of a *paint*.

tinting strength [pai.] *See* **staining power**.

tip The thin end of a tapered *shingle*, laid at the upper end.

titanium dioxide or **titanium white** TiO_2 [pai.] An outstandingly opaque *white pigment*, the strongest known. It is used in cosmetics with zinc oxide, and in paints, sometimes with precipitated barium carbonate or sulphate.

title The right of ownership to a property, shown on legal documents called title deeds.

toat [joi.] The handle of a bench *plane*. Many planes have none.

toe (1) [joi.] The lower part of the *shutting stile* of a door.

(2) [carp.] That part of the foot of a *rafter* which is not resting on the *wall plate* but is within the building. *See also* C.

toe board A *scaffold board* set on edge at the side of a scaffolding to prevent tools dropping off the *scaffold* and doing damage below.

toe nailing [carp.] *Skew nailing*.

toggle bolt [joi.] A proprietary device which enables a strong fixing to be made to thin board such as plasterboard or hardboard.

tolerance [mech.] The allowable range of dimensions of a part, whether hand- or machine-made. It is usually expressed as in the following example of common building bricks: $214·5 \times 103 \times 65$ mm with tolerances on length of $\pm 2·5$ mm, on width and depth of ± 2 mm; meaning that the dimension of length may be from 212 to 217 mm (also written $214·5 \pm 2·5$ mm, the usual method). Concrete walls and floors can, with care, be built to within 3 mm of the dimensions shown on the drawings, but variations of 2·5 cm are often allowed.

tommy bar [mech.] A loose bar inserted into a hole in a capstan or in a box spanner to provide the leverage for turning it.

toner [pai.] A pure organic dye without *extender*, usually of strong *colour*.

tongue [carp.] *See* **cross tongue**, **straight tongue**.

tongue-and-groove joint [carp.] A joint between the edges of boards to form a smooth wall, floor, or roof surface which is relatively air-tight. The tongue in one board fits the groove of its neighbour. *See* **matchboard.**

Tonk strips [joi.] *Sherardized* steel strips housed into the inner faces of the sides of an adjustable book-case. These strips are slotted for small metal tongues which, inserted into the slots, carry the shelves.

ton slates Large *random slates* sold by weight.

tool [mech.] An object held in the hand, used for working metal, timber, earth, masonry, mortar, and so on, such as the file, saw, shovel, hammer and trowel. However, since lathes and other metal-and wood-working machines have come to be called *machine tools* (*C*) the first are sometimes now called hand tools. For illustrations *see* **carpenter, house painter, mason, plasterer, plumber.**

tooled ashlar *Batted surface.*

tooling (1) *See* **batting.**

(2) (USA) The shaping and compressing of a mortar joint in *jointing* with any tool other than a trowel. *Concave* and *vee-joints* are both tooled in this sense.

tool pad or **tool holder** or **pad** [joi.] A combination tool consisting of a handle with a screw clamp for holding one of several small tools, such as gimlets, saws, awls, and screwdrivers of various sorts.

tooth [pai.] The surface roughness of a paint *film* due to the coarseness or abrasiveness of its *pigment*. Such a surface is very suitable for rubbing and gives good adhesion to paints put on to it, but is not a good top coat.

toothed plate or **bulldog plate** [tim.] A *connector.*

toother A *stretcher* which projects in *toothing.*

toothing or **indenting** Leaving stretchers projecting 6 cm (2¼ in.) at the end of a wall to bond with future work. Bricks project like teeth from alternate courses.

top beam [carp.] A *collar beam.*

top course tiles The *tiles* of the *course* next the *ridge*, shorter than those below to maintain the same *gauge.*

top cut [carp.] (USA) A *plumb cut* at the top of a *rafter.*

top-hung window [joi.] A window hinged at its top edge, opening outwards and held by a *casement stay.*

top lighting Lighting from overhead by *skylight, borrowed light*, or artificial light.

top log [tim.] The highest log from a tree trunk.

topman A *demolisher.*

topping or **jointing** or **breasting** [carp.] The first operation in putting a cutting edge on a handsaw or *circular saw*, the making level of the tips of all the teeth with a file. The later operations are *shaping, setting*, and sharpening. *See* **stoning,** also *C.*

topping coat [pla.] A *floating coat.*

torching or **tiering** or **rendering** Pointing the underside of *slates*, where the *heads* bed on the upper edge of the *battens*, with *haired mortar* (*shouldering* or half torching). Full torching includes, in addition, the plastering of the whole underside of the slate seen between battens. It has been common practice on unboarded, unfelted roofs, in spite of the fact that moisture is sucked on to the battens by capillary movement through the mortar. The battens therefore decay quickly and the practice is now condemned unless the torching is between slates or tiles only and does not touch the battens.

torn grain [tim.] *Chipped grain.*

torus roll A wooden horizontal *roll*, covered usually with lead sheet, at the intersection between the two slopes at each side of a *mansard roof.*

touch A plumber's term for tallow.

touch dry [pai.] A stage in *drying*, when very slight pressure with the fingers leaves no mark and does not show stickiness.

toughened glass *See* **safety glass.**

tough-rubber sheathing or **TRS** [elec.] A cable insulation which used to be called cab-tyre sheathing because it is made of rubber like motor-tyre rubber.

tower bolt A massive, steel *barrel bolt.*

town gas The gas provided in the piped supply of a town or village, usually sold by the *therm*, now called *manufactured gas.*

town planning The coordination by *town planners*, who are usually architects or municipal engineers, of the interests of the town, represented by the views of economists, doctors, sociologists, and so on. In USA town planning is a short-term process, city planning the long-term process of continuously developing the master plan. *See* **planning.**

TRADA The Timber Research and Development Association has been in existence since 1933. One of its most interesting achievements was the completion in 1969, at 2 per cent inside the Ministry's cost limits, of its design for a four-storey development of 48 dwellings with a structure entirely of timber for Wycombe Rural District Council. The necessary fire resistance and sound absorption were achieved by mineral wool quilt within the walls, 2·5 cm (1 in.) of plasterboard on wall surfaces and ceilings, and a 'double' plywood floor with a layer of sand between the two sheets of plywood. The thermal insulation was much better than brickwork and the weight was a quarter that of brickwork, thus reducing foundation costs. Some wall components were prefabricated.

trade A building trade is the same as a building *craft. See* **tradesman, sequence.**

tradesman or **craftsman** A man who has been an *apprentice* for some years in a building trade and has therefore enough skill to be considered a *journeyman* at his trade. He may be a carpenter-and-joiner, *bricklayer, mason, slater-and-tiler, plumber, electrician, house* *painter, glazier, floor-and-wall tiler, plasterer, paperhanger, steeple-jack, hot-water fitter,* etc. *See* **skilled man, semi-skilled man, labourer.**

trade union or **trades union** A voluntary organization of people who work in the same trade, such as *carpenters, bricklayers,* or *plumbers,* (craft union), or of people who work in the same industry, such as miners (industrial union). Both types of union aim at improving the working conditions of their members.

trammel [pla.] A device for drawing ellipses and for plastering elliptical surfaces.

translucent glass *Obscured glass.*

transmittance *See* **U-value.**

transom or **transome** (1) A horizontal beam, particularly the stone or timber bar separating the *lights* of a window, or separating a door from a fanlight over it.

   (2) (USA) A *fanlight.*

trap (1) [plu.] A U-shaped bend in a pipe, so shaped as to contain sufficient water to seal the air downstream of the U from that above the U.

   (2) A *scaffold board* which overhangs its support and is therefore dangerous to tread on.

traverse A *dressing iron. See also* C.

tread The level part of a step, or its length. *See* **flight** (illus.).

treadmill The only source of power on building sites in the Middle Ages, introduced into British prisons in 1817 as a wooden wheel 5 m (17 ft) wide of 1·5 m (5 ft) dia., with twenty-four equidistant treads round the rim, on which the prisoners had to climb.

treenail or **trenail** [carp.] A hardwood pin driven into a hole bored across a *mortise-and-tenon joint* or other joints in *carpentry.* A tree-nail is sometimes called a *drawbore* pin, but this term is best reserved for the steel pin which pulls the holes into line before the treenail is inserted. The word *dowel* is used more for *joinery,* treenail for *carpentry. See also* C.

trellis window A *lattice window.*

trench (1) *See* **creep trench.**

   (2) [joi.] A groove across a member, a *housing.*

trestle A support for *scaffold boards,* used by *plasterers* or *painters* when working on a ceiling. It consists of two broad ladders on four spreading legs about 1·5 m long, usually hinged together at the top. Two of these are needed to carry the staging of *scaffold boards. See also* C.

trim (1) or (USA) **casing** [joi.] *Architraves, skirtings, dado,* or picture rails, etc. made, if of wood, by linear machining. *See* **metal trim.**

   (2) *See* **trimming.**

trimmed joist or (USA) **tailpiece** [carp.] A *common joist* which has been cut short (trimmed) at an opening and is carried by a *trimmer joist.*

trimmer [carp.] (1) A *trimmer joist*.

(2) (USA) A *trimming joist*.

trimmer arch A brick arch which carries a fireplace in a wood floor and spans from chimney back to *trimmer joist*.

trimmer joist or (USA) **header joist** [carp.] A short *joist* which encloses one side of a rectangular hole in a wood floor. Another trimmer joist may form the side opposite and parallel to it, and *trimming joists* form the sides parallel to the *common joists*. The trimmer joists are carried on the trimming joists by *tusk tenons* through them. The trimmer joists carry the common joists which are cut off by the hole.

trimming Framing round or otherwise strengthening an opening through a floor, roof, or wall, whether of timber or other material. *See below*.

trimming joist [carp.] A heavy joist parallel to the common joists. One is fixed each side of a hole in a wooden floor in place of a *common joist*. It carries one end of the *trimmer joists*, and thus the loads of the common joists carried by the trimmer joists.

trimming machine or **mitring machine** or **guillotine** or **bench trimmer** [joi.] A lever- or pedal-operated machine with a heavy sharp blade for cutting the ends of *mouldings* or timbers at any desired angle. (For mitring, the angle is 45°.)

trimming piece A *camber slip*.

triple course Three rows of *shingles* laid together at eaves.

trough gutter A *box gutter*.

trowel Wood-handled steel-bladed tools of many different shapes and sizes used by *bricklayers*, *masons*, and *plasterers*. *Compare* **float**.

trowelled face [pla.] A plaster or mortar surface finished with a trowel. Trowelling cement mortar must be done with care, as over-trowelling brings the cement to the surface and often leads to *crazing*.

trowelled stucco [pla.] *Stucco* which, after ruling off and *scouring*, is trowelled with a *laying trowel*.

trowel man A *finisher*.

trunk or **trunking** Large sheet-iron or wood pipe, generally for ventilation.

trunk lift *See* **freight elevator**.

truss A frame, generally nowadays of steel (but also sometimes of timber, concrete, or *light alloy*) to carry a roof or other load, built up wholly from members in tension and compression. Steel trusses generally weigh 10 to 15 kg/m² (2 to 3 lb/ft²) of floor area for spans of 12 m (40 ft) or less. For every 3 m (10 ft) increase in span, another 2 to 5 kg/m² (0·5 to 1 lb/ft²) should be added. Bracing against wind adds a further 2 to 5 kg/m². Roofing with *slates* instead of the usual asbestos, steel, or aluminium sheets increases the truss weight by 15%. Trusses are usually placed about 3 m apart, but their spacing is fixed by the design of the *purlin*.

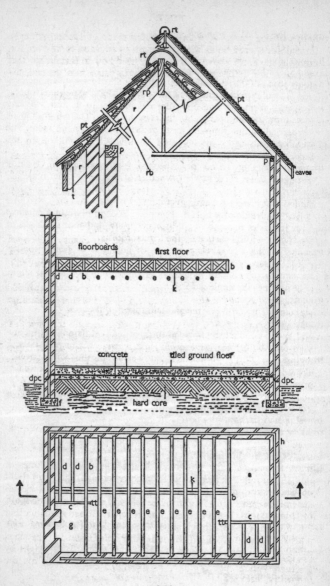

floorboards   first floor

concrete   tiled ground floor

hard core

trussed partition [carp.] A *framed partition* made of timber, strongly framed like a *truss*, so as to carry weight in addition to its own, for example, the floors above it. It is rarely used now in Britain but was very common until cheap building blocks came into general use about 1920.

trussed purlin A *purlin* reinforced by a camber rod beneath it like a *trussed beam* (*C*).

trussed rafter roof (1) A *collar-beam* roof.

(2) A *scissors truss*.

trussed roof A roof carried on *trusses*.

trying plane or **try plane** or **truing plane** [carp.] A *bench plane* about 56 cm (22 in.) long used after the *jack plane*. It is usually 6·3 cm (2½ in.) wide with a 5·7 cm (2¼ in.) *cutting iron*.

trylon A tall slim pylon tapering to a point built at the New York World's Fair 1939. Compare *skylon*.

try plane [carp.] A *trying plane*.

try square [carp.] A *square*. *See* **mason's tools** (illus.).

tubular [plu.] A pipe *fitting* formed from pipe, for example, a *barrel nipple* or *connector* or *bend*. These are generally of *dead-mild steel* (*C*), as opposed to most fittings which are of *malleable cast iron* (*C*).

tubular saw [mech.] A *hole saw*.

tubular scaffolder A *skilled man* who erects steel or *light alloy* scaffolding, joined with clips, fish plates, nuts and bolts, and so on. If he also puts up suspended scaffolds he may be called a *rigger* (C).

tubular scaffolding *Scaffolding* of 5 cm (2 in.) outside dia. steel or *light-alloy* tube. Steel is much stronger, about three times as heavy, and half the cost of light alloy per metre. Both are made of tubes with wall thickness of 4·5 to 4·9 mm. Light alloy has the advantage that it does not rust. *See* BSCP 97, Metal Scaffolding.

tuck A recess in a horizontal mortar *joint* made in *tuck pointing*. It is filled with a line of *lime putty* which projects about 3 mm (⅛ in.) and makes an outstanding white line.

Plan and section of a house carcase, omitting doors, windows, and plaster; and floorboards in plan.

a = opening trimmed for stair	k = herring-bone strutting
b = trimming joist	p = wall plate
c = trimmer joist	pt = plain tiles (resting on battens)
d = trimmed joists	r = rafter
dpc = damp-proof course	rb = roof boarding (covered with sarking felt)
e = common joists	
f = footing	rp = ridge piece (or ridge)
g = opening trimmed for fireplace	rt = ridge tile (set in mortar)
h = cavity wall	t = tilting fillet
j = ceiling joist	tt = tusk tenon

tuck in That part of a *bitumen felt* roofing, *skirting*, or *cover flashing* which is bent into a chase in the wall.

tuck pointing A decorative and protective method of *pointing* old brickwork. The joints are first filled flush and then grooved to form a **tuck**. The white *lime putty* in the tuck emphasizes the joint strongly.

tumbler [joi.] The part of a *lock* which holds the bolt in place until the key is turned.

tumbler switch [elec.] A simple lever-operated switch.

tumbling bay [plu.] A *back drop*.

tumbling in or **tumbling courses** Sloping *courses* of brickwork meeting horizontal courses as at the sloping top of a *buttress* (*C*) or as a *coping* to a *gable wall*.

tung oil or **wood oil** or **China wood oil** [pai.] An oil from the seeds of certain tropical trees (aleurites), which also grow in China and Japan. The oil dries quickly but frosts unless properly heat treated. It has excellent water resistance. *See* **frosting**.

tupper A *bricklayer's labourer* in the north of England.

turn button [joi.] A simple catch for a cupboard door consisting of a metal or wood piece held pivoted on the frame outside the door by a screw.

turner A *tradesman* who works on wood or metal in a lathe. *See* **wood turner**.

turning [joi.] Making an object in a lathe.

turning bar A *chimney bar*.

turning piece or **trimming piece** A *camber slip*.

turning saw [joi.] A *bow saw*.

turnpin or **turning pin** [plu.] A *boxwood tampin*.

turnscrew [joi.] An ironmonger's term for screwdriver.

turpentine or **spirits of turpentine** [pai.] A valuable *solvent* obtained by distilling the *oleo-resin* of the pine tree. Resin is leached from the living tree, as in the Landes of France, or given off when the wood is heated.

turpentine substitute [pai.] *White spirit*.

turret step A triangular stone step from which a *spiral stair* is built up. The central *solid newel* consists of the rounded ends of the turret steps laid on top of each other.

tusk nailing [carp.] *Skew nailing*.

tusks or **tusses** Stones projecting as *toothing*.

tusk tenon or **keyed mortise and tenon joint** [carp.] A small *tenon* usually formed at the end of a *trimmer joist* to fix it to a *trimming joist*. The tenon passes through a mortise in the trimming joist, and is itself mortised and clamped to the trimming joist by a wedge-shaped key.

twin cable [elec.] or **duplex cable** Two insulated conductors laid in a common insulating covering.

twin-shaft paddle mixer or (USA) **twin pug** [c.e.] A *pug mill* with two horizontal shafts turning in opposite directions.

twin tenon or **divided tenon** [carp.] A *tenon* from which the central part has been cut away, leaving its top and bottom.

twist [tim.] Spiral *warp*.

twist drill [mech.] A hardened steel *bit* with helical cutting edges, used in electric drills, *hand drills*, or *breast drills* for drilling metal or wood. The cutting edges slope at about 30° to the axis of the drill. In the *auger bit* the twist slopes at about 45° to the axis of the drill.

twisted fibres [tim.] *Interlocked grain.*

twist gimlet [joi.] A simple *gimlet* with a helical groove by which the wood cuttings are removed.

twitcher [pla.] An *angle trowel*.

two-bolt lock [joi.] A lock which is turned by a key, combined with a *latch* operated by a knob on a spindle.

two-coat work (1) [pla.] Plastering with a *floating coat* and a *finishing coat*, now very common.

(2) Paint or asphalt applied in two layers. *See* **three-coat work.**

two-foot rule A *fourfold rule* 2 ft (61 cm) long.

two-handed saw [tim.] A large hand-operated saw worked by two men, one at each end, for example, a felling saw, or a saw for cross-cutting logs in the forest.

two-light frame [joi.] A window with one *mullion* dividing it into two *lights*.

two-pipe system [plu.] (1) A soil and waste system comprising two independent pipes for fluid, namely a soil pipe conveying soil directly to the drain and a waste pipe conveying waste water to it through a trapped gulley. The system may also need ventilating pipes. *See* **one-pipe, single-stack.**

(2) A heating circuit with a flow pipe as well as a return connected to each radiator. The furthest radiator is therefore very little cooler than the one nearest the boiler. *See* **one-pipe.**

Tyrolean finish [pla.] Plaster thrown on to a wall and left rough. The description is usually given to plaster thrown on by a hand-operated machine. The plaster is porous, as is generally desirable for a rain-proof surface, and it sticks well to the backing.

U

UCATT Union of Construction, Allied Trades and Technicians. *See also* **Association of Building Technicians.**

U-gauge or **water gauge** [plu.] A glass U-tube half filled with water, one end being connected by rubber flexible tube to a system of drains or gas pipes under test. It quickly shows whether they are gas-tight. *See* **air test.**

ultramarine [pai.] A complex sulpho-silicate of aluminium and sodium with good *alkali resistance*, a precious blue *pigment* got from crushed *lapis lazuli* in the past, first synthesized commercially in France in 1828. Since then the artificial pigment has been more used than the mineral. The latter is still obtainable but costs nearly as much as gold.

umber [pai.] Raw umber is a *pigment* consisting of natural hydrated iron oxide with some oxide of manganese in it. When calcined it is called burnt umber and is a rich, deep reddish brown. *See* **sienna.**

unbuttoning Demolition of buildings, particularly of steel frames by breaking off the rivet heads.

uncoursed A description which may be applied to *random* or *snecked-rubble* walls. Both may also be coursed. In USA broken-range ashlar is English uncoursed rubble.

undercloak (1) A course of *slates* or *plain tiles*, at *eaves* or verges, placed under the surface tiles. *See* **double eaves course.**

(2) A row of *shingles* laid up the *verge* of a roof with their butts overhanging the gable. The ordinary shingles are laid over these and are thus given a slight slope towards the roof.

(3) In *flexible-metal roofing* the lower layer of metal sheet at a *roll* or *drip* or *seam* which is covered by the *overcloak*.

undercoat (1) [pai.] Any coat applied to a surface after *priming* and before the finish coat. (An undercoat can be also a *priming coat*.) It is matt with a high content of *pigment* and *extender* and a *colour* approaching or helping that of the *finish* coat. It should increase the thickness and therefore the protectiveness of a *paint system*. *Varnish* undercoats have less oil per unit of *resin* than varnish finishes and are usually *short oil varnishes*.

(2) [pla.] A *floating coat* or *rendering* or *pricking-up coat*, in any case not a *finishing coat*.

undercuring [tim.] Insufficient hardening of a *glue* due to low temperature or too short a hardening period.

undercut tenon [carp.] A *tenon* with its shoulder cut slightly off square to ensure that it bears on the mortised piece.

under-eaves course A course of *eaves tiles* in the *eaves course*.

underfloor heating Electric heating cables or hot water pipes, sunk near the upper surface of a concrete floor slab or in the screed over it, provide a warm floor that heats the room without the presence of ugly radiators. Expert design and execution are essential. For example if the floor temperature exceeds 27° C. (81° F.), failures with floor gluing can be expected that may cause vinyl tiles to curl. The method was used by the Romans in their hypocausts, except that these were heated by flue gases.

underlay (1) or **isolating membrane** A layer of *sheathing felt* or asphalt separate from the surface asphalt, laid so as to allow vibration or thermal movement of the surface and not to restrict it.

(2) A layer of *building paper* or *inodorous felt* fixed below *flexible-metal roofing* to allow movement.

(3) *See* **double-skin roof.**

(4) Hardboard, plywood, etc., fixed over a rough floor to make a smooth surface suitable for laying lino, cork tiles, parquet flooring, etc.

underlayment A thin (2 mm or $\frac{1}{16}$ in.) specially formulated fine-grained mortar put over a *screed* or concrete floor to make it smooth enough for a pvc (vinyl) floor surfacing, especially the flexible vinyl sheet.

underlining felt *Sarking felt.*

underpitch groin or **Welsh groin** An intersection of two cylindrical *vaults* of different *rise. See* **groin.**

under-ridge tiles or **under-tiles** Special *plain tiles* laid in the top *course* of a roof below the *ridge tiles*. They are 23 cm (9 in.) instead of 27 cm (10½ in.) long. *Compare* **eaves tiles.**

undersize slates In Scotland, *slates* smaller than 36 × 20 cm (14 × 8 in.).

under-tile (1) or **tegula** The lower tile of *Spanish* or *Italian tiling*.

(2) An *under-ridge* tile.

undertone [pai.] (1) The *colour* obtained when a coloured *pigment* is reduced with a large proportion of *white pigment*.

(2) The colour seen when a coloured pigment is spread on glass and viewed with light passing through it.

undressed timber or **unwrought timber** Sawn but not planed timber.

uneven grain or **uneven texture** [tim.] A texture with considerable contrast between *springwood* and *summerwood*.

unframed door [joi.] A *batten door* or *ledged and braced* door.

ungauged lime plaster [pla.] *Plaster* made with *lime*, sand, and water only. Only those limes which have a compressive strength of at least 689 kN/m² (100 psi) at twenty-eight days are recommended. Otherwise *Portland cement* or *gypsum plaster* should be *gauged* with it to give early strength.

Unified thread [mech.] A screw *thread* accepted by the *BSI* and by Canadian and US authorities, now replacing the Whitworth thread in Britain.

union or **cap and lining** or **plumber's union** A pipe fitting that can be screwed at one end to steel pipe and is soldered to lead at the other by its brass lining or sleeve. It is easily joined or disconnected by hand to or from the steel pipe.

union bend, union cock, etc. [plu.] A bend, cock, etc., with a *union* at one end.

unit [elec.] An electrical unit is the energy used when one kilowatt flows for one hour, that is, one kilowatt-hour, kW.

unit air conditioner Usually a steel box (the unit) measuring about $60 \times 30 \times 30$ cm ($2 \times 1 \times 1$ ft) containing an electrically driven fan, a compressor, an evaporator and a condenser, through which air is blown, filtered and cooled. The box can be placed in a window or in a hole through an outside wall or through a roof. Sometimes the unit is 'split' with the evaporator (the cooler), the air fan and the air filter in one box and the noisier compressor, condenser, and condenser fan in another box away from the living room. Often used in hot countries, these units have the advantage that they can be installed very quickly in any room that has an electrical supply and an outside wall or roof, but they are noisy and a central *air conditioning* system is preferable because the noisy machines are then away from living rooms.

unit heater [plu.] (1) A fixed space heater used in large rooms such as workshops, hung at about 2·5 m (8 ft) above floor level. Air is blown over its finned tubes by an electric fan. It can be supplied with hot water or steam by pipework like a radiator or can be gas-fired. Because of the forced flow of air over the surface, its heat output is several times more than a radiator of the same size. Although it is a warm air heater, this description is usually reserved for those that are ducted, at least for the supply into the room.

(2) A **fan convector.**

unit of bond The smallest length of a brick *course* which repeats itself. In *Flemish bond* it is $1\frac{1}{2}$ bricks long; in *English bond* $1\frac{1}{2}$ bricks thick, it is 1 brick long.

universal plane or **combination plane** [joi.] A hand-operated metal *plane* with many *cutting irons*, often about fifty, each of different shape, to cut different *mouldings*, *tongues* or grooves, *rebates*, *beads*, etc.

unsound [pla.] A description of slaked *limes*, *cements*, *plasters*, and *mortars* that contain particles which may expand (*blowing*).

unsound knot or **rotten knot** [tim.] A *knot* which is softer in whole or in part than the wood round it.

untrimmed floor [carp.] A floor carried on *common joists* only.

unwrought timber or **undressed timber** Sawn but not planed timber.

upset [tim.] A tear across the fibres of wood caused by a shock, often during felling. *See also C.*

upstand (1) or **upturn** That part of a felt or *flexible-metal* flashing, or roof covering which turns up beside a wall without being tucked into it, and is covered usually with a *stepped flashing*.

(2) or **upstand beam** A beam in a concrete floor which rises out of the floor like a wall, instead of, as usual, projecting below it.

urea-formaldehyde glue [tim.] A *synthetic resin* glue made by chemically condensing urea with formalin. An *accelerator* must be added to the liquid *resin* to make the *glue* harden. The two can be mixed in a pot or applied to each face of the joint in *separate application*. With some slight chemical modification these glues are used in baking enamels. Urea glues can be used cold, but if heated they harden more quickly.

usable life The *pot life* of a glue.

U-tie A *wall tie* used in USA, made of heavy wire bent into a top-hat shape.

utility A *service* such as water, gas, electricity, telephone, sewage disposal, etc., which is available to all.

U-value or **thermal transmittance** or **air-to-air heat-transmission co-efficient.** A figure determined by experiment for a certain wall, roof, or floor in a certain situation, which tells how many watts (Btu/hour) will pass through one m² (ft²) of the wall when the temperature of the air on one side is 1° C. (1° F.) higher than on the other side. In an exposed situation, the same wall will have a higher U-value than in a sheltered situation. The U-value recommended in good Continental domestic practice for the outer wall or roof-plus-ceiling is about 0·5 Watt/m² deg. C. (0·09 Btu/ft²h deg. F.) but British recommended values are twice as high, implying heavy heat losses from British houses. The more extreme the climate, the lower should the U-value be.

V

vacuum-cleaning plant or **centralized vacuum-cleaning plant** Vacuum-cleaning plant, permanently installed in a building, consists of an exhauster fan and filter, usually in the *basement*, connected to all floors through a network of smooth-bore pipe (usually cold-drawn, hard-steel tube). The pipe diameter is fixed by the air velocity (15 to 24 m/sec (50 to 80 ft/sec)) and by the quantity (2·2 m³ (80 ft³)) per minute for each operator cleaning bare floor. Such high velocities can only be obtained by a fan working to the considerable suction of 17 kN/m² (2½ psi or 127 mm of mercury). For carpet cleaning, the air consumption is about 1·1 m³ (40 ft³) per minute. Hose outlets are fitted at a spacing which enables all the floor to be cleaned by hoses from 6 to 12 m (20 to 40 ft) long.

vacuum heating A steam-heating system for buildings in which a vacuum pump is connected to the return main. It removes condensate and air from the radiators and returns the water to the boiler feed tank (USA).

valley The intersection between two sloping surfaces of a roof, towards which water flows, the opposite of a *hip*. With *thatch*, *plain tiles*, or *slates*, the valley can be continuous with the slopes, that is there need be no sharp angle and no *valley gutter*. *See* **laced valley, swept valley, valley tile**.

valley board [carp.] A board about 28 × 2·5 cm (11 × 1 in.) fixed on and parallel to the *valley rafter*. It is used for supporting slates or tiles in a *laced valley* or *swept valley*.

valley gutter A *gutter* lined with *flexible metal* in a *valley*, for example, a *secret gutter* or a box gutter. It may also be of concrete, precast, or cast in place.

valley jack (USA) A *jack rafter* which fits on to a *valley rafter* or *valley board*.

valley rafter The *rafter* which lies along the line of a *valley* and carries a *valley gutter* or a *valley board*. The *jack rafters* meet on it.

valley shingle A *shingle* laid next to the *valley*, cut so that its *grain* is parallel to the valley.

valley tile A specially large *tile*, concave upwards, shaped for forming a *valley* without *flexible metal* and without lacing or sweeping the tiles. Burnt clay valley tiles may be less well burnt than the *plain tiles* which course with them owing to the maker's fear of the tile warping when it is hard burnt. They may therefore not weather well.

value–cost contract [q.s.] A *cost-reimbursement contract* in which the *contractor* receives a larger fee when his final costs are low than when his final costs are high.

valve [mech.] A device to open or close a flow (stop valve or stopcock) or to regulate a flow (discharge valve).

vapour barrier An airtight skin consisting of rubber-like paint, or metal, or *roofing-felt* sheets, bonded together to prevent moisture from the warm damp air in a building passing into and condensing within a cold, insulating wall or ceiling material. It is therefore placed on the inner, warm face of the insulation and thus prevents condensation. *See* **temperature gradient** (*C*).

vapour heating A steam heating system in which the condensate returns to the boiler by gravity. The pipes are at a pressure near atmospheric (USA). *Compare* **vacuum heating**.

variation order A written order from the *building owner* (represented by the consulting *architect* or engineer) authorizing an increase of the work above the amount shown in the *contract*.

varnish [pai.] A *resin*, asphalt, or pitch dissolved in oil or spirit which dries in air to a brilliant, thin, transparent, protective film. Varnishes are usually subdivided into *spirit varnish* and oil varnish, the latter drying by oxidation of *drying oil*. Varnish may be mixed with paint, put on over it to increase its *gloss*, or put on unpainted wood. *See* **lacquer**.

vault (1) An arched masonry roof, generally with a smooth curved *soffit*, or a room or passage with an arched masonry roof.

(2) A room below ground, of massive construction, for keeping valuables safe from fire and thieves.

vault light A *pavement light*.

V-brick A *perforated brick* 23×23 cm (9×9 in.) in plan and 7·6 cm (3 in.) deep, with vertical cavities formed in it to make a one-brick wall equivalent in dryness and warmth to the 28 cm (11 in.) *cavity wall*. It was designed by the Building Research Station in 1958 but did not quickly become popular although it is easier to build than a cavity wall.

vee-beam sheeting *Asbestos-cement sheeting* in which the corrugations are formed by three flat surfaces instead of by curves.

vee gutter A *valley gutter*.

vee joint (1) [joi.] A small chamfer on the face edge of *matchboards* made so that the two boards, when put together, form a V at their junction which masks shrinkage.

(2) A slight concave horizontal < formed in a mortar joint by *tooling*.

vee roof The shape formed by two *lean-to roofs* which meet at a *valley*.

vee tool [joi.] A *parting tool*.

vegetable glue [tim.] Starch glue such as *cassava* or protein glue like *soya glue* or that from ground nuts or rape seed.

vehicle [pai.] The entire liquid part of the *paint* in the container, consisting of *binder* and *thinner* but no *pigment*.

veneer (1) [tim.] A thin layer of wood of uniform thickness which may have been *sliced*, *rotary cut*, cut *half-round*, or sawn. It is used either as a facing, for its beauty of *figure*, to stronger, less beautiful wood or for its strength in building up *plywood* or *coreboard*.

(2) (mainly USA) A layer of brick or marble or other facing outside a wall. The veneer looks well and resists the weather. *See* **brick veneer, veneered wall.**

veneer cutter [tim.] A *tradesman*, who may decide on the method of cutting *veneer* from a log, whether by *slicing* or *rotary cutting*, but who in any case can set a knife and operate the machine. If he works a slicing machine, he is called a veneer slicer.

veneered stock [tim.] An early name for *plywood*.

veneered wall A wall having a *facing* which is attached to the backing but, not being bonded to it, it cannot resist load equally with the backing. A veneered *brick* facing is attached to the backing across the cavity by *wall ties*, which may be of wire only 2·5 mm (0·1 in.) dia. If it is desired to count the strength of the facing with the backing, the two must be bonded without a cavity, by bricks. 'Veneer' is used particularly in Australia and North America for the 10 cm (4 in.) leaf of brickwork which sometimes covers *frame construction*. *See* **brick veneer, faced wall, veneer tie.**

veneering [tim.] Fixing a decorative *veneer* to a structural backing of wood, *plywood*, *hardboard*, etc.

veneer tie (USA) A *wall tie* for holding a *veneered wall* to its wood backing.

Venetian mosaic [plu.] *Terrazzo.*

Venetian red [pai.] Red iron oxide (Fe_2O_3), a *pigment* from Italy.

vent An outlet for air, a ventilating *duct*, *expansion pipe*, etc.

ventilating bead [joi.] A *deep bead.*

ventilating brick An *air brick.*

ventilating jack (USA) A sheet metal hood over the inlet to a ventilating pipe to increase the down-draught.

ventilation *See* **air change.**

ventilation pipe or **vent stack** or **vent pipe** or (USA) **continuous vent** A *soil drain* must be ventilated at its upper end by a pipe of at least 7·6 cm (3 in.) dia., which is often a continuation of the *soil pipe*. *Waste pipes* also may need ventilating. *See* **anti-siphon pipe, fresh-air inlet, loop vent.**

ventilator or **ventlight** Any means of ventilating a room, such as a *night vent.*

vent pipe An *anti-siphon pipe* or a *ventilation pipe.*

vent stack A *ventilation pipe.*

verdigris Green basic acetate of copper formed as a protective *patina* over copper exposed to the air. It may be of any colour from brown to black in cities.

verge The edge of a sloping roof which overhangs a *gable*, sometimes including the bricks which cope the gable wall. *See also* C.

verge board or **verge rafter** A *barge board.*

verge fillet A *batten* fixed on a *gable* wall to the ends of the roof battens, as a neat finish beyond which roofing *shingles* overhang.

verge tile A *tile-and-a-half tile* used at the *verge* of a roof in alternate courses.

vermiculite A mica, found in USA and South Africa, which when heated expands (exfoliates) into worm-like threads to form a very light insulating aggregate. It weighs, loose, 65 to 130 kg/m³ (4 to 8 lb/ft³) and has a thermal *conductivity* or K-value of 0·06 to 0·07 W/m deg. C. (0·4 to 0·5 Btu in./ft²h deg. F.). A concrete or floor screed can be made of it that weighs 340 to 720 kg/m³ (21 to 45 lb/ft³) with a K-value from 0·08 to 0·2 W/m deg. C. (0·54 to 1·4 Btu in./ft²h deg. F.) but it is weak (0·9 N/mm² or 135 psi) though this may not always matter because its high insulation value so greatly improves fire resistance. *See* **mineral wool.**

vermiculite–gypsum plaster [pla.] For a 4-hour *fire grading* which ordinarily involves a 15 cm (6 in.) concrete slab with simple plaster ceiling, a 10 cm (4 in.) slab can be substituted, having a vermiculite–gypsum plaster ceiling 19 mm (¾ in.) thick. This reduces the slab-and-plaster weight to 274 kg/m² (56 lb/ft²) from 380 kg/m² (80 lb/ft²). The *pre-mixed plaster* is usually a 1·5 to 1 mix which weighs only 670 kg/m³ (42 lb/ft³).

vermilion [pai.] A brilliant red, slightly orange-coloured *pigment* composed of mercuric sulphide (HgS), which was used in antiquity first in China, but is now too expensive except for use in very small quantities. The mineral is called cinnabar.

vertical grain [tim.] The *edge grain* of *quarter-sawn* wood.

vertical sash [joi.] (USA) A *sash window.*

vertical shingling or **weather shingling** or **hanging shingling** *Shingles* hung on a wall, like *tile hanging.*

vertical spindle moulder [tim.] A *spindle moulder.*

vertical tiling *See* tile hanging.

vestibule An antechamber just within the entrance to a building or before the entrance to a large room. *See* **tambour.**

vice or (USA) **vise** A screwed metal or timber clamp fixed to a workbench and used for holding timber or metal which is being worked. *See* **chop.**

vinyl flooring Vinyl polymers are many but this usually means PVC, either vinyl asbestos tiles to BS 3260 or flexible vinyl sheet or tiles or backed vinyl sheet to BS 3261.

vision-proof glass *See* obscured glass.

vitreous enamel *See* enamel.

vitrified brick or **pipe** Bricks or *ceramic* pipe which have been surface glazed by beginning to melt in the kiln or have been glazed by salt or other material being thrown into the furnace. *Hard-burnt* bricks may be vitrified.

vitrified clay pipe The correct description of *glazed ware pipe.*

void former In the original *hollow-tile floor* (C), lines of hollow clay blocks are placed on the decking between the reinforcing bars before concreting. These reduce the weight of the concrete floor by eliminating part of it that has no structural function. Clay blocks have disadvantages of weight, breakability, high absorption of water, laborious unloading, etc., and so have in some contracts been replaced by lightweight blocks of expanded plastics whose one apparent disadvantage is that they must be fixed to the decking to prevent them floating during concreting. They can be made much bigger than clay blocks, up to 2·4 m (8 ft) long, 30 cm wide and 30 cm thick. Unloading is consequently easier but, during placing, the blocks must be anchored by nailing them to the decking.

volatile [pai.] Literally that which flies away. It therefore describes well those liquids which boil at ordinary temperatures or at temperatures below the boiling point of water (100° C.).

volt [elec.] The unit of electrical pressure, related to the units of flow (amperes) and power (watts) in the following way: watts = volts × amperes.

voltage [elec.] Electrical pressure expressed in volts. The conventional descriptions are: high voltage more than 650 volts; medium voltage, from 250 to 650 volts; low voltage from 50 volts DC or 30 volts AC to 250 volts; extra low voltage, below 50 volts DC or 30 volts AC.

volume yield (1) The volume of *lime putty* of a stated consistency obtained from a stated weight of *quicklime*.

(2) The volume of *concrete* of a certain mix obtained from unit weight of *cement*.

voussoir An *arch-stone* in a stone arch or an *arch-brick* in a brick arch.

W

wagtail [joi.] A *parting slip*.

wainscot or **wainscoting** [joi.] Wood panelling on boards up to *dado* height in a room.

wainscot oak [tim.] *Quarter-sawn* oak used for panelling. That used in Britain is usually imported from central Europe.

waist The narrow part of an object, in particular the least thickness of a reinforced-concrete *stair* slab. *See* **flight** (illus.).

walking line A setting-out line at 45 cm (18 in.) from the centre line of the handrail of a stair. Along this line the *going* of the *winders* is the same as that of the *fliers*.

walk-up or **walk-up apartment house** (USA) A block of *flats* of four storeys or less, without a lift.

walkway A permanent gangway with handrails to give safe access along a roof.

wall *See* **bond, loadbearing wall** (*C*), **panel wall, partitions, party wall.**

wall anchor or **joist anchor** [carp.] A steel strap screwed to the end of every second or third *common joist* and built into the brickwork to ensure that the joists give lateral support to the wall. It is fixed to the same joists as those to which the *strap anchors* are fixed.

wallboard *Building boards* made for surfacing, rather than for insulating ceilings and walls. Wallboards include *asbestos-cement sheet, asbestos wallboard, plywood, gypsum plasterboard,* and glossy *laminated plastics* glued to a backing of *hardboard* or plywood. Fibre wallboards were made in England first as a substitute for *matchboard*. They give some insulation, are fairly cheap, and much quicker to put up than matchboard. *See also* **compressed straw slabs, wood-wool slab.**

wall box or **beam box** A cast-iron box built into brickwork to provide a bearing for a timber beam or joist. *See* **wall hanger.**

wall column A steel stanchion or reinforced-concrete column partly within the thickness of a wall.

wall hanger A cast-iron or bent-steel *stirrup strap*, built into a wall to carry the end of a wood joist. They are generally used instead of wall boxes in modern buildings.

wall hook A spike or heavy nail driven into the mortar joint of a wall to carry a pipe or timber. Its head is specially shaped for the purpose. *See* **wall plate, fasteners.**

wall-hung boiler [plu.] A gas boiler mounted on a wall, of very light weight because of its low water content. Almost all are provided with *balanced flue*.

walling mason or **waller** A *mason* who sets stones in walls and bridges, cutting them to shape with a *scutch*.

wall joint A mortar *joint* parallel to the face of the wall.

wall panel *See* **panel wall.**

wallpaper Decorated printed paper sold in rolls for sticking on the plaster of walls. A *sealer* should be put on new plaster to prevent it discolouring paper. Often lining paper is used. A roll (or piece) of paper in Britain is 10 m × 53 cm (33 ft × 21 in.), in France 8·2 m × 46 cm (27 ft × 18 in.), in USA 7·3 m × 46 cm (24 ft × 18 in.).

wall piece [carp.] *See* **wall plate.**

wall plate (1) A horizontal timber along the top of a wall at *eaves* level. It carries the *rafters* or *joists*.

(2) or **wall piece** A vertical timber in *raking shoring* held on to the shored wall by *wall hooks* and by short *needles* which pass through it into the wall. The shores bear on the wall plate under the needles.

wall plug *See* **plug.**

wall string [joi.] The *string* on the side of a *stair* next the wall, as opposed to the *outer string*.

wall tie (1) or **tie iron** A piece of twisted bronze or galvanized steel plate or wire built into the *bed joint*, across the cavity of a *cavity wall*, to hold the two leaves together. In USA wall ties are less carefully designed for their function of keeping water from the inner leaf than in Britain, and often consist of a straight bar which would not be acceptable in the British climate. *See* **bonding brick, butterfly wall tie, prefabricated tie, rectangular tie, veneer tie.**

(2) (USA) Reinforcement placed in the bed joint of a brick or glass block wall parallel to its length.

wall tile [pla.] A tile of burnt clay glazed or unglazed, *terra cotta*, faience, glass, concrete, *asbestos-cement*, or *plastics*, used for putting a decorative or smooth face on a wall, by sticking the tile on to the wall. For internal walls the joints are later filled with *Keene's* or similar *gypsum plasters*. Tiles are usually 15 or 10 cm (6 or 4 in.) square and 1 cm (⅜ in.) thick. *See* BSCP 212 Wall Tiling.

walnut (Juglans regia) or **European walnut** [tim.] A decorative hardwood, mainly from southern Europe. The English variety has a finer *figure* and colour but is rare. Walnut is a wood with a grey background and dark, sometimes ruddy streaks, used for carving, turnery, and *veneers*, in which its *burrs* and *crotches* are valued.

wane or **waney edge** [tim.] Bark or the rounded surface under the bark, seen on *square-sawn timber*.

ware pipes Pipes of *glazed ware*, properly called vitrified clay.

Warerite A *laminated plastic*.

warm air heating [plu.] A *central heating* method introduced in Britain in the 1950s with small so-called selective heaters of about 6 kilowatts capacity. They are now made as large as any other unit and are heated by any fuel. All types use a *heat exchanger* and an electric fan.

warning pipe [plu.] An overflow pipe from a *cistern*, which discharges overflow water into the open so as to show up this plumbing fault.

warp [tim.] Any distortion of timber during seasoning such as *cup*, *bow*, *spring*, and *twist*, always caused by changing *moisture content*.

Warrington hammer A *joiner's hammer*.

wash (USA) A weathered slope or *weathering*.

washability [pai.] The ability of a coat of *water paint* to withstand washing without damage, or the ease with which dirt can be removed from a coat.

washboard (USA) A *skirting board*.

washer (1) [plu.] A flat ring made of rubber or plastics or leather or fibrous composition, which is held by a nut to the underside of the *jumper* of a water tap to make it watertight.

(2) *See* (*C*).

washleather glazing Bedding of glass in washleather instead of putty, commonly used for swing doors; the fixing is by *glazing bead*.

waste (1) (USA) Building rubbish.

(2) All dirty water except rainwater, and *soil*. It usually means water from basin, bath, and kitchen, but includes industrial wastes.

(3) [q.s.] The space in the description column of *dimensions paper*, in which collections and calculations are entered.

waste-disposal unit [plu.] A small electrically driven (0·5 hp) rubbish grinder placed near a kitchen sink, into which all kitchen waste except newspapers, tins and bottles can be thrown. The rubbish is disposed of down the drain, as in the *Garchey sink*.

waste mould [pla.] A mould for fibrous plaster used once only and then destroyed.

waste pipe [plu.] A pipe to carry water away from a basin, bath, or sink. It is trapped at its exit from the basin and the *trap* is fixed with a screw cap which can be removed to clean it if it should get stopped.

waste preventer or **waste water preventer** [plu.] A flushing *cistern*.

waster (1) A *facing brick* with minor defects which allow it to be used as a backing brick.

(2) A *mason's* chisel, either one with claw cutting edge for removing waste stone or a 19 mm (¾ in.) wide chisel. *See* **mason's tools** (illus.).

waste table A *water table*.

wasting Removing excess stone with a *waster* before dressing a stone block. *See* **picking**.

water bar A galvanized steel bar 25 × 6 mm (1 × ¼ in.), bedded on edge in a groove filled with white lead, on the stone or concrete *sill* of an opening. It fits a corresponding groove under the sill of the fixed wooden frame. It prevents draughts and water passing into the gap between sills during storms.

water-burnt lime [pla.] When *semi-* or *non-hydraulic limes* are slaked without excess of water they are likely to form grains of a very slowly reacting *lime*. These grains combine with water, swell after the plaster has been applied to the wall, and cause *blowing*.

water channel A *condensation groove* in a *patent glazing* bar.

water check A kerb standing above a roof covered with *bitumen felt*.

water-checked casements [joi.] A *casement* having *sill* and *meeting stile* grooved to break any possible *capillary* path for water.

water content *See* **moisture content**.

water gauge [plu.] A *U-gauge*.

water joint A *saddle joint*.

water level An instrument for setting out levels on a building site. It consists of a rubber tube connecting two vertical glass tubes filled with water. The level of the water in one tube is the same as that in the other if there are no air locks in the rubber tube.

water paint or **washable distemper** Any paint or *distemper* which can be thinned by water, such as oil-bound or *emulsion paints*, though the *binder* is insoluble in water. Ordinarily, however, distempers contain water-soluble binders like *size*.

water pipe Water supply pipes are usually of dead mild steel or cast iron, or, for small diameters, also copper or plastics. Large iron or steel mains are lined internally with 3 to 6 mm ($\frac{1}{8}$ to $\frac{1}{4}$ in.) of bitumen or with up to 25 mm (1 in.) of cement mortar for protection. The outsides when buried are covered with asphalt. *See* **fittings, light-gauge copper**.

waterproof cement *Water-repellent cement*.

waterproofing Asphalt *tanking* or other waterproof skins of bituminous material over a wall are the only certain methods of waterproofing walls and floors, apart from providing an air gap between the wall and the source of damp. *Integral waterproofers* are occasionally effective, but most concretes can be made waterproof by careful grading, mixing, and placing without any admixture. *See* **surface waterproofer**.

waterproof paper *Building paper*.

water-repellent cement A cement which is *hydraulic* but has higher resistance to water penetration than ordinary *Portland cement*.

water retentivity A property of *mortar* which prevents it losing water rapidly to bricks with high *absorption rate*. It also prevents water coming to the surface when the mortar touches *vitrified* or other bricks with low absorption. It is thus a property which allows mortar to develop good bond with every sort of brick as well as high strength. *See* **masonry cement**.

water seal [plu.] The *seal* in the *trap* of a drain.

water seasoning [tim.] Soaking timber for fourteen days and then air-drying it.

watershot walling *Dry walling* with stones laid sloping so that water falling on them pours to the outside of the wall. Walling is often built like this in the English Lake District where the rainfall is about 3·7 m (150 in.) per year.

water softener [plu.] A chemical plant for treating water in such a way that soap lathers easily in it. It removes from the water the calcium and magnesium salts called hardness, which produce *furring* in pipes, kettles, and boilers. The best known types are *base exchange* softeners and the *soda-lime process*. For domestic use, some chemical manufacturers now produce a low-cost water softener consisting merely of a metal basket that hangs in the cold water tank, and contains crystals of low solubility. The crystals have to be replaced every six months. They discourage *furring* but do not remove scale that is already deposited.

water spotting [pai.] Pale spots on a paint film caused by drops of water on the surface. They may or may not be permanent.

water stain [tim.] (1) A discoloration (which may improve the *figure* of the wood) caused by *converted timber* getting wet.

(2) Colouring matter dissolved in water and applied to wood to accentuate the grain. *See* **spirit stain.**

water table or **watershed** or **offshoot** or **canting strip** A board or masonry projection fixed to the foot of a wall (particularly if it is weatherboarded) to shoot water away from it. *See* **earth table.**

water test [plu.] A severe *drain test*, used ordinarily only for new drains. They are stopped at the lower end and filled with water for an hour to a maximum head which may be up to 2·1 m (7 ft). If no fall occurs during the hour, the drain is accepted.

water-waste preventer [plu.] The *cistern* which flushes a WC.

watt [elec.] The power obtained from 1 ampere flowing at a pressure of 1 *volt*, one thousandth of a kilowatt, 1 joule per second.

wavy grain [tim.] A curly attractive *grain* often seen in *birch*, *mahogany*, and sycamore.

wayleave A permission to pass over land, which may sometimes include leave to lay cables, pipes, etc.

weather The amount measured along the slope, by which a *shingle* overlaps the next shingle but one below it. It is the same as the *lap* of a *centre-nailed* slate.

weather bar (1) A term occasionally used for *water bar*.

(2) A water and draught excluder for an inward-opening *casement*. *See* **weather strip.**

weather-board [joi.] A *weather moulding*, but *see below*.

weather-boarding or **weather boards** or (USA) **siding** [tim.] Horizontal boards nailed on edge over the outside of light buildings. The boards generally overlap each other, either with or without a rebate at the lower edge of the upper board, which helps to keep out rain and wind. *Clapboard* is not *rebated*. *See* **shiplap boards.**

weather check A *drip* or *throat*.

weathercock or **weathervane** A pivoted ornamental *finial* which turns with the wind to show which way it is blowing.

weathered *See* **weathering.**

weathered pointing *Weather-struck joints*.

weather fillet A *cement fillet*.

weathering (1) or (USA) **wash** A slight slope to throw off rainwater.

(2) A change in colour of the surface of a building material after exposure to rain and sun.

(3) [tim.] The mechanical and chemical break-up of a wood surface exposed to rain and sun. It is not *decay*.

(4) *See* **double-skin roof**.

weather joint A *weather-struck joint*.

weather moulding (1) or **weather board** [joi.] A *moulding* housed into the bottom *rail* of an outside door to throw water off the *threshold*.

(2) A stone *drip*.

weather-proof glues These include phenolic, resorcinol and perhaps melamine resins. Some others are, however, at least weather resistant, even if not quite weather-proof.

weather shingling *Vertical shingling*.

weather slating *Slate hanging* (vertical slating).

weather strip or **wind stop** or (USA) **air lock** A piece of metal, wood, rubber, or other material which stops the draught passing the joints of a closed door or window.

weather-struck joint or **struck joint** or **struck-joint pointing** A mortar *joint* smoothed off by pressing the trowel in at the upper edge so as to throw rain out to the face of the *brick*. This work can be done as the walls go up and is therefore more durable than *pointing*.

weather tiling *Tile hanging* (vertical tiling).

weathervane A *weathercock*.

weaving or (USA) **Boston hip** Laying *shingles* on adjoining surfaces of a wall or roof so that shingles on each face lap each other alternately. This provides a weathered angle at a *ridge*, *hip*, or corner of a wall.

webbing [pai.] The wrinkling in *gas-checking*, which may or may not be desirable.

wedge A tapered piece of wood, used in timbering or centering such as a *folding wedge*. *See also C*.

wedge coping A *feather-edged coping*.

weephole (1) [joi.] A small hole in a wood *sill* which allows condensation water to escape outwards.

(2) Small gaps left in the cross joints at the foot of a *cavity wall* to allow water to escape.

weight box [joi.] The space for the sash weights in a *cased frame*.

weighting out [q.s.] Multiplying out to obtain quantities in tons or other weight unit, for the *extended prices*.

well (1) The space (horizontal distance) between the *flights* of a *stair*.

(2) or **wellhole** or **liftshaft** An open space passing through one or more floors for a lift.

Welsh arch or (USA) **jack arch** A small opening, less than about

30 cm (12 in.) span, bridged by a *stretcher* cut to a wedge shape, resting on two *corbelled* bricks or stones of matching shape.

Welsh groin An *underpitch groin*.

welt A *seam* in *flexible-metal roofing*.

welted drip *Roofing felt* turned down at an *eave* or *verge* to make a *drip*, folded back on the roof and continuously sealed.

welted nosing A junction of *flexible-metal roofing* between a vertical sheet and a horizontal sheet. They are folded together and dressed down at the top of the vertical surface.

welting strip A strip of *flexible metal* with one edge screwed or nailed to the roof and the other bent round to hold the lower edge of a vertical sheet such as a *dormer cheek*.

western framing [carp.] *Platform framing*.

western method (USA) Bricklaying by *stringing mortar* and *shoved joints*.

western red cedar [tim.] (Thuja plicata) Also called British Columbia cedar, giant cedar, and Pacific red cedar. A straight-grained, coarse, soft, weak *softwood*, very useful for making *shingles* and cheap *joinery*. Shingles made of it have a reddish-brown colour which weathers to grey. The wood contains an aromatic oil which repels insects. It splits easily and shingles were originally made from it by splitting, but are now usually sawn.

Westmorland slates Thick heavy *random slates* which a man can walk on without breaking. They are from 6 to 16 mm ($\frac{1}{4}$ to $\frac{5}{8}$ in.) thick, of a pleasing green colour with a chipped edge. This gives a satisfactory rough texture to a roof, which cannot be obtained with thin slates.

wet on wet [pai.] Painting with special paints in which one coat is sprayed on to another before the first has dried.

wet rot [tim.] *Decay* of timber caused by fungi which flourish in alternate wet and dry conditions. *See* **moisture content.**

wet sprinkler *See* **sprinkler.**

wet time The smaller-than-usual payment which men receive when they cannot work on a site because of bad weather, although they have reported for work.

wet venting [plu.] The use of the lower length of a *ventilation pipe*, in a waste-pipe or soil-waste-pipe system, as a waste pipe.

wheelbarrow A wood, steel, or *light-alloy* container with a single steel or rubber-tyred wheel in front and two hand-holds behind, by which it is lifted and pushed forward. It carries 60 to 70 litres (2 to $2\frac{1}{2}$ ft^3) of material, of which the man lifts about a quarter.

wheeling step or **wheel step** Scots for *winder*.

whetstone A stone used for sharpening cutting tools. It is sometimes a gritty slate containing garnet or a quartz rock with cavities which are abrasive.

whip A steel-wire *bond* (C) used by a *timber scaffolder*.

white cement *Portland cement* which has been selected and ground without contamination by iron (the green colouring matter of ordinary cement) or to which *white pigment* has been added. It can be used as cheap, durable but dirty white paint for masonry.

white coat [pla.] A *finishing coat.*

white lead [pai.] An opaque but not very brilliant *white pigment* much used now in the best *undercoats* for exterior work. Although poisonous, it was used as a cosmetic in ancient Greece. It consists of basic lead carbonate ($2PbCO_3.Pb(OH)_2$). The description white lead is sometimes used for basic lead sulphate. *See* **lead paint, titanium white.**

white lead putty The best *putty*, useful for plugging gaps that are narrow or exposed to the weather, for example under a *water bar.*

white lime *See* **high-calcium lime.**

whitening in the grain [pai.] A streaky white unpleasant appearance which is sometimes seen in varnished or polished woods with *coarse texture*, whether filled or not.

white pigments [pai.] The commonest white *pigments* are *white lead, zinc oxide, antimony oxide, basic lead sulphate, leaded zinc oxide,* and *titanium white. See whiting.*

white spirit or **turpentine substitute** [pai.] A thinner for *oil paint*, distilled from petroleum at about 150° to 200° C., used in Britain because it is much cheaper than *turpentine* and has similar properties.

whitewash [pai.] *See* **limewash, whiting.**

whitewood [tim.] (Picea abies and Abies alba) Also called **white deal, yellow pine, white pine, common spruce, Norway spruce.** A soft, light, general purpose timber described under *silver fir.*

whiting or **Paris white** [pai.] Crushed chalk used for making *distempers, glazier's putty*, and as an *extender*, probably the cheapest *white pigment.*

Whitworth screw thread [mech.] A British *screw thread*, used in building and suitable for heavy engineering work. *See* **Unified thread.**

whole-brick wall A wall whose thickness is the length of one *brick*, about 23 cm (9 in.) in Britain.

whole timber [tim.] A *baulk* which is square, usually about 30 × 30 cm (12 × 12 in.).

wide-ringed timber or **coarse-** or **open-grained timber** Timber with *annual rings* which are far apart. It has grown quickly and is known as coarse growth. In *softwoods, narrow-ringed timber* is stronger, but not always in *hardwoods.*

wiggle nail [carp.] A *corrugated fastener.*

Winchester cutting The intersection between *tile hanging* and a *verge*. The two tiles at the ends of each course are splay cut so that each cut tile has one nail hole left for fixing. The topmost tile meets the overhanging verge at right angles.

wind or **winding** [tim.] The twist of converted timber. *See* **warp.**

wind beam A *collar beam. See also* **C.**

winder or **wheel step** A *tread* of triangular or wedge shape, changing the direction of a *stair*. *See* **balanced step.**

wind filling *See* **beam filling.**

winding [tim.] *See* **wind.**

winding stair (1) A *spiral stair*.

(2) A circular or elliptical *geometrical stair*.

winding strips or **winding sticks** Two short straight edges about 45 cm (18 in.) long and 5 cm (2 in.) wide, tapering in thickness from 8 to 3 mm ($\frac{5}{16}$ to $\frac{1}{8}$ in.). They are used for setting out a plane surface on wood or stone by *boning* (*C*).

window A glazed opening in a wall to let in light and usually also air. In a roof it is called a *skylight* or *lantern*. Windows are usually either side-hinged on vertical hinges (*casements*) or they slide up and down (*sash windows*). Other less usual windows are *sliding sashes*, pivoted sashes, *night vents*, *hopper windows*. *See also* **dead light, opening light.**

window back [joi.] The panelling, *matchboarding*, or other *joinery* between the floor and the *window* on the inner face of a window with a *lifting shutter*. The window back hides the lifting shutter.

window bar [joi.] A *glazing bar*.

window bead A *guard bead*.

window board or **elbowboard** or (USA) **stool** [joi.] A horizontal board fixed like a shelf at *sill* level inside a window. It was always of wood but may now be of *pressed steel* 1·5 mm (0·06 in.) thick.

window efficiency ratio The *daylight factor*.

window frame [joi.] The part of a *window* surrounding the *casements* or *sashes*, in which they hinge or slide as the case may be.

window glass *See* **sheet glass.**

window lock [joi.] A *sash fastener*.

window stile [joi.] A *pulley stile*.

window stool [joi.] (USA) A *window board*.

wind shake [tim.] *Ring shake*.

wind stop A *weather strip*.

wing (1) Part of a building projecting from one side of the body of the building.

(2) A heavy slate which forms one side of a *slate ridge* and is covered by a slate roll.

wing compasses [joi.] *Quadrant dividers* adjusted by a *wing nut* on the quadrant stay.

winglight [joi.] A *flanking window* with its sill above doorsill level.

wing nut or **thumb screw** [mech.] A nut which can be turned by hand without a spanner. Wings outside it can be gripped with the fingers.

wiped joint [plu.] A joint made with *plumber's solder* between lead pipes of which one is widened with a *boxwood tampin* and the other tapered down to be inserted in it. The *solder* is poured on and moulded round the pipes with a cotton (moleskin) wiping cloth, so that if the more-fluid solder drops out of the bottom of the joint it is wiped back to the top. The *blow lamp* now used was not available in ancient times and the joints wiped 2000 years ago were heated over an open fire, the solder being held in a ladle.

wiping cloth *See* **plumber's tools** (illus.) and **wiped joint.**

wire comb [pla.] a *drag.*

wirecut bricks Bricks which are shaped before burning by extrusion (squeezing through a hole 114×76 mm ($4\frac{1}{2} \times 3$ in.)). A long bar of clay of this size is thus formed which is cut into bricks by a set of wires 23 cm (9 in.) apart attached to a frame. They are cheaper and less dense than *pressed bricks*, and have no *frog.*

wired glass *Sheet* or *plate glass* in which wire mesh is used as reinforcement to hold the pieces together if it breaks. It is used for *skylights* or for glass screens in gardens.

wire gauge [mech.] A method of defining wire diameter by a number which stated originally the number of passes through different, increasingly smaller dies, to make the wire. The number for large wire is therefore smaller than that for thin wire. The commonest gauge used in Britain is the *Standard Wire Gauge* or S W G but many other gauges exist in Britain and U S A. The Paris wire gauge increases with the wire diameter, unlike those of English-speaking countries. *See* **cold drawing** (*C*).

wire nail A *nail* made by cutting and shaping a piece of round or elliptical steel wire. (Brass wire is used for escutcheon pins, and other expensive metals for *roofing nails.*) Steel nails in exposed places are galvanized or *sherardized.* The commonest nails are round wire nails which are stocked in sizes from $25 \times 1 \cdot 6$ mm to $150 \times 5 \cdot 9$ mm, oval wire brads from 19 to 100 mm long, and clout nails, usually galvanized, from $19 \times 2 \cdot 6$ mm to $50 \times 3 \cdot 7$ mm.

wire scratcher [pla.] A *devil float.*

wiring regulations [elec.] In Britain wiring regulations are issued mainly by the Institution of Electrical Engineers in consultation with the Department of Trade and Industry but also by the D T I's Factory Inspectorate. Many electrical codes and standards are issued by the *B S I.*

withdrawal load [carp.] *Wire nails* driven across the grain are considered to have a safe pulling resistance of about 4·3 kg per cm length per 2·5 mm dia.

withe (1) or **mid feather** (Britain) A half-brick wall such as a *partition* between chimney *flues.*

    (2) or **wythe** or **tier** (USA) One leaf of a *cavity wall* or *hollow wall* or a half-brick thickness of a wall bonded into solid brickwork.

withies or **osiers** Flexible sticks like those used for basket-making, cut about every two years from willow trees. Several are twisted into a cable for tying *reed* on to rafters in thatching. *See* **runner.**

wobble saw [tim.] A *drunken saw.*

wood [tim.] *Standing timber, round timber, lumber, plywood, block-board, laminated wood,* etc.

wood-block floor layer *See* **parquet floor layer.**

wood-block paving or **wood flooring** Many sorts of wood-block floor exist, for example, end-grain blocks about 2·5 cm (1 in.) cube glued down in decorative patterns, *parquet floor, plywood parquet, wood mosaic,* etc. These floors are warm and hard-wearing.

wood brick A *fixing brick.*

wood-casing system or **casing** or **moulding system** [elec.] An old method of interior wiring in grooves in wood strips which are later covered by strips called capping.

wood chipboard [tim.] The correct term for resin-bonded *chipboard.* Chipboard itself is a material like cardboard, of which grocers' cartons are made.

wood-cutting machinist [tim.] A craftsman who sets and operates every type of *circular saw* and *band saw.*

wood element [tim.] A wood cell, botanically called xylem.

wood flour [tim.] Fine sawdust, sometimes used as an *extender* for glues. It is also used in explosives, *plastic wood,* and so on.

wood lath [pla.] *See* **lath.**

wood mosaic A floor of wood blocks $11 \times 2·5 \times 1$ cm ($4\frac{1}{2} \times 1 \times \frac{3}{8}$ in.) deep, arranged in squares in the workshop on strong paper, like *Roman mosaic,* each block being surrounded with an *oxychloride cement* which sticks fast to any *sub-floor* including concrete, tile, or metal. It is a decorative hard-wearing floor, which because of its numerous joints is also flexible.

wood ray [tim.] A term preferred by some people to the expression *medullary ray,* since the rays grow from the outer living wood towards the inner dead medulla or pith.

wood roll A piece of wood, usually round-topped, fixed on to roof boarding to enable *flexible-metal roofing* sheets to be lapped over it. *See* **hollow roll.**

wood scaffolder *See* **timber scaffolder.**

wood screw [joi.] An ordinary *screw* for fixing into wood.

wood slip A *fixing fillet.*

wood turner [tim.] A *tradesman* who works to drawings, setting the wood in a lathe, and marking the parts to be cut. He works the lathe, cutting the wood with *gouges* and *chisels.*

woodwool nailing slab A *woodwool slab* with a 6×2 cm ($2\frac{1}{2} \times \frac{3}{4}$ in.) timber let into the face along the centre of the slab.

woodwool slab Slabs 2·5 to 10 cm (1 to 4 in.) thick, made of long-fibre wood shavings compressed and bound together with cement. This fire-resisting insulating material of light weight and low cost was first made in Austria in 1914. Most boards are 60 cm (2 ft) wide but lengths can be up to 3·8 m (12 ft). *See* **reinforced woodwool**.

woodworking machinist [tim.] In Britain, a *tradesman* with considerable experience, capable of setting and operating every woodworking machine except, possibly a *spindle moulder*, which is a specialist's machine.

workability (1) [tim.] The labour of working hard dense timber with twisted grain may be five times as much as for soft, light timber with straight grain. The following table gives the rough order of workability of some well-known timbers, the least workable first.

6. Lignum vitae

5. Ebony, greenheart

4. English oak

3. Ash, beech, makore, Japanese oak

2. Pitchpine, larch, redwood

1. Spruce, poplar

(2) For the *workability* of concrete *see* (*C*).

(3) [pla.] The flow properties of a mix, its cohesiveness, and its moisture retention against the suction of the background.

work or **working edge** [joi.] (USA) The *face edge* of timber.

work or **working face** [joi.] (USA) The *face side* of timber.

working [tim.] Swelling and shrinkage of timber as the relative humidity of the air becomes higher (in winter) or lower (in summer).

working life [tim.] The *pot life* of a glue.

working up [q.s.] *Squaring* quantities and adding them.

worm hole [tim.] Any hole bored by insects such as *beetles* in timber, whether *shot hole*, *powder-post* damage, or *pinhole*.

wreath [joi.] That part of a handrail which is curved both in plan and in elevation. It occurs in every *geometrical stair*.

wreath piece or **wreathed string** [joi.] The curved part of the *outer string* below a *wreath*.

wrecking (USA) Demolition.

wrecking bar A *pinch bar*.

wrench [mech.] A spanner, usually adjustable. *See also* **pipe wrench**.

wrinkle [tim.] An unglued area caused by failure of a *veneer* to slip into place.

wrinkle finish or **wrinkling** [pai.] A *finish* with intentional wrinkles made during drying, usually by *stoving*. *See* webbing, gas-checking.

writer (1) or **signwriter** or **glass painter** or **letterer** or **advertisement painter** [pai.] A *tradesman* who paints signs for shop fronts either in the workshop or on the site. He uses a palette, *sable pencils*, and other artists' tools, sketching the outline first in chalk. He can make his own stencils and may also do gilding and silvering.

(2) *See* **sable writer**.

writing short [q.s.] Placing *items* out of their proper category in a *bill of quantities*; they are often small items, for example *shoes* (1) (priced each) with *downpipes* (priced per metre length).

wrot [joi.] Abbreviation for *wrought timber*, common in English *bills of quantities*.

wrought nail A wrought iron *nail* with a head forged to a rounded 'rose' shape. It can be bent over and clenched like a wire nail but is now less used than *wire nails*.

wrought timber or **wrot timber** or **surfaced timber** [joi.] Timber which has been planed on one or more surfaces.

wye or **Y** or **(USA) yoke** [plu.] A branch pipe leading off a straight main *run* usually at 45° to the run.

wythe *See* **withe.**

X

X-mark [joi.] *See* **face mark**.

xylem [tim.] The botanical name for wood.

xylol [pai.] An aromatic hydrocarbon distilled from coal tar at about 140° C. It is a *solvent* for *synthetic resins* and gums.

Z

zax or **sax** or **slater's axe** or **slate knife** or **chopper** or **whittle** A straight blade like a butcher's chopper with a point projecting from the back for punching holes in *slates*. *See* **plasterer's tools** (illus.).

Z-bar or **zee-bar** (USA) A Z-shaped bar used as a *wall tie*.

zee American and Australian pronunciation of Z, pronounced 'zed' in Britain.

zein [pai.] A protein from maize. When dissolved in alcohol it gives a tough *film* which has partly replaced *shellac* in USA.

zeolites Minerals which are used in the *base exchange* process of *water softening*.

zig-zag rule or **folding rule** A rule made of wood or metal pieces 15 cm (6 in.) long, pivoted together at each end, not hinged like the *fourfold rule*. It is a more recent invention and is usually longer. *Compare* **push-pull rule**.

zinc A whitish-blue metal used in *flexible-metal roofing* and as a protective coating to corrugated steel sheet in *galvanized iron* (*C*) and to smaller articles in *sherardizing*. Zinc is attacked by tar or by copper, iron, or steel in contact with it and it is therefore in roofing fixed direct to wood. British sizes of zinc sheet are up to 2·4 by 0·9 m (8 by 3 ft), in thicknesses from 0·6 to 1·0 mm. Zinc gauge thicknesses are not the same as *standard wire gauge*. *See also* **metal coating**.

zinc chromes [pai.] Bright yellow *pigments* which may contain alkali chromates (potassium chromate) but are mostly zinc chromate ($ZnCrO_4$) combined with some zinc hydroxide. *Lead chromes* are affected by hydrogen sulphide (H_2S).

zinc dust [pai.] Powdered zinc which is used in *priming* paints for use on galvanized iron.

zinc drier [pai.] The most important zinc *drier* is zinc naphthenate, which is used to prevent *wrinkling* in *stoving*. It is also a preservative for wood or rope or sash cord.

zinc oxide or **zinc white** or **Chinese white** [pai.] (ZnO) A permanent *white pigment* which prevents *chalking* in other pigments but may cause *feeding* in an unsuitable medium. It is not poisonous. *See* **leaded zinc oxide**.

zoning [t.p.] The reservation under the master plan of certain areas of land for certain uses and the enforcement of these uses by restrictions on building types, heights, and sizes. In this way certain areas can be kept for light industry, others for heavy industry, dwellings, offices, shops, and so on, each area being called a zone. *See* **Local Authority**.